Performance-Based Fire Engineering of Structures

Performance-Based Fire Engineering of Structures

YONG WANG, IAN BURGESS,
FRANTIŠEK WALD AND MARTIN GILLIE

CRC Press
Taylor & Francis Group
Boca Raton London New York

CRC Press is an imprint of the
Taylor & Francis Group, an **informa** business

A SPON PRESS BOOK

CRC Press
Taylor & Francis Group
6000 Broken Sound Parkway NW, Suite 300
Boca Raton, FL 33487-2742

First issued in paperback 2017

ISBN 13: 978-0-415-55733-7 (hbk)
ISBN 13: 978-1-138-07492-7 (pbk)

Library of Congress Cataloging-in-Publication Data

Performance-based fire engineering of structures / authors, Yong Wang ... [et al.].
    p. cm.
  Includes bibliographical references and index.
  ISBN 978-0-415-55733-7 (alk. paper)
    1. Building, Fireproof. 2. Buildings--Performance. I. Wang, Y. C. (Yong C.), 1964-

TH1065.P47 2012
693.8'2--dc23
                                                  2011052160

Visit the Taylor & Francis Web site at
http://www.taylorandfrancis.com

and the CRC Press Web site at
http://www.crcpress.com

Contents

Preface

Performance-based fire engineering of structures, based on the application of scientific and engineering principles for achieving structural safety in fire, has now firmly taken root in many parts of the world, particularly in the United Kingdom and in Europe. Its fundamental aims are to enable improvement of structural safety in fire, to increase design flexibility and to reduce the cost of fire protection to structures. Performance-based fire engineering of structures is now increasingly being practised in the whole range of structural engineering projects, from the iconic to the run of the mill, to the benefit of the client, the design team and society in general.

Undertaking performance-based fire engineering of structures demands an understanding of the fundamental scientific and engineering principles that control the behaviour of structures affected by extreme local heating. In particular, it is necessary to identify the required performance of a structure when a fire occurs, and the structural characteristics needed to achieve this required performance, based on the principles of fire science, materials science, heat transfer and structural mechanics. It should be appreciated that there are different levels of requirement for structural performance in fire, ranging from compliance with simple prescriptive rules, through ensuring fire safety with reduced levels of fire protection costs, to improving structural safety in fire under exceptional circumstances (for example, by controlling fire-induced disproportionate collapse). In the year that included the 10th anniversary of 9/11, the last point is especially important because it was the fire-induced total collapse of the World Trade Center buildings on September 11, 2001 that brought structural fire safety most dramatically to the attention of engineers, architects, building owners, government and others involved in building construction, creating a demand for authoritative information on this subject. The event was also the main trigger for the massive recent growth in research activity intended to further the understanding of the causes and mechanisms of fire-induced progressive collapse and the development of methods intended to improve the robustness of structures under fire action. The intensity of research in the last 10 years has been such that there has been a vast amount of progress in this area, although many wheels have also been reinvented.

Against this background, we felt that an authoritative book was necessary to synthesize, in a systematic manner, the areas where recent progress has taken place and its implications. Our intention is to inform those who have an interest in the area, including structural/fire protection engineers, researchers, educators/students and regulatory bodies. We feel that the few existing books on the topic have served a useful but different purpose, mainly concerned with dissemination of the now almost standard knowledge of structural fire engineering that is encapsulated in established codes of practice such as the Eurocodes. Although for completeness this book also includes aspects of this set of "standard" knowledge on structural fire engineering, its main emphasis is on issues that have become more important in the last 10 years, including the integrity of fire-resistant compartmentation; fire-induced progressive failure and provisions for robustness; joint (connection) behaviour and its critical role in providing structural robustness; global modelling of structural behaviour in fire; accurate information on material properties at elevated temperatures; and moving fires.

We are among the small number of internationally highly active researchers in the field of structural fire engineering, and this book reflects our recent research endeavours. We were the core members of the working group on fire resistance in the European Network project "COST C26 Urban Habitat under Catastrophic Events" and were instrumental in originating the current European Network project "COST TU0904: Integrated Fire Engineering and Response". The idea for writing this book originated during one of the COST C26 core group meetings, which determined the format of outputs of the project. The main purpose of the COST networks is to engage all those interested in the subject matter in the European Union to disseminate state-of-the-art knowledge across the European Union. Such networks generate large numbers of wide-ranging outputs, to which we have made significant contributions. However, because of the inevitable problems of dissemination of these outputs, and because of their rather specialised nature, it became clear that there was a need to produce an authoritative and focused book on the topic, with the whole structural fire engineering community in mind.

We are all based in Europe. Although we have extensive interactions with other international researchers, it is inevitable that this book, in particular the sections on codified design practice, focuses mainly on Europe as its context. However, we believe that current European codes on structural and fire engineering have the soundest research foundation at this time and will be considered as relevant and informative by readers in other parts of the world.

Writing a book requires a significant amount of time and effort, and we are grateful to our respective institutions for giving us the freedom to pursue this "self-interest" project. Coauthoring a book allows the book to benefit from the pooled expertise of all the authors, and we hope that readers will recognize its benefits. Coauthoring was also meant to share the considerable time and effort in delivering the book, and this was so; however, the coordination involved delays in delivering the manuscript to the publisher due to the inevitable differences in the priority of work of the different coauthors during different periods. Because of this need for extreme patience, we would like to thank our publishers, Taylor & Francis, and in particular the commissioning editor, Siobhan Poole. Chapter 11 of this book, on case studies, was based on evidence contributed by Dr. Florian Block. We thank him for his most valuable contribution, which will enable readers to appreciate how some of the principles of performance-based fire engineering of structures are already being applied in practice. We are grateful to the many colleagues with whom we have interacted over the years who have helped us develop an understanding of structural fire engineering. With regard to this book, particular thanks are due John Carpenter, Rachel Yin, Angus Law, Dougal Drysdale, Charlotte Röben Guillermo Rein, Stephen Welch, Jaroslav Procházka and Petr Kuklík, who all contributed by offering advice, providing references and correcting errors, both typographical and conceptual.

Y. C. Wang, I. W. Burgess, F. Wald and M. Gillie

Notations

The following list of notations is provided primarily for Chapter 6, where many notations are used following the Eurocode system. In the other chapters, the meaning of each notation is given following its first appearance.

LATIN UPPERCASE LETTERS

A_i	the area of the cross section with a temperature θ_i; the elemental area
A_m	the surface area of a member per unit length
A_m/V	the section factor for unprotected steel members
A_r	the area of the residual timber or timber-based cross section
A_s	the area of the reinforcement
C	concrete grades
E_a	the modulus of elasticity of steel for normal temperature designs
$E_{a,\theta}$	the modulus of elasticity of steel at elevated temperature θ_a
E_c	the elastic modulus of the concrete at ambient temperature
E_d	the design value of the relevant effects of actions from the fundamental combination
$E_{fi,d}$	the design effect of actions for the fire situation, determined in accordance with EN 1991-1-2, including the effects of thermal expansions and deformations
$(EI)_{fi,eff,z}$	the effective bending stiffness at fire situation to z axis
$(EI)_z$	the bending stiffness of the section to z axis
F	the resultant of internal forces in compression
G_k	the characteristic value of a permanent action
I_z	the second moment of area of the section to z axis
L	the system length of a column in the relevant storey
$M_{b,fi,t,Rd}$	the design buckling resistance moment at time t
$M_{el,Rd}$	the elastic moment resistance of the gross cross section for normal temperature design
$M_{fi,Rd+}$	the sagging moment resistance in fire situation
$M_{fi,t,Rd}$	the design moment resistance at time t
$M_{2,fi}$	second-order moment for fire conditions

$M_{pl,Rd}$	the plastic moment resistance of the gross cross section for normal temperature design
M_{Rd}	the moment resistance of cross section for normal temperature design
$M_{0Rd,fi}$	first-order moment resistance for fire conditions
$N_{b,fi,t,Rd}$	the design buckling resistance compression member at time t
$N_{fi,cr,z}$	the Euler critical force in the fire situation to z axis
$N_{fi,pl,Rd}$	the design value of the plastic resistance to axial compression in the fire situation
$N_{fi,\theta,Rd}$	the design resistance of a tension member at uniform temperature θ_a
$N_{fi,Rd,z}$	the design buckling resistance in the fire situation to z axis
$N_{pl,Rd}$	the plastic design resistance of the cross section for normal temperature design
N_{Rd}	the design resistance of the cross section for normal temperature design
$P_{fi,Rd}$	the design shear resistance in the fire situation of a welded headed stud
O	the opening factor [m$^{1/2}$]
Q_l	the representative value of the variable action
$Q_{k,l}$	the principal variable load
$R_{fi,d,0}$	the value of the design resistance in the fire situation $R_{fi,d,t}$ for time $t = 0$
$R_{fi,d,t}$	the design resistance in the fire situation
T	the resultant of internal forces in tension
V	the volume of a member per unit length
$V_{fi,t,Rd}$	the design shear resistance at time t
V_{Rd}	the shear resistance of the gross cross section for normal temperature design, according to EN 1993-1-1 2005
$W_{pl,y}$	plastic section modulus
$X_{d,fi}$	design values of mechanical material properties in the fire situation
X_k	the characteristic values of a strength or deformation property for normal temperature design

LATIN LOWERCASE LETTERS

a_{fi}	the addend thickness, width of the side members or the end and edge distance to fasteners of side timber and timber-based members
a_z	the reduced thickness of cross section; absorptivity of flames
a_{500}	the position of the isotherm 500°C
b	the thickness; width
b_{eff}	the effective width of the concrete slab
b_{fi}	the effective thickness
b_{st}	the width of the internal steel plate in timber connection
c	factor ($c \approx 10$) depending on the curvature distribution; the specific heat
c_a	the specific heat of steel
c_p	the temperature-independent specific heat of the fire protection material
d	the height
$d_{char,0}$	design charring depth for one-dimensional charring

d_{ef}	the effective charring depth
d_f	the thickness of the fire protection material
d_{fi}	the effective height
d_i	the cross-sectional dimension of member face i
d_p	the thickness of fire protection material
$e_{\Delta\theta}$	the eccentricity due to variation of temperature across masonry
$f_{ay,\theta}$	maximum stress level or effective yield strength of structural steel in the fire situation
$f_{ay,\theta cr}$	strength of steel at critical temperature θ_{cr}
f_b	the characteristic unit strength for masonry
f_{ck}	the characteristic cylinder compressive strength of the concrete
$f_{ck,fi\theta}$	the characteristic value of compressive strength of concrete at temperature θ for a specific strain
$f_{d\theta}$	the design compressive strength of masonry at temperature θ
f_k	the characteristic strength of timber at ambient temperature
$f_{o,\theta}$	the 0.2% proof strength at elevated temperature
f_y	the yield strength at ambient temperature
$f_{y,\theta}$	the effective yield strength of steel at elevated temperature θ_a
$f_{y,i}$	the nominal yield strength for the elemental area A_i
f_u	the ultimate tensile strength of the material of the stud but not greater than 500 N/mm^2
g	the permanent action
h	the height; the depth
h_c	the slab depth
h_{eff}	the effective height for a composite slab with a profiled steel sheet
h_p	the wood-based panel thicknesses
$h_{net,d}$	the design value of the net heat flux per unit area
k_θ	the reduction factor for a strength or deformation property dependent on the material temperature
k_σ	the buckling reduction factor
$k_{c,\theta}$	the reduction factors of compressed strength of concrete
$k_{c,\theta,M}$	the reduction factor for concrete at particular point M
$k_{E,\theta}$	the reduction factor for modulus of elasticity at the temperature θ
k_{fi}	the coefficient for transferring the characteristic value of timber characteristic strength to average one
k_{flux}	the coefficient taking into account increased heat flux through the fastener
$k_{max,\theta}$	the reduction factors for studs et elevated temperature
$k_{mod,fi}$	the modification factor for duration of fire load and moisture content of timber
$k_{o,\theta}$	the strength reduction factor for the 0.2 proof strength at fire
$k_{o,\theta,max}$	the strength reduction factor for the 0.2 proof strength at the element temperature
k_{sh}	correction factor for the shadow effect
$k_{y,\theta}$	the reduction factor for the yield strength of steel dependent on temperature
l_{fi}	the buckling length of a column for the fire design situation
$l_{fi}t$	he buckling length at fire

ℓ_θ	the buckling length of steel to concrete composite column at its elevated temperature θ
n	the number of zones
p	the perimeter of the fire exposed residual cross section of timber
q	the variable action
$1/r$	is the curvature
t	the plate thickness; the time in fire exposure
$t_{d,fi}$	the fire resistance period of the unprotected timber connection
t_F	thickness of a wall for a period of fire resistance
$t_{fi,d}$	time of fire classification
t_{req}	the required standard fire resistance period
w	the width of the zone
z_i	the distance from the plastic neutral axis to the centroid of the elemental area A_i

GREEK UPPERCASE LETTERS

Δt	the time interval
Δl	the temperature-induced expansion
$\Delta\theta_{g,t}$	the increase of the ambient gas temperature during the time interval Δt
Φ	the capacity reduction factor of the masonry wall; the configuration factor

GREEK LOWERCASE LETTERS

α	the ratio of the applied design load on the masonry wall to the design resistance of the masonry wall; the convective heat transfer coefficient
α_θ	the ratio of reduction factor of elasticity modulus and reduction factor of 0.2% proof strength at elevated temperature
α_t	the coefficient of thermal expansion of masonry
β_0	the design charring rate for one-dimensional charring under standard fire exposure of timber and timber-based elements
$\beta_{0,\rho,t}$	the charring rate for other characteristic densities ρ_k and panel thickness t_p
β_n	design notional charring rate under standard fire exposure of timber and timber-based elements
γ_G	the partial safety factor for permanent actions
γ_{M2}	the partial safety factor for normal temperature
$\gamma_{M,fi}$	the partial safety factor for the relevant material property for the fire situation
$\gamma_{M,fi,v}$	the partial safety factor of connectors for fire situation
$\gamma_{M,fiv}$	the partial factor for shear connectors at fire situation
$\gamma_{Q,I}$	the partial safety factor for variable action I
ε	the reduction factor for steel yield strength
ε_f	the emissivity of a flame; the emissivity of an opening
ε_m	the surface emissivity of the component

η_{fi}	the reduction factor for design load level in the fire situation
θ	the temperature in degrees centigrade
θ_a	the steel temperature
$\theta_{a,cr}$	critical temperature of steel
θ_{al}	the aluminium temperature
$\theta_{g,t}$	the ambient gas temperature at time t
θ_i	the temperature in the elemental area A_i
θ_{web}	the average temperature in the web of the section
κ_1	the adaptation factor for non-uniform temperature across the cross section
κ_2	the adaptation factor for non-uniform temperature along the beam
λ	the thermal conductivity
$\bar{\lambda}$	the relative slenderness subjected to buckling at ambient temperature
$\bar{\lambda}_\theta$	the relative slenderness for the temperature θ_a
λ_a	the thermal conductivity of steel ambient temperature
$\bar{\lambda}_{LT}$	the relative slenderness subjected to lateral-torsional buckling in normal temperature design
$\bar{\lambda}_{LT,\theta,com}$	the relative slenderness subjected to lateral-torsional buckling for the element temperature θ
$\bar{\lambda}_{z,\theta}$	the relative slenderness to z axis for the temperature θ_a
μ	the Poisson's ratio
μ_0	the degree of utilization at time $t = 0$
ξ	the reduction factor for unfavourable permanent actions G
ρ_a	the density of steel
ρ_p	the density of the fire protection material
ρ_k	the characteristic density
σ	the Stefan Boltzmann constant [$5.67 \cdot 10^{-8}\,\text{W/m}^2\text{K}^4$]
χ_{fi}	the reduction factor for flexural buckling in the fire design situation
$\chi_{LT,fi}$	the reduction factor for lateral-torsional buckling in the fire design situation
$\chi_{z,fi}$	the reduction factor for flexural buckling about the z axis in the fire design situation
χ_z	the reduction coefficient for buckling to z axis
ψ_{fi}	the combination factor for frequent values, given by either $\psi_{1,1}$ or $\psi_{2,1}$

Chapter 1

Introduction to fire safety engineering and the role of structural fire engineering

1.1 INTRODUCTION TO FIRE ENGINEERING

For many types of infrastructure, such as buildings and tunnels, ensuring fire safety is one of the most fundamental requirements of design, construction and operation. The principal aim of fire safety precautions is to minimise the loss of life, but in many cases, fire safety provision will also have to take into consideration other requirements, such as minimising financial loss incurred as a result of damage to property and contents, business continuity and the environment. It is essential that, at the beginning of the design stage, the requirements for fire safety are clearly defined, taking into consideration the risk posed by fire and the level of acceptable risk.

Providing means achieving fire safety encompasses many different aspects, including fire management to minimise occurrence of ignition and control of combustible materials; fire detection, providing means of warning to occupants and adequate means of escape to aid evacuation; controlling building materials to limit fire growth, propagation and spread; fire containment to limit the extent of fire damage, firefighting and rescue. This book addresses issues of structural performance, which is related to fire containment.

Design for fire safety may adopt one of two generic approaches: (1) following the simple "deemed-to-satisfy" prescriptive rules, such as those laid down in government regulations for life safety or in insurance company specifications for property protection; (2) employing the performance-based fire engineering approach. This book is focused on performance-based fire engineering of structures.

Performance based fire engineering is about specifying performance requirements and then developing solutions based on sound understanding of the underlying scientific and engineering principles. This contrasts with the prescriptive approach that tends to be based on experiences developed through many years as specific responses to past fire incidents. It is important to note that, since the prescriptive approach is simple to implement and has generally resulted in achieving the desired level of fire safety, it should be the first choice when specifying a fire safety strategy. Performance-based

fire engineering design should be exploited only if its use can be justified because it offers flexibility in design, reduced construction cost and
improved safety. Because of the level of flexibility and complexity involved
in implementing a performance-based fire engineering approach, there is a
much greater requirement to understand the fundamental principles of this
approach. The aim of this book is to help the reader to develop a thorough
understanding of the principles of performance-based approaches to the
fire safety engineering of structures.

In many other engineering disciplines such as structural engineering, the
performance-based approach has long been taken for granted. However,
since specifying fire safety precautions has been dominated by statutory
requirements and regulations, the development of the scientific and engineering basis of fire safety started relatively late. The adoption in practice
of this approach to specifying fire safety precautions has developed only
since the 1980s. However, the pace of research and development in the field
has increased to the extent that performance-based fire engineering is now
routinely practised on a significant number of projects in many countries,
by an increasing number of professionally qualified fire engineers. Projects
that have benefited from performance-based fire engineering design include
some iconic buildings, such as the China Central Television (CCTV) building (Figure 1.1; Luo et al. 2005), in which a fire engineering approach was
required to give assurance that the prescriptive fire protection approach was
adequate. In more conventional buildings (Figure 1.2; http://www.mace.
manchester.ac.uk/project/research/structures/strucfire/) the application of a
performance-based fire engineering approach has produced significant sav-
ings in construction cost while giving assurance of the level of safety.

The prescriptive treatment of structural performance in fire is rather
simplistic; therefore, adopting the performance-based fire engineering

(a) Building

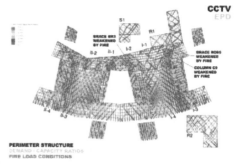

(b) Structural analysis model based on
incorporating fire-weakened members

Figure 1.1 Fire Engineered CCTV Headquarter Building, Beijing, China (Luo et al 2005).

(a) Nuffield Hospital, Leeds

(b) Fire protection scheme

Figure 1.2 Fire Engineered Nuffield Hospital, Leeds, United Kingdom. (From http://www.mace.manchester.ac.uk/project/research/structures/strucfire/, accessed on 24 May 2010. With permission.)

approach provides excellent potential for improving structural fire safety design from the points of view both of reducing the cost of construction and of improving structural safety. The aim of this book is to help the reader develop a thorough understanding of structural performance in fire so that the benefits of performance-based fire engineering of structures are fully and safely exploited.

1.2 ROLES OF STRUCTURAL FIRE RESISTANCE

It is important to understand the overarching requirements for structural performance in the case of fire. For example, the building regulations for England and Wales (Department of Communities and Local Government [DCLG] 2006) states: "The building shall be designed and constructed so that, in the event of fire, its stability will be maintained for a reasonable period." This defines the fundamental performance requirement for the structure of a building in fire. How this functional requirement is met may be interpreted in different ways. In the prescriptive approach for structural resistance in fire, such as Approved Document B (ADB) to the building regulations for England and Wales (DCLG 2006), the interpretation places emphasis on the elements of the structure, stating that the load-bearing elements of the structure of the building should be capable of withstanding the effects of fire for an appropriate period without loss of stability. One may deduce two implicit assumptions from this approach:

1. *The necessity condition:* Loss of any element in fire results in the structure losing its stability;
2. *The adequacy condition:* If each element possesses sufficient resistance in fire, this ensures that the structure has sufficient load-bearing capacity.

Figure 1.3 Fire-damaged Broadgate building, London. (From Steel Construction Industry Forum (SCIF), *Investigation of Broadgate Phase 8 Fire*, Steel Construction Institute, Ascot, UK, 1991. With permission.)

Neither of the above two assumptions is entirely correct. With regard to the necessity condition 1, it should be understood that a structure usually has sufficient redundancy to allow a number of alternative load paths. The loss of one load path rarely leads to overall loss of stability. This can be illustrated using one of the well-known examples of fire-damaged structures. Figure 1.3 shows a photo taken from the fire-damaged Broadgate building in London, which suffered a severe fire in 1990 (Steel Construction Industry Forum [SCIF] 1991). In this building, the steel columns were not protected at the time of the fire and clearly suffered extensive damage, to the extent that they would have been considered to have lost their load-bearing capacity in a prescriptive approach. However, since the flooring system was able to redistribute the loads using an alternative load path, the structure was able to maintain its stability during the entire period of the fire.

To disprove the adequacy condition 2, it is necessary to consider the conditions under which the capacity of a load-bearing element is assessed. The element-based approach has led to the development of a set of procedures for this purpose, based on standard fire resistance tests. These procedures are typically based on the loading and boundary conditions of the element at ambient temperature. However, because the loading and boundary conditions of the load-bearing element under fire attack may differ considerably from those at ambient temperature, adequate load-bearing capacity of the elements of the structure does not mean that it

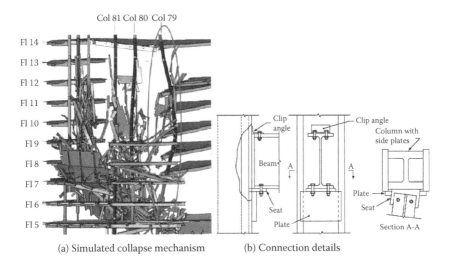

Col 81 Col 80 Col 79

Fl 14
Fl 13
Fl 12
Fl 11
Fl 10
Fl 9
Fl 8
Fl 7
Fl 6
Fl 5

(a) Simulated collapse mechanism   (b) Connection details

Figure 1.4 Possible collapse mechanism of World Trade Center Building 7. (From National Institute of Standards and Technology (NIST), *Federal Building and Fire Safety Investigation of the World Trade Center Disaster: Structural Response and Probable Collapse Sequence of World Trade Center Building 7*, NIST Report NCSTAR 1-9(2), National Institute of Standards and Technology, Gaithersburg, MD, USA, 2008.)

will maintain its stability in fire. The case of World Trade Center Building 7 (WTC7) may be used to illustrate this point. Figure 1.4, taken from the National Institute of Standards and Technology (NIST) investigation report (NIST 2008), shows the damaged structure prior to total collapse of the building. The inset figure shows how a main girder was connected to Column 79. In this case, each individual element of the structure would have been assessed as retaining sufficient load-bearing capacity. However, the main girder would have been assessed based on a simply supported boundary condition, without any out-of-plane force, and the connection would have been assumed as resisting only vertical shear. In reality, during the fire attack the main girder was subjected to a high out-of-plane force due to restraint to thermal expansion of the connected secondary beams, which may have fractured the seating bolts. This could have caused the main girder to become detached from the column. In the NIST investigation, this is considered to have been the triggering event that led to progressive collapse of the building.

What the examples demonstrate is that there are major shortcomings associated with the implicit assumptions of the element-based prescriptive approach. Of course, dealing with a whole structure can be difficult, and it may be necessary to consider it on an elemental basis. However, should this be the case, the definition of elemental loading and boundary conditions should take into consideration the expected interaction of the element with the wider structure.

The overall aim of this book is to facilitate this alternative interpretation by explaining the real performance of structures in fire. This has important implications for structural fire safety design. On the one hand, the cost of fire protection may be reduced if it can be demonstrated that a deterioration of some structural elements can be tolerated without compromising fire safety. On the other hand, as demonstrated by the behaviour of WTC7, the whole-structure approach can be used to overcome the severe shortcomings of the element-based approach to fire safety, ultimately leading to safer structures in the event of fire.

1.3 THE PROCESS OF PERFORMANCE-BASED FIRE ENGINEERING OF STRUCTURES

The previous section set out the fundamental requirements for structural fire safety and contrasted the two different interpretations: the prescriptive approach based on elements of the structure and the performance-based approach considering the structure as a whole. These two approaches are applied differently. After many years of development and refinement, the prescriptive approach has evolved into a set of simple rules that can easily be implemented in practice. For example, fire safety design of a steel-framed structure is reduced to finding the required thickness of fire protection to be applied to the structural elements, according to the type of fire protection and the fire resistance requirement of the structure (Association for Specialist Fire Protection [ASFP] 2002). Ease of implementation is its main attraction. Such prescriptive rules have been implemented in the vast majority of structures that have been shown to perform satisfactorily in the event of fire. Thus, it may be accepted that the prescriptive rules are largely satisfactory in ensuring safety. However, in some cases the prescriptive approach can be restrictive, uneconomical and even unsafe. In addition, the prescriptive approach requires little understanding by the engineer of how a structure behaves in fire. These shortcomings may be overcome by implementing a performance-based approach, which will by necessity be more complex. Therefore, the choice of which approach to implement depends on the skills available and the potential benefits to be gained. Typically, simplified approaches will be used; these may or may not be purely prescriptive. Where such approaches become unrealistic, restrictive or uneconomical, the performance-based approach should be considered as an alternative.

When prescriptive rules are followed, it is usually deemed by regulatory authorities that fire safety design of the structure fulfils the legal requirements. Checking the application of these rules does not require the designers or the building control officers to have much understanding of structural performance in fire. When performance-based structural fire engineering is adopted, the designer has to understand the performance of the structure in a range of fire scenarios. This book aims to help those who have an interest in performance-based structural fire engineering to develop a thorough understanding of the different aspects of the subject.

Performance-based structural fire engineering includes consideration of fire severity and behaviour, heat transfer as the mechanism of temperature development in the structure, and a thermostructural assessment to check that the residual load-bearing capacity of the structure is not exceeded by the applied load at the time of the fire. Linking these three general steps requires information on the thermal and mechanical properties of materials. This book attempts to deal with the key aspects of the process by explaining the behaviour of the fire, the structure and its materials, and the different options for quantifying these effects.

1.4 INTRODUCTION TO THE BOOK

Two events, described in more detail in Chapter 2, have played a transformational role in the development of performance-based structural fire engineering. The first is the structural fire engineering research programme of the mid-1990s, in which instrumented and controlled structural fire tests were carried out in a realistic full-scale structure at Cardington, Bedfordshire, England. The second is the collapse of buildings of the World Trade Center (WTC) in New York in fires caused by terrorist attacks on September 11, 2001. The Cardington research programme was the catalyst in accelerating the take-up of structural fire engineering in practical design as a method of eliminating unnecessary fire protection to steel structures. The WTC building collapses made clear the need for more research studies to be undertaken to understand the vulnerability of building structures under fire attack. These events have brought the subject of performance-based structural fire engineering to prominence, and there is now a great demand for information and knowledge on the subject.

Interest in this subject has created a huge surge in output. Academic papers in the area are now being published in journals at an accelerating rate. Fire resistance is also becoming a major theme in many structural engineering conferences, where a few years ago it would have been a peripheral subject attracting no more than a handful of participants. There is now even a biennial international conference series devoted entirely to the topic, Structures in Fire (SiF), at which attendance over the first six has grown from around 30 (Copenhagen 2000) to 220 (Michigan 2010). Although this explosion in output makes information easily accessible to anyone who is interested, it creates a problem for people who have a great interest in the subject but lack a deep enough knowledge to assess critically the true value of the output. We consider that it is highly desirable that there should be an authoritative reference book to summarize the up-to-date knowledge on the subject and to explain thoroughly the associated topics in a systematic way.

A number of books have already been published that are relevant to performance-based structural fire engineering. However, among these books some concentrate on "design" aspects, mainly concerned with dissemination of the now-almost-routine knowledge of structural fire engineering encapsulated in established codes of practice such as the structural Eurocodes. These do not address in any depth the following issues, which have become important in recent years:

- Global structural behaviour and modelling,
- Progressive collapse of structures in fire and the importance of connection robustness,
- Integrity of compartmentation in fire,
- Structural fire engineering under realistic fire conditions and its implications for basic input data such as material properties.

A previous book by the first author (Wang 2002) differed from the "design" texts in that it was intended to provide fundamental underpinning to fire-resistant design methods. Knowledge of these topics at the time of writing of that book was either non-existent because the importance of some topics had not been fully appreciated (e.g. progressive collapse/connection behaviour) or was very limited because research was at an early stage. It is the intention of this book to provide up-to-date, authoritative, fundamental information on the subjects that relate to performance-based structural fire engineering. In particular, it attempts to pay attention to those aspects that affect whole-structure behaviour, structural robustness and integrity under realistic fire conditions.

The book has 11 chapters: Chapter 2 reviews a few events that have defined the major phases of development in performance-based structural fire engineering. This includes the development of the suite of European standards (Eurocodes) on fire resistance of structures, as well as a few well-known structural fires: Broadgate (an accidental fire in London), Cardington (a structural fire research programme), WTC (fires caused by terrorist action on September 11, 2001) and the Windsor Tower (an accidental fire in Madrid). Through a review of these major structural fires, this chapter identifies their most important implications for performance-based fire engineering of structures, some aspects of which have now been researched thoroughly and are influencing everyday practice and some of which still require extensive research studies.

Performance-based fire engineering of structures deals with structural performance at high temperatures. This requires an understanding of the fire behaviour and the transfer of heat from the fire to the structure to raise its temperatures, as well as knowledge of the thermal and mechanical properties of the structural materials and the thermal properties of the fire protection materials. Chapters 3 to 5 deal with these subjects. Chapter 3 covers fire dynamics, Chapter 4 heat transfer and Chapter 5 material properties. In addition to covering some of the "standard" knowledge within this field, which is available in previous books, this book incorporates recent developments in these topics. In particular, Chapter 3, on fire dynamics, includes travelling fires in large spaces and fires involving more than one fire-resistant design compartment. It also provides explanations of how these may affect performance-based structural fire engineering design. Appropriate use of material properties is vital in achieving accurate and realistic solutions. Because the majority of material property models have been developed within the bounds of standard fire exposure and element-based structural behaviour, their applicability to whole-structure performance under realistic fire conditions requires careful consideration. Chapter 5, in addition to presenting generally well-known material models, provides an introduction to new material property models and their implications for performance-based design. These include load-induced transient thermal strain in concrete and intumescent coating performance under realistic fire exposure.

As mentioned, the main focus of this book is on the fundamental aspects of performance-based structural fire engineering. For completeness, and for readers who are interested in applying structural fire engineering in relatively simple situations, Chapter 6 presents the essential information to enable simple fire resistance calculations to be carried out for structural elements using the four most common construction materials: steel, concrete, timber and masonry.

Chapters 7 to 11 are the heart of the book, dealing with the essential subjects of performance-based structural fire engineering, and their content represents the real change from the currently accepted knowledge and methods. In the new context, it is essential to take into consideration structural interactions, which makes modelling of whole-structure behaviour inevitable. Chapter 7 introduces the unique features of thermostructural simulation, in particular large-deflection behaviour, and provides detailed guidance on modelling techniques to enable faithful representation of structural behaviour in fire. Benchmark examples are provided to help readers to verify and validate their numerical simulations.

A key component in control of the risk of fire integrity failure and disproportionate collapse, which are covered in Chapters 9 and 10, respectively, is in ensuring that the structural joints are robust. This requires that the joints do not fracture, even if distortions are very high. Until very recently, joints had been assumed to be non-critical in fire since their temperatures tend to be lower than those of the surrounding structural components. This view is not valid because joints can be subjected to additional normal forces that would not have been considered in their ambient-temperature design. Joint behaviour in fire is complex, and it is only very recently that some fundamental understanding of joint behaviour in fire has been achieved. Chapter 8 presents details of the current knowledge of joint behaviour, including detailed modelling, integration of joint models into global structural modelling, and guidance on joint details to improve structural robustness in fire.

Provision of adequate structural load-bearing capacity is one of the three fundamental aspects of providing sufficient fire resistance, the other two being adequate insulation and compartment integrity. Insulation performance is principally about heat transfer across separating elements by conduction and can now be quantified with sufficient accuracy. Integrity (which is referred to as fire compartment integrity and should not be confused with structural integrity) is about preventing fire spread through openings in the building. Chapter 9 explains the requirement of maintaining fire compartment integrity and its interaction with structural performance. Despite progress in fire engineering, practical checking of fire compartment integrity is still based on the results of standard fire resistance testing. Application of performance-based structural fire engineering design conducted mainly based on load-bearing capacity may make some of the construction details inadequate for the maintenance of fire integrity. In particular, large structural deformations may be tolerated to the extent that fissures appear in slabs. This requires careful consideration to ensure that fire compartment integrity is not compromised.

Since the WTC building collapses, fire-induced progressive structural collapse has become a significant research topic. Progressive collapse may be proportionate or disproportionate. The ability of a structure to avoid disproportionate collapse is termed *structural robustness* and is one of the key requirements of structural design against exceptional hazard loadings. Chapter 10 discusses how fire-induced disproportionate collapse can occur and how performance-based design principles may be used to reduce the risk of disproportionate collapse. In particular, failure of fire compartment integrity, leading to multicompartment fires, increases the risk of fire-induced disproportionate collapse considerably. Chapter 10 provides some recommendations for controlling this risk.

Performance-based fire engineering is no longer solely a research tool. It is now being practised by a number of specialist engineering consultancies to improve structural fire safety and to reduce the cost of fire protection. Chapter 11 presents a case study of practical applications of performance-based fire engineering of structures, explaining the decision-making processes and the detailed solution methods employed. It is to be hoped that the successful applications of performance-based structural fire engineering principles can motivate more engineers to take advantage of the developments in this area.

Chapter 2

Recent major structural fire events and their implications

2.1 INTRODUCTION

Humankind has learnt from fire events to develop practices that mitigate the dangers of fire. The great fires of London (2–5 September 1666), Moscow (14–18 September 1812), Chicago (8–10 October 1871), and others have laid the foundations of regulations on fire safety. Within the narrower context of structural performance in fire, a few recent fires have been influential and may be claimed to have influenced strongly the development of best practice in modern structural fire engineering. This chapter describes these fires and explains their impacts and influences on the development of modern structural fire engineering practice. These fires are the Broadgate fire in London in 1990 (Steel Construction Industry Forum [SCIF] 1991); the Cardington structural fire research programme in the mid-1990s (Newman et al. 2006); the World Trade Center fires on September 11, 2001 (Federal Emergency Management Agency [FEMA] 2002; National Institute of Standards and Technology [NIST] 2005, 2008); and the Windsor Tower fire in Madrid in 2005 (Intemac 2005).

Before these fires, traditional best practice for structures in fire was based on idealized structural element behaviour with the design fire having almost no relevance to real fire behaviour. In fact, these principles are still being used and will continue to be used in the future because they are simple to apply and are known by all those concerned. It may also be said that such traditional practices have proved their worth in that their applications have not resulted in extensive fire-related structural damage. However, this "good" record may be partially caused by chance in that other fire precaution measures have managed to prevent fires from reaching the dangerous stage at which they affect structural performance and partially due to overprotection of structures against fire. The Cardington structural fire research programme, which resulted from the Broadgate fire accident, successfully challenged the issue of overspecification of fire protection and has resulted in practices that enable the cost of fire protection to be reduced. However, complete collapse of the World Trade Center structures, particularly Building 7, suggests that merely following the conventional element-based structural fire safety practice may not be sufficient to ensure structural safety in fire.

This chapter describes these recent fires and explains their implications for design of structures against fire. This forms the contextual background for the other chapters of this book.

2.2 BROADGATE FIRE, LONDON, 1990

2.2.1 The fire and observations

The Broadgate fire occurred during construction of a 14-storey steel-framed building (SCIF 1991). The structure consisted of steel columns supporting floors constructed using composite long-span lattice trusses, 13.5 m in span, and composite beams forming the floor slabs. Fire design was according to the conventional method at the time. Fire protection was intended to provide 90 minutes of standard fire resistance to the steel structure, but the fire protection had not been applied to all columns as the building was still under construction. No other active fire protection system (including the sprinkler system) was operational. This lack of active fire protection measures allowed the fire to develop unhindered to very high temperatures, which severely damaged the structure.

The fire started inside a site hut on the first floor of the building. Even though combustible materials were limited, it is estimated that the fire temperature reached over 1000°C in some locations. However, the maximum steel temperatures were estimated to be below 600°C.

2.2.2 Implications

Two implications can be deduced from the observations of the structural damages suffered by the Broadgate building. On the one hand, since the applied load at the time of fire was very low (because there was virtually no imposed load), the critical steel temperatures would have been higher than the temperatures experienced by the steelwork. According to the element-based fire design approaches, there should not have been any structural element failure. However, the structure suffered extensive structural element failure. For example, some of the steel beams suffered distortion and local buckling of their bottom flanges and webs near their supports. Some of the columns suffered extensive damage and localised failure (Figure 2.1a). The reduction in load-carrying capacity of the columns resulted in the floor slab losing vertical support and suffering extensive vertical deformation; for example, the maximum permanent slab vertical deflections were as high as 600 mm (Figure 2.1b). This structural element damage was a result of

(a) Buckled column and deformed beams   (b) Floor deformation

Figure 2.1 Damages from the Broadgate fire. (From Steel Construction Industry Forum (SCIF), *Investigation of Broadgate Phase 8 Fire*, Steel Construction Institute, Ascot, UK, 1991. With permission.)

restrained thermal expansion (Wang et al. 1995), which would not have been considered in the element-based fire safety design approach.

On the other hand, despite these structural element failures, the Broadgate building remained stable, and the floors slab retained their integrity, preventing fire spread to the upper floors. This can be attributed to the ability of the structure to redistribute loads from the failed structural elements to other parts of the structure.

Following the fire, all of the structure, except for an area approximately 40 by 20 m, was deemed reusable without any structural repair. The damaged area of the structure was replaced at a cost of less than £2 million, out of direct fire loss of over £25 million. The structural repairs were completed in 30 days.

The ability of the Broadgate building to remain stable without collapse after multiple structural element failures became the catalyst for the U.K. and European steel industries to carry out the influential Cardington structural fire research programme, which enabled the behaviour to be reliably observed and quantified so that the economic benefits of reducing fire protection costs for steel structures can be realised.

2.3 CARDINGTON FIRE RESEARCH PROGRAMME, 1994–2003

2.3.1 Fire tests

At the time of the Broadgate fire, the U.K. and European steel industries were engaged in research studies to base structural fire safety design of steel framed buildings on a more scientific footing. At this time, the first-ever formal structural fire engineering code of practice, BS 5950 Part 8 for fire resistant design of steel structures (British Standards Institution [BSI]

1990b), was published. At the same time, the suite of fire-engineering-based structural Eurocodes were at an advanced stage of preparation. However, all of these codes of practice are based on isolated structural elements. It would have been difficult to use the element-based approach of these codes of practice to help the steel construction industry to achieve a step change in substantially reducing the cost of fire protection. For example, the existing steel structural fire protection guide (Association for Specialist Fire Protection [ASFP] 2002) was based on a steel beam limiting temperature of 620°C and a steel column limiting temperature of 550°C, which imply a load ratio of 0.6. Both BS 5950 Part 8 and the equivalent Eurocode EN 1993-1-2 (Committee of European Normalization [CEN] 1993–2005b) allow the steel limiting temperature to increase if the load ratio is lower. However, even at the lowest realistic load ratio (about 0.2), the steel limiting temperature could only be increased by a maximum of about 150°C. This increase in steel-limiting temperature is useful but is insignificant to allow the steel industry to reduce the cost of fire protection substantially because the cost of fire protection mainly derives from the labour cost required for application, with the cost of the protection material a relatively small part of the total. Increasing the limiting temperature of steel by about 150°C is unlikely to be sufficient to eliminate the necessity for applying fire protection.

Before the Broadgate fire, the potential of using alternative load-carrying mechanisms, and hence allowing some steel members to fail in fire, had not been considered as an option because the understanding of structural behaviour in fire had been built on structural element performance in isolation. The Broadgate fire provided a vision based on a different way of thinking, and the Cardington structural fire research programme and subsequent related research activities provided tools to realize the potential of this vision.

The Cardington structural fire research programme was ambitious, aiming at understanding the performance in fire of full-scale, realistic structures. It was fortunate that the principal partner of the research programme, the United Kingdom's Building Research Establishment (BRE), had in its possession one of the two former airship hangers (Figure 2.2a), each covering a space approximately 240 m long, 85 m wide and 54 m high, located at Cardington in Bedfordshire.

Initially, an eight-storey steel-framed composite structure was constructed in the hanger. Later, with input from the concrete and timber industries, a seven-storey reinforced concrete structure and a six-storey timber-framed structure were also constructed and tested in fire. Figures 2.2b–2.2d show these three structures. This chapter only discusses the findings from the steel-framed structure because this was the most thoroughly researched. More detailed information on the Cardington fire tests can be obtained from Timber Frame 2000 (TF 2000) for the timber structure;

(a) BRE Cardington hanger   (b) Timber-framed structure

(c) Concrete-framed structure   (d) Steel/concrete composite structure

Figure 2.2 BRE Cardington laboratory and full-scale test structures.

Chana and Price (2003) and Bailey (2002a) for the concrete-framed structure; and Newman et al. (2006), Wang (2002), and Wald et al. (2006) for the steel/concrete composite structure.

2.3.2 Main results from the tests on the steel-framed structure

The steel building was completed in 1994 (Lennon 1996). It was of steel-framed construction, using concrete slabs cast on steel decking, which were in composite action with the steel beams. It had eight storeys (overall height 33 m) and five bays (5 × 9 m = 45 m) wide by three bays (6 + 9 + 6 = 21 m)

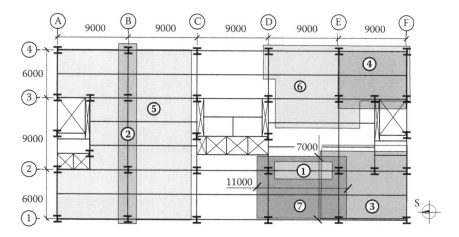

Figure 2.3 The Cardington fire tests on steel structure.

deep (Figure 2.3). The structure was designed as non-sway with a braced central lift shaft and two braced end staircases providing the necessary resistance to horizontal loads. The main steel frame was designed for gravity loads, and the connections, which consisted of flexible end-plates for beam-column connections and fin plates for beam-beam connections, were designed to transmit only vertical shear loads. The building simulated a typical medium-rise commercial office building in the Bedford area, and all the elements were designed according to British standards and checked for compliance with the provisions of the Eurocodes.

The building was designed for a permanent floor load of 3.65 kN/m² and an imposed load of 3.5 kN/m². The floor construction used steel decking with a lightweight in situ concrete composite floor cast onto it, incorporating an anticrack mesh of 142 mm²/m in both directions. The floor slab had an overall depth of 130 mm, and the steel decking had a trough depth of 60 mm. Seven large-scale fire tests at various positions within the experimental building were conducted; see Figure 2.3 for their locations on the plan and Table 2.1 for their essential details.

After each fire test, the structure remained standing, without any apparent danger of total collapse. Table 2.2 summarises the main test results.

According to the structural element-based fire-resistant design approach, under the applied mechanical loading given in Table 2.2 the maximum steel limiting temperature would be less than 720°C. This temperature was exceeded in every test. It was inevitable that the steelwork suffered local damage, such as flange/web local buckling and distortion of the beam (Figure 2.4a) and severe distortion of the column (Figure 2.4b). However, in all cases, the structure was able to redistribute loads from these failed

Table 2.1 Fire tests on steel structure in Cardington laboratory

No.	Test	Size, m × m	Area, m²	Fire	Mechanical (ULS), %
		Fire compartment		Load	
1	Restrained beam	8 × 3	24	Gas	30
2	Restrained frame	21 × 2,5	53	Gas	30
3	Corner compartment	10 × 7	70	Wood: 45 kg/m²	30
4	Corner compartment	9 × 6	54	Wood: 45 kg/m²	30
5	Large compartment	21 × 18	342	Wood: 40 kg/m²	30
6	Office demonstrational	18 × 9	136	Wood: 46 kg/m²	30
7	Internal compartment	11 × 7	77	Wood: 40 kg/m²	56

structural members to other parts of the structure, so that it remained safely standing during and after the fire.

2.3.3 Implications

The main objective of the Cardington tests on the steel-framed structure was to use the results to develop methods of substantially reducing the cost of fire protection to steelwork in fire. This objective can be considered to have been successfully met. The more general objective was to develop a better understanding of composite structure behaviour in fire. Whilst some of the extensive follow-on research studies have indeed greatly enriched understanding of structural behaviour in fire, they have largely been used

Table 2.2 Summary of results from fire tests on steel structure in Cardington laboratory

No.	Organization	Level	Time to maximum temperature	Gas	Steel	Maximum	Permanent
				Reached temperature, °C		Measured deformations	
1	BS	7	170	913	875	232	113
2	BS	4	125	820	800	445	265
3	BS	?	75	1020	950	325	425
4	BRE + SCI	3	114	1000	903	269	160
5	BRE	3	70	—	691	557	481
6	BS	2	40	1150	1060	610	—
7	CTU	4	55	1108	1088	>1000	925

BRE, Building Research Establishment, United Kingdom; BS, British Steel (now Tata Steel), United Kingdom; SCI, Steel Construction Institute, United Kingdom; CTU, E.U. collaborative research led by Czech Technical University in Prague.

(a) Severely distorted beams   (b) Squashed column

Figure 2.4 Damages to steel from fire in Cardington steel-framed structure. (From Newman, G.M., Robinson, J.T., and Bailey, C.G., *Fire Safe Design: A New Approach to Multi-Storey Steel-Framed Buildings*, 2nd ed., SCI Publication P288, Steel Construction Institute, Ascot, UK, 2006. With permission.)

in the context of reducing the cost of fire protection. A few of the economically less beneficial observations would not have been given prominence had fire-induced progressive collapse, such as that described in the next section of this chapter in relation to the World Trade Center buildings, not become such an important issue in structural fire safety design.

2.3.3.1 Reduction of fire protection

Among the many observations and conclusions that emerged from the Cardington structural fire test programme and related research studies, the ability of the composite floor slabs to develop tensile membrane action (TMA) in the wake of a yield line mechanism (Figure 2.5b), at large deflections (see Figure 2.5a) to bridge over the unprotected, fire-damaged composite beams was identified as a key effect to provide sufficient fire resistance without protecting many of the steel downstand beams. Although the theories of TMA were originally developed in the 1960s (Park 1964a, 1964b; Hayes 1968), it is this economic benefit of utilising TMA in fire-resistant design of steel-framed buildings that has been enthusiastically taken up by engineers. Among a large number of recent studies related to TMA in composite floor slabs under fire exposure, the work of Bailey and Moore (2000) has been the most influential due to its being taken up by practitioners in real projects, and the application of TMA to reduce the cost of fire protection is

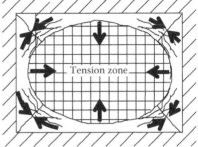

(a) Large deflection of Cardington floor slab
(Note near zero deflection around the edges)

(b) Tensile membrane action mechanism

Figure 2.5 Tensile membrane action in composite floor slab.

now widespread. Figure 2.6 shows two examples of the application of this method, one in the United Kingdom and one in China.

2.3.3.2 Fire-induced structural collapse

The Cardington structural fire tests not only demonstrated that it was feasible to reduce the cost of fire protection substantially by making use of alternative load-carrying mechanisms but also revealed some potential dangers of fire for structures that may initiate disproportionate collapse. It should be pointed out that there was no sign of disproportionate collapse of the Cardington test structures after any of the various fire tests. Nevertheless, the following

(a) Nuffield Hospital, Leeds, UK
(www.structuralfiresafety.com)

(b) Tongji University, China
(courtesy of Prof. G. Q. Li)

Figure 2.6 Examples of application of tensile membrane action to reduce fire protection. (Figure 2.6a from http://www.mace.manchester.ac.uk/project/research/structures/strucfire/. With permission. Figure 2.6b courtesy of Professor G. Q. Li.)

(a) Floor cracks in the steel-framed
structure (Newman et al. 2006)

(b) Fractured connection in the steel-
framed structure (Newman et al. 2006)

(c) Thermal bowing in the concrete
structure (Bailey 2002a)

Figure 2.7 Important damages to different Cardington structures after fire tests.
(Figures 2.7a and 2.7b from Newman, G.M., Robinson, J.T., and Bailey, C.G.,
Fire Safe Design: A New Approach to Multi-Storey Steel-Framed Buildings, 2nd
ed., SCI Publication P288, Steel Construction Institute, Ascot, UK, 2006.
With permission. Figure 2.7c from Bailey, C.G., *Design of Steel Structures
with Composite Slabs at the Fire Limit State*, Final Report prepared for the
Department of the Environment, Transport and the Regions, and the Steel
Construction Institute, Report No. 81415, Building Research Establishment,
Garston, Watford, UK, 2000. With permission.)

observed phenomena from the Cardington tests do have the potential to cause
disproportionate structural collapse if they occur in other circumstances.

Figure 2.7a shows a large crack in the floor of the Cardington steel-
framed structure after a test. Integrity failure of the floor under very large
deflections and rotations (which are necessary for TMA to develop) may
lead to fire spread from one floor to another. Figure 2.7b shows connection
fracture that occurred due to the large tensile forces that developed in the

attached beam during cooling. Connection fracture occurred because the tensile force in the connection was not considered in design. Fracture of the connections to the main beams may cause these beams to detach from the remaining structure, which may lead to disproportionate collapse. Figure 2.7c shows the reinforced concrete building columns being pushed outwards by the thermal expansion of the connected beams during heating, after a fire test (Bailey 2002a). Although this could also have happened to steel-framed structures, the consequences for concrete-framed structure would be more severe because if a concrete column is not designed to resist this horizontal force, extensive cracking could develop in the column under bending, leading to its premature failure.

2.4 WORLD TRADE CENTER COLLAPSES, 11 SEPTEMBER 2001

The Cardington structural fire research programme helped to find means of substantially eliminating fire protection to steel structures while still meeting the regulatory safety demands of the time. However, the Cardington structural fire research programme also revealed a few modes of structural member failure that had not previously been considered, because no code rules existed, in conventional structural fire resistance design. It is no exaggeration to say that had the World Trade Center buildings not collapsed on 11 September 2001, fire-induced structural collapse would not have been treated as a credible design requirement, and there would have been no enthusiasm for continued research on structural robustness in fire.

Vigorous research studies are now being conducted to understand the mechanisms of the World Trade Center building collapses and to develop means of reducing the risk associated with disproportionate collapse. This section briefly explains possible triggering mechanisms of the collapses of the World Trade Center Buildings 1, 2 and 7 based on interpretation of the FEMA (FEMA 2002) and NIST (NIST 2005, 2008) reports. Chapter 10 of this book examines means of controlling fire-induced structural collapse in more detail.

2.4.1 World Trade Center Buildings 1 and 2

Under the expected design fire condition, the floor trusses in WTC buildings 1 and 2 ("the Twin Towers") would have been designed to resist the applied vertical load in bending, with the floor system providing lateral support to the columns so that their buckling length would be equal to or less than one floor height. However, deterioration and subsequent impact by flying debris from the airplane would have caused nearly all of the fire protection on the floor trusses to be substantially damaged or removed

Figure 2.8 Damaged sprayed fire protection material. (From National Institute of Standards and Technology (NIST), *Federal Building and Fire Safety Investigation of the World Trade Center Disaster: Final Report of the National Construction Safety Team on the Collapses of the World Trade Center Towers (Draft)*, NIST NCSTAR 1, National Institute of Standards and Technology, Gaithersburg, MD, USA, 2005.)

(Figure 2.8), so that the floor trusses could have experienced much higher temperatures than the limiting temperatures allowed for under the design assumption. Consequently, the floor trusses would have to develop catenary action (Figure 2.9a) to resist the applied floor load. Due to this development of catenary action, the role of the floor trusses in the vertical load-carrying system would change from being beneficial to being doubly detrimental: (1) The floor trusses applied additional horizontal pulling forces on the columns; and (2) the floor trusses no longer provided lateral support to the columns, thus increasing their buckling lengths to a few floor heights. This role reversal led to dramatic reduction in the load-carrying capacity of the system of columns. Subsequent numerical simulation (Figure 2.9b; Usmani et al. 2003) have been able to re-create this triggering mechanism. Once the columns across the floors directly involved in fire buckled, it would not be possible to arrest total collapse due to the very large dynamic forces produced once column buckling was initiated.

Two issues are therefore crucial to understanding the causes of progressive collapse of these two buildings, yet they do not form part of any regulatory requirement of structural fire safety design. Their implications should be addressed in performance-based fire engineering design of structures. These two issues are

• Durability of fire protection materials;
• Changes in structural loading and support conditions during the course of fire exposure.

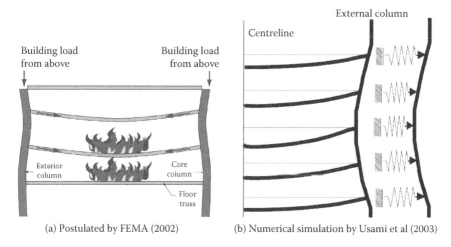

(a) Postulated by FEMA (2002)   (b) Numerical simulation by Usami et al (2003)

Figure 2.9 Possible trigger to progressive collapse of WTC Buildings 1 and 2. (Figure 2.9a is from http://www.fema.gov/news/event.fema?id=127. Figure 2.9b from Usmani, A.S., Chung Y.C., and Torero, J.L., How Did the WTC Towers Collapse: A New Theory, *Fire Safe Journal*, 38, pp. 501–533, 2003. With permission.)

2.4.2 World Trade Center Building 7

The probable initiating event to World Trade Center Building 7 collapse was attributed to failure of the connection of a primary beam to a key column, Column 79. Figure 2.10a shows a typical floor plan. The connections to Column 79 used seated connections with a top clip, whose details are shown in Figure 2.10b.

The connection would have been designed to resist gravity load. The sprayed fire protection material, as shown in Figure 2.11, was considered intact and well maintained after renovation in 1989. Also, the fire temperatures were relatively low. Therefore, it would not have been possible for the steel structure to fail under the assumed load-carrying mechanism of bending in the beam and vertical shear in the connection.

The probable cause of the primary beam's connection failure to Column 79 was attributed to a lateral force in the beam resulting in horizontal shear failure of the connection. The primary beam (between Column 76 and Column 79) was non-composite and supported a number of secondary beams. During heating, these secondary beams expanded, but the expansions were restrained by the primary beam, resulting in axial compression forces being generated in them; these forces would be quite large due to the large sizes and long spans of these secondary beams. Because of a lack of shear connector between the primary beam

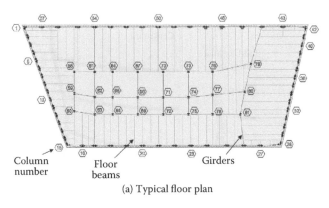

Column number   Floor beams   Girders

(a) Typical floor plan

(b) Details of seated connection with top clip (STC) with indication of forces

Figure 2.10 Key construction details around Column 79. (From National Institute of Standards and Technology (NIST), *Federal Building and Fire Safety Investigation of the World Trade Center Disaster: Structural Response and Probable Collapse Sequence of World Trade Center Building 7*, NIST Report NCSTAR 1-9(2), National Institute of Standards and Technology, Gaithersburg, MD, USA, 2008.)

and the floor slab, the compression forces from the secondary beams were transmitted to the bolts in the seated connections at the ends of the primary beam (shown by the arrows in Figure 2.10b). These positioning bolts had very little shear resistance. They would easily have been fractured by the forces from the secondary beams. Once the bolts were fractured, the primary beam would be detached from, and hence lose support to, Column 79. Repeated occurrence of this sequence on a number of lower floors would result in Column 79 being unsupported over these

Figure 2.11 Probable state of sprayed fire protection in WTC Building 7 on the day of fire attack. (From National Institute of Standards and Technology (NIST), *Federal Building and Fire Safety Investigation of the World Trade Center Disaster: Structural Response and Probable Collapse Sequence of World Trade Center Building 7*, NIST Report NCSTAR 1-9(2), National Institute of Standards and Technology, Gaithersburg, MD, USA, 2008.)

floors, and Column 79 would eventually collapse (Figure 2.12), bringing down these floors and causing the cascading progressive collapse as the unheated floors below were not able to sustain the dynamic forces generated by the falling weights of the structure above.

2.5 WINDSOR TOWER FIRE, MADRID, 12 FEBRUARY 2005

The Windsor Tower (or Edificio Windsor) was a 32-storey, 106-m high building with a reinforced concrete (RC) central core (Intemac 2005). A typical floor was constructed of two-way-spanning 280-mm deep waffle slabs supported by the concrete core, internal RC columns with additional 360-mm deep steel I-beams and steel perimeter columns. Originally, the perimeter columns and the internal steel beams were left unprotected in accordance with the Spanish building code at the time of construction. The building (see Figure 2.13a) featured two heavily reinforced concrete transfer levels (technical floors) between the 2nd and 3rd floors and between the 16th and 17th floors, respectively. The original cladding system had been fixed to the steel perimeter columns and the floor slabs. The building had been refurbished and reclad, and the reinstatement was nearly complete when the fire occurred.

The Windsor Tower's original structural design complied with the Spanish building codes in the early 1970s. At the time of original construction, the

Col 81   Col 80   Col 79

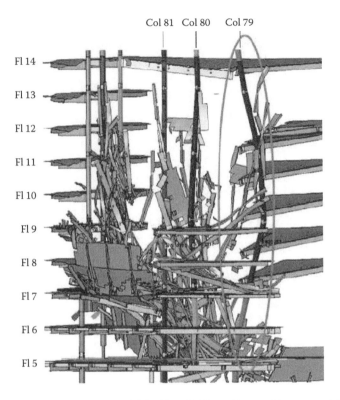

Figure 2.12 Simulated failure of WTC Building 7. (From National Institute of Standards and Technology (NIST), *Federal Building and Fire Safety Investigation of the World Trade Center Disaster: Structural Response and Probable Collapse Sequence of World Trade Center Building 7*, NIST Report NCSTAR 1-9(2), National Institute of Standards and Technology, Gaithersburg, MD, USA, 2008.)

Spanish codes did not require fire protection to steelwork or sprinkler fire protection to the building. The steelwork was in the process of being protected, although floors above level 19 (and floor 9) had not been protected. Some vertical fire stopping had not been installed. In particular, the gaps between the new cladding and the floor slabs were not fire-stopped. In fact, these weak links in the fire protection of the building were being rectified in the refurbishment project at the time of the fire. Since the building adopted an open-plan office concept, the fire compartmentation could only be floor by floor, a footprint of about 40 × 25 m. However, vertical compartmentation was clearly incomplete due to the lack of fire stopping in floor openings and between the new cladding and the floor slabs.

The fire is believed to have been caused by an electrical short circuit. It was reported that the fire started at 2300 on the 21st floor. Within 1 hr,

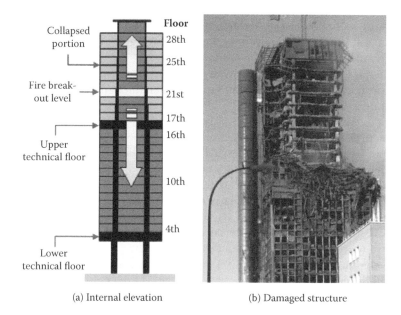

Floor

Collapsed portion ← 28th

25th

Fire break-out level → 21st

17th
16th

Upper technical floor →

10th

4th

Lower technical floor →

(a) Internal elevation   (b) Damaged structure

Figure 2.13 Madrid Tower. (From Instituto Tecnico De Materiales Y Construcciones (Intemac), *Fire in the Windsor Building, Madrid: Survey of the Fire Resistance and Residual Bearing Capacity of the Structure after the Fire*, Intemac, Spain, 2005. With permission.)

all floors above the 21st floor were on fire. In the following hours, the fire gradually spread downwards to the lower technical floor at level 3.

It is believed that the multiple-floor fire, along with simultaneous buckling of the unprotected steel perimeter columns on several floors, triggered the collapse of the floor slabs above the 17th floor (Figure 2.13b).

Whilst progressive collapse of a large part of the structure may be directly attributed to lack of fire stopping between the floors and the unprotected steel perimeter columns, it illustrates the danger of fire spread and the lack of stability of the vertical perimeter column system.

2.6 SUMMARY AND CONTEXT OF THIS BOOK

As explained in this chapter, recent developments in structural fire engineering practice may be roughly divided into three phases. They are as follows:

Phase 1: Element-based structural behaviour in fire: This has enabled a slightly more realistic consideration of fire behaviour and structural member behaviour than in the prescriptive treatment based on standard fire resistance tests.

Phase 2: Exploitation of alternative load paths in whole-structure behaviour in fire to enable reductions in the cost of fire protection while maintaining the required level of structural fire safety implicit in the regulatory fire resistance requirements.

Phase 3: Understanding the causes of fire-induced disproportionate structural collapse and developing means of engineering robust structures under fire attack.

Structural fire engineering research has developed to such an extent that rational treatment of element-based structural behaviour in fire, under well-defined loading and boundary conditions, is now part of the codified standard knowledge of structural fire engineering. Chapters 3–6 of this book deal with these topics. Chapter 3 deals with fire dynamics relevant to structural fire engineering; Chapter 4 introduces relevant heat transfer topics; Chapter 5 provides material properties at high temperatures; and Chapter 6 presents element-based design methods.

Developments in Phases 2 and 3 deal with variable loading and boundary conditions of structural members; structural fire engineers increasingly resort to numerical modelling of part- or whole-structural behaviour in fire. Chapter 7 is devoted to effective modelling of structural behaviour in fire, considering material and geometrical non-linearities and complex interactions between different structural members in fire.

Exploiting whole-structure behaviour to reduce the cost of structural fire protection is now firmly established in much of structural fire engineering design. Chapter 11 of this book provides a practical example of performance-based structural fire engineering using this technique.

In contrast, fire-induced disproportionate structural collapse still requires substantial research effort. Understanding fire-induced disproportionate structural collapse inevitably involves dealing with the two interrelated topics: fire integrity and structural robustness. The former is concerned with preventing fire spread from one fire-resistant compartment to another, and the latter is concerned with limiting any fire-induced structural collapse to remain proportionate to the initiating event. Chapters 9 and 10 present our thinking on these two issues. Joint (connection) behaviour in fire underlines every aspect of whole-structure behaviour, and Chapter 8 of this book is devoted to this topic.

Chapter 3

Introduction to enclosure fire dynamics

3.1 INTRODUCTION

To design structures to resist fire loading, an assessment of the likely gas temperatures to which they may be subjected is required. This is turn requires consideration of the behaviour of fires that may occur within structures. Since modelling burning in compartment fires presents considerable challenges to fire scientists, for structural fire engineering application, grossly simplifying assumptions have to be made to enable usable gas temperature-time relationships to be obtained. These relationships are then used as the thermal boundary conditions of the structure in subsequent calculations. This chapter discusses the various phenomena that govern the behaviour of fires that may occur in buildings and presents the models that are available to represent such fires. The models are presented broadly in the order in which they were developed, with emphasis placed on the more recently developed models of fires because these are most applicable to performance-based design.

3.2 STANDARD FIRES

The Standard Fire Curve is the best-known and most widely used method of estimating temperatures in building fires. It assumes that the temperature in a fire compartment is uniform, and that it increases indefinitely with time according to a logarithmic relationship (Figure 3.1). The Standard Fire curve has been incorporated (with minor differences) into a number of design standards worldwide. In Eurocode 1 (Committee of European Normalization [CEN] 1991–2002), the gas temperature (θ in degrees centigrade) at time t in minutes is given as

$$\theta = 20 + 345 \big[\log(8t + 1) \big] \tag{3.1}$$

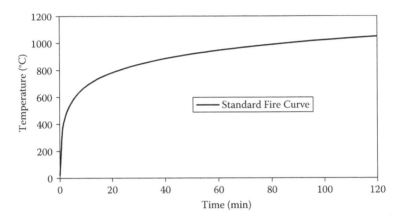

Figure 3.1 The Standard Fire temperature-time curve as defined in Eurocode EN 1991-1-2. (From Committee of European Normalization (CEN), *EN 1991-1-2-2002, Eurocode 1: Actions of Structures, Part 1–2: Actions of Structures Exposed to Fire*, CEN, Brussels, 1991–2002.)

This form of temperature-time relationship was originally derived from estimates of temperatures in tests undertaken early in the 20th century and has been shown to have only a very limited similarity to the temperatures in real compartment fires. Notable shortcomings in the Standard Fire curve include the lack of a cooling branch and no consideration of any of the physical parameters that govern fire behaviour, such as fuel load or available ventilation. Thus, it is often referred to as a "nominal" temperature-time curve. Other nominal fire models are available for particular cases, such as the hydrocarbon curve and the external fire exposure curve in Eurocode EN 1991-1-2 (CEN 1991–2002). These can be similarly criticised.

Since the mid-1990s, performance-based methods have been introduced to structural fire engineering. Adopting the Standard Fire curve in performance-based design is difficult to justify on scientific grounds due to its lack of similarity with real fires. Despite this, its use remains widespread, mainly due to its long history and the familiarity with using the standard fire condition by all concerned with fire safety design, particularly control authorities.

For performance-based fire engineering of structures, more appropriate fire models should be used. These are discussed in the remainder of this chapter.

3.3 FIRES IN SMALL COMPARTMENTS

Most building structures are divided into a number of compartments, such as offices, meeting rooms and the like. From a fire safety perspective, these divisions are significant for two reasons. Firstly, they provide a means of

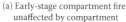
(a) Early-stage compartment fire unaffected by compartment

(b) Pre-flashover fire

(c) Flashover results in fully involved burning

Figure 3.2 Growth of a compartment fire.

preventing fire spread and hence often allow a structure to be designed on the assumption that a fire will only occur in a single compartment. Secondly, a fire occurring within a compartment (a "compartment fire") will develop in a different manner from a fire occurring in an unrestricted space. Historically, research into the behaviour of compartment fires has focused on small compartments and assumed a uniform distribution of fuel over the floor of the compartment. This work has informed most of the design fires currently used in structural design and is discussed in this section. Section 3.4 considers the behaviour of fires in the large compartments that are increasingly present in modern structures.

An uninterrupted compartment fire goes through a number of stages. Immediately after ignition, the behaviour of a fire in a small compartment will be unaffected by the compartment boundaries. However, hot gases will not be free to escape from the compartment but, due to buoyancy effects, will begin to accumulate rapidly in a layer under the ceiling. As the fire grows in size and this layer of gases develops, a point is reached at which the downward radiation and convection from the smoke layer and flames become sufficiently intense to ignite objects distant from the seat of the fire. Assuming an adequate supply of air, this results in ignition of all combustible materials in the compartment. The transition from localised to fully involved burning tends to be rapid and is known as "flashover". Figure 3.2 shows the development of a compartment fire.

After flashover, the speed of burning, as measured by the rate of heat release, within the compartment is governed either by supply of oxygen through openings such as doors and windows or by the amount of fuel available. The former case is known as a ventilation-controlled fire and the latter a fuel-controlled fire. The peak heat release rates in fuel-controlled fires will depend on the type and arrangement of the fuel.

3.3.1 Heat release rate

The heat released in the preflashover phase of a compartment fire is generally small in comparison to that released postflashover and is therefore normally ignored in structural fire engineering applications. Consequently, fire models for use in structural engineering generally only consider fully developed fires.

The heat release rate in fully developed building compartment fires is a dominant factor in determining the gas temperatures that will occur. Thus, some estimate of this rate is needed for all performance-based structural fire engineering. The most widely quoted relationship between the rate at which fuel is consumed (kilograms per second) and the available ventilation is that of Kawagoe (1958), reported, for example, by Drysdale (1998) as

$$\dot{m} = KA_v\sqrt{H} \tag{3.2}$$

where A_v is the area of the vertical ventilation opening (square meters) and H its height (m). For a ventilation-controlled fire, assuming a stoichiometric ratio of air supply to complete burning of wood, an approximate value of $0.09 \text{ kg}^{-1}\text{m}^{-5/2}$ (DiNenno 2002) may be used for the constant K. However, in reality, the value of K can vary over a large range. By multiplying the rate of mass loss by the heat of combustion of the fuel h_c, an estimate of the rate of heat release within a compartment fire is obtained as

$$\dot{q} = KA_v\sqrt{H}h_c \tag{3.3}$$

For structural fire engineering applications, the fuel load in a compartment is converted, based on equivalent calorific value, to an equivalent wood fuel load for which the value of h_c is approximately 18 MJ/kg.

Kawagoe derived Equation (3.2) from fire tests in small compartments, so the application of Equations (3.2) and (3.3) to typical fire compartments, which are comparatively large, can be questioned. Various more refined relationships similar to Kawagoe's are available that take account of factors such as the observed dependence of heat release rate on the total internal area of a compartment and on radiation reflected from compartment boundaries. Fuller discussions of these and other effects are given by DiNenno (2002) and Drysdale (1998). However, Equation (3.2) does enable the most influential factor that governs compartment fire behaviour to be identified. This is the ventilation factor $A_v\sqrt{H}$. As shown in this chapter, this factor features in all approximate models of compartment fire where ventilation control is assumed.

3.3.2 Gas temperatures: Pettersson's method

Recognising that the Standard Fire curve was not physically reasonable, researchers in Sweden in the 1970s (Pettersson et al. 1976) developed a method of predicting compartment fire gas temperatures by considering

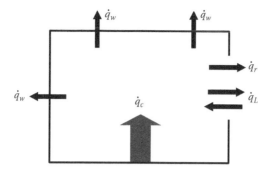

Figure 3.3 Pettersson et al.'s heat balance model for a compartment fire. (From Pettersson, O., Magnuson, S.E., and Thor, J., *Fire Engineering Design of Structures*, Publication 50, Swedish Institute of Steel Construction, Stockholm, 1976.)

the heat balance in a fire compartment based on Kawagoe's equation for heat release rate. By assuming that the temperature within a fire compartment is uniform, all available fuel is burnt within the compartment, and the thermal properties of the compartment walls are uniform, the heat balance can be expressed as

$$\dot{q}_c = \dot{q}_L + \dot{q}_W + \dot{q}_R \tag{3.4}$$

where $\dot{q}_c$ is the rate of heat release due to combustion, $\dot{q}_L$ is the rate of heat loss due to the replacement of hot gases by cold, $\dot{q}_W$ is the rate of heat loss through the compartment boundaries and $\dot{q}_R$ is the rate of heat loss due to radiation through the compartment openings (Figure 3.3). By evaluating these terms, it is possible to arrive at a differential equation that relates the temperature within a compartment to the fuel load, the available ventilation and the thermal properties of the compartment walls. The solution to this equation cannot be expressed explicitly, so compartment temperatures based on Pettersson's approach are normally presented graphically for various fire loads and ventilation conditions (Pettersson et al. 1976; more readily available in Drysdale 1998). This description of a compartment fire is sometimes referred to as a "natural fire" to distinguish it from the Standard Fire curve.

3.3.3 Gas temperatures: Eurocode parametric curves

Pettersson's model forms the basis of the Eurocode 1 (CEN 1991–2002) parametric fire model, which fits an equation to the graphical temperature-time curves presented by Pettersson. The input variables are very similar and take account of fire load, ventilation conditions and the thermal properties of the compartment.

The method is based on the expression

$$\theta = 20 + 1325(1 - 0.324e^{-0.2t^*} - 0.204e^{-1.7t^*} - 0.427e^{-19t^*}) \tag{3.5}$$

for the heating phase of a fire, in which θ is the compartment gas temperature (degrees centigrade), and t^* is a non-physical time parameter, related to the physical time t through

$$t^* = t\Gamma \tag{3.6}$$

where t is given in hours, and Γ is a dimensionless quantity determined by the various input data for the fire compartment being considered, which depends on whether the fire is predicted to be ventilation or fuel controlled. This in turn is governed by the time t_{max} at which the peak gas temperature occurs and cooling begins. For ventilation-controlled fires, t_{max} is determined by

$$t_{max} = 0.2 \times 10^{-3} q_{td} \frac{A_t}{A_v\sqrt{h_{eq}}} \tag{3.7}$$

where q_{td} is the design fire load in the compartment in terms of megajoules per square meter of the *total* compartment internal surface area A, A_v is the area of ventilation openings and h_{eq} is the weighted average height of the ventilation openings. However, if

$$t_{lim} > 0.2 \times 10^{-3} q_{td} \frac{A_t}{A_v\sqrt{h_{eq}}} \tag{3.8}$$

then, the fire is fuel controlled, and

$$t_{max} = t_{lim} \tag{3.9}$$

Values of t_{lim} depend on whether a fire is considered to grow at a slow, medium or fast rate. Guidance is given in Eurocode 1 (CEN 1991–2002) on growth rates for different types of fire and examples are repeated in Table 3.1.

If a fire is ventilation controlled and in a compartment with boundaries made of uniform, thermally thick materials, Γ is given by

$$\Gamma = 8.41 \times 10^8 \left(\frac{A_v}{A_t}\right)^2 \left(\frac{h_{eq}}{\rho c \lambda}\right) \tag{3.10}$$

Table 3.1 Examples of fire growth rates for use with Eurocode parametric fires

Occupancy type	Growth rate	t_{lim}, h
Public space	Slow	0.417
Classroom, office, hotel bedroom	Medium	0.333
Library, cinema, shopping centre	Fast	0.250

Source: Committee of European Normalization (CEN) (1991–2002), *EN 1991-1-2-2002, Eurocode 1:Actions of Structures, Part 1–2: Actions of Structures Exposed to Fire*, CEN, Brussels.

where λ, ρ and c are, respectively, the conductivity, density and specific heat capacity of the compartment boundaries.

In the fuel-controlled case,

$$\Gamma = 8.41 \times 10^{8} \left(\frac{0.1 \times 10^{-3} q_{td}}{t_{lim}} \right)^{2} \frac{1}{\rho c \lambda} \tag{3.11}$$

which is an expression that does not include the opening factor of the compartment, indicating that the temperature-time curve is not governed by the availability of ventilation to the fire. An effect of the different definitions of Γ for fuel- and ventilation-controlled fires is to introduce a potential step change in the predicted temperature-time curves when small changes are made to the input data (Franssen and Real, 2010). This is illustrated in the example that follows.

The cooling phase of a parametric fire is assumed to be linear with time and is given by

$$\theta = \theta_{max} - 625(t^{*} - t_{max}^{*}x) \qquad \text{for } t_{max}^{*} \leq 0.5$$

$$\theta = \theta_{max} - 250(3 - t_{max}^{*})(t^{*} - t_{max}^{*}x) \qquad \text{for } 0.5 < t_{max}^{*} < 2 \tag{3.12}$$

$$\theta = \theta_{max} - 250(t^{*} - t_{max}^{*}x) \qquad \text{for } t_{max}^{*} \geq 2$$

where $x = 1$ for $t_{max} > t_{lim}$, and $x = t_{lim}\Gamma/t_{max}^{*}$ for $t_{max} - t_{lim}$.

Complete details of the Eurocode EN 1991-1-2 parametric fire curves are given in the code and are also presented by Franssen and Zaharia (2005) in a clearer manner. The previous description highlights the key points but does not repeat a number of restrictions and details that are present in the Eurocode. Some of these, such as limits on the thermal properties of compartment walls, are based on physical considerations; others, such as the

Table 3.2 Indicative thermal properties of generic fire protection materials

Generic material	Density, kg/m³	Thermal conductivity, W/(m.K)	Specific heat, J/(kg.K)	Moisture content, % by wt
Sprayed mineral fibre	250–350	0.1	1050	1.0
Vermiculite slabs	300	0.15	1200	7.0
Vermiculite/gypsum slabs	800	0.15	1200	15.0
Gypsum plaster	800	0.2	1700	20.0
Mineral fibre sheets	500	0.25	1500	2.0
Aerated concrete	600	0.3	1200	2.5
Lightweight concrete	600	0.8	1200	2.5
Normal weight concrete	2200	1.7	1200	1.5

Source: Lawson, R.M., and Newman, G.M. (1996), *Structural Fire Design to EC3 & EC4, and Comparison with BS 5950*, Technical Report, SCI Publication 159, Steel Construction Institute, Ascot, UK.

maximum compartment size to which the parametric fire curves may be applied (500 m²), appear less easily justifiable.

In Chapter 5, temperature-dependent thermal properties of a number of construction materials are given. For the purpose of quantifying compartment fire temperature-time relationships, where the effects of compartment construction materials are relatively small and simplicity is desirable, it may be assumed that the material thermal properties are temperature independent, and the indicative values in Table 3.2 may be used.

Whilst being simple, the accuracy of the Eurocode parametric approach is considered acceptable in comparison with test results (Lennon and Moore 2003) and computational fluid dynamics (CFD) simulations (Pope and Bailey 2006). Its simplicity has allowed it to be used successfully in a wide range of structural fire calculations. The Eurocode parametric fire curves allow for the effects of the key phenomena that govern compartment fire behaviour (fuel load, ventilation and compartment characteristics), and they include a cooling phase. However, one key shortcoming is the restriction on the size of fire compartment to which they are applicable. Many modern office spaces are considerably larger than the 500-m² limit that is specified in the code. Use of the parametric curves in larger spaces is sometimes allowed (British Standards Institution [BSI] 2007), but this will often result in an unrealistically onerous design fire. Conversely, the effects of localized cooling, which may be structurally more severe than assuming a uniform fire compartment temperature (Röben 2010), cannot be taken into consideration if a Eurocode parametric fire is used. The travelling nature of fires in large compartments also cannot be captured. Section 3.4 of this chapter addresses these two issues.

Example 3.1

Produce a Eurocode parametric temperature-time curve for a compartment with the following properties:

Length	4.75 m
Breadth	3.5 m
Height	2.45 m
Boundary material	Normal-weight concrete (ρ = 2300 kg/m³, λ = 1.5 W/mK, c = 1000 J/kgK)
Fire load	32 kg/m² wood over the floor area with heat of combustion 18 MJ/kg
Ventilation	One door 1 by 2 m high, one window 2.35 by 1.18 m high
Use	Living room

Calculations:

$$A_v = (1 \times 2) + (2.35 \times 1.18) = 4.77 \text{ m}^2$$

$$A_t = (3.5 \times 4.75 \times 2) + (3.5 \times 2.45 \times 2) + (4.75 \times 2.45 \times 2) = 73.68 \text{ m}^2$$

$$q_{td} = (32 \times 18) \times (4.75 \times 3.5)/A_t = 130 \text{ MJ/m}^2$$

$$h_{eq} = [(1 \times 2 \times 2) + (2.35 \times 1.18 \times 1.18)]/A_v = 1.52 \text{ m}$$

Use as a living room gives t_{lim} = 0.333 hr and

$$0.2 \times 10^{-3} q_{td} \frac{A_t}{A_v \sqrt{h_{eq}}} = 0.2 \times 10^{-3} \times 130 \times \frac{73.68}{4.77 \times \sqrt{1.52}} = 0.326 < t_{lim}$$

Therefore, t_{max} = 0.333 hr = 20 min and the fire is fuel-controlled. This means that

$$\Gamma = 8.41 \times 10^8 \left(\frac{0.1 \times 10^{-3} q_{td}}{t_{lim}} \right)^2 \frac{1}{\rho c \lambda}$$

$$= 8.41 \times 10^8 \left(\frac{0.1 \times 10^{-3} \times 130}{0.333} \right)^2 \frac{1}{2300 \times 1000 \times 1.5} = 0.3709$$

which allows the temperature-time curve of the heating phase of the fire to be determined using Equations (3.5) and (3.6) with suitably small time increments. The peak temperature occurs at the end of the heating phase (20 min) and is 647°C. This time also indicates the start of the cooling phase of the fire, which is linear according to Equations (3.12) with $t_{max}^* = 0.333 \times 0.3709 = 0.124$.

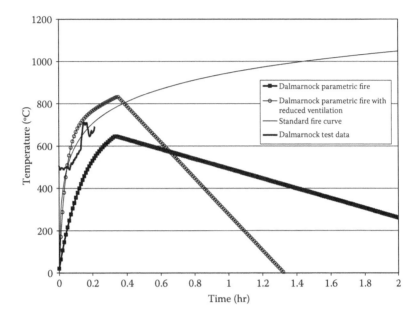

Figure 3.4 Predicted temperatures within a compartment for two ventilation conditions using the Eurocode parametric approach. The Standard Fire curve is
shown for comparison as well as experimental data from a fire in a similar
compartment.

Example 3.1 is an approximation of the Dalmarnock Test 1 fire undertaken in 2007 (Abecassis-Empis et al. 2008), which burned real living room
furniture in a very heavily instrumented compartment. Data from this test
(limited to 0.2 hr due to firefighting) is shown in Figure 3.4, which also
shows the Parametric Fire calculated previously, the Standard Fire curve
and a Parametric Fire for the example but with ventilation restricted just
sufficiently to result in a ventilation-controlled fire. This last case results if
the input data provided are changed so that the ventilation conditions are
as follows:

Ventilation   One door 0.92 by 2 m high, one window 2.22 by
1.18 m high.

The following changes then occur in the calculations:

$A_v = (0.92 \times 2) + (2.22 \times 1.18) = 4.46$ m²

$h_{eq} = [(0.92 \times 2 \times 2) + (2.22 \times 1.18 \times 1.18)]/A_v = 1.52$ m which leads to

$$0.2 \times 10^{-3} q_{td} \frac{A_t}{A_v \sqrt{h_{eq}}} = 0.2 \times 10^{-3} \times 130 \times \frac{73.68}{4.46 \times \sqrt{1.52}} = 0.348 > t_{lim}$$

Thus, t_{max} is now greater than t_{lim}, and the fire is ventilation controlled. This means

$$\Gamma = 8.41 \times 10^{8} \left(\frac{A_v}{A_t} \right)^2 \left(\frac{h_{eq}}{\rho c \lambda} \right)$$

$$= 8.41 \times 10^{8} \left(\frac{4.46}{73.68} \right)^2 \left(\frac{1.52}{2300 \times 1000 \times 1.5} \right)$$

$$= 1.358$$

which again allows the temperature-time curve to be calculated according to Equations (3.5) and (3.6). It is noticeable that a small change in input parameters has resulted in a large change in the predicted temperature-time curve as a result of the different ways in which the value of Γ is calculated for fuel- and ventilation-controlled conditions. It is difficult to reconcile this step change with the behaviour of real fires.

3.3.4 Decay

The decay period for compartment fires has received less attention than the growth and postflashover periods. However, from a structural engineering perspective, cooling is an important part of fire behaviour due to the very large plastic strains that may develop in structural elements during heating. On cooling, these unrecoverable strains can produce substantial tensile forces, which may result in tensile failure of connections or other components, as was observed, for example, in the Cardington fire tests (British Steel 1998).

   If combustion is completed naturally, it is probable that compartment gas temperatures will fall rapidly because of the low thermal inertia of gases and as a result of the continued supply of cool air from ventilation openings. However, this does not imply that structural members will no longer be subject to large heat fluxes because other material within a fire compartment will probably remain hot and continue to radiate strongly. Alternatively, very rapid cooling might result from firefighting with water. Details of the behaviour will probably be highly compartment dependent, and at present there appears to be no simplified quantitative description of such behaviour other than the assumption of linear cooling given in Eurocode EN 1991-1-2 (CEN 1991–2002).

3.4 FIRES IN LARGE COMPARTMENTS AND TRAVELLING FIRES

3.4.1 Horizontally travelling fires

Experimental and analytical evidence increasingly suggests that in larger fire compartments, such as typical office spaces, the assumption of uniform temperatures at any level within a compartment is not valid. Instead, fires in larger compartments will tend to travel within the compartment as fuel is consumed at a rate governed by the available ventilation. This causes variations in gas temperatures within such compartments that are not present in older descriptions of compartment fires such as those discussed above.

Clear experimental evidence of localised burning was presented by Cooke (1998), who undertook a number of fire tests with uniform fire loads of wooden cribs in a long, narrow compartment ($4.5 \times 8.75 \times 2.75$ m high) in which ventilation was provided at one end. The results showed a clear progression over time of localised high temperatures within the compartment. Peak values occurred near the source of ventilation early in the fire and then progressed away from the opening as fuel was consumed. This progression occurred even when ignition was distant from the source of ventilation. In this case, initial superficial burning took place advancing towards the source of ventilation. When fuel in this region ignited, a fully developed fire occurred, which then travelled away from the ventilation source. The progression of peak temperatures from the front to the rear of the compartment took 20–30 min, with higher levels of ventilation resulting in more rapid-fire spread.

Similarly, Welch et al. (2007) reported on the analysis of a series of fire tests undertaken in a $12 \times 12 \times 3$ m high compartment in which a combination of wood and plastic fuel was burnt. Ventilation was provided along either one or two walls of the compartment. The tests were heavily instrumented, which allowed both temperature and heat flux maps to be produced after various periods of burning. Figure 3.5 shows the manner in which a local area of peak temperature traversed the compartment from near the ventilation source to points distant from it over time. Although the differences in temperature were modest, when converted to incident heat fluxes on the compartment ceiling (Figure 3.6), very significant differences arose as a result of the dependence of radiation on the fourth power of absolute temperature.

Correspondingly, the effect of non-uniform gas temperatures on structural temperatures has been shown to be significant in work by Gillie and Stratford (2007), who reported on temperatures in a concrete slab above a fire compartment. Lower-surface concrete temperatures varied by as much as 400°C, despite a compartment size of only 3.5 by 4.5 m (Figure 3.7).

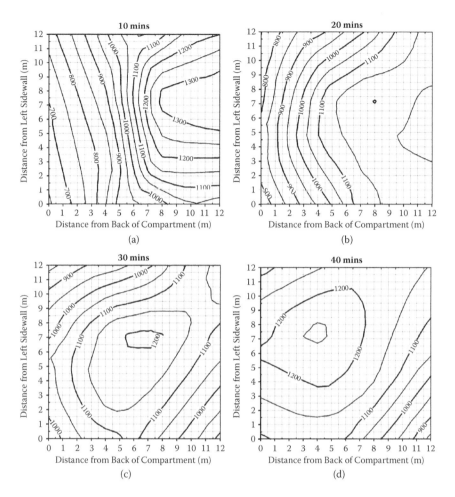

Figure 3.5 Measured gas temperature maps for a 12 ′ 12 ′ 3 m high compartment just under the ceiling at various times after ignition. Ventilation was provided from the right of each plot. (After Welch, S., Jowsey, A., Deeny, S., Morgan, R., and Torero, J.L., BRE Large Compartment Fire Tests—Characterising Post-Flashover Fires for Model Validation, *Fire Safety Journal*, 42, pp. 548–567, 2007, by permission of Elsevier.)

These results have all been recorded in compartments that are small in comparison to many modern office spaces; the effects of travelling fires will likely be more significant in even larger compartments. This suggests that travelling fires should be considered in structural design for large spaces, even though this is currently unusual.

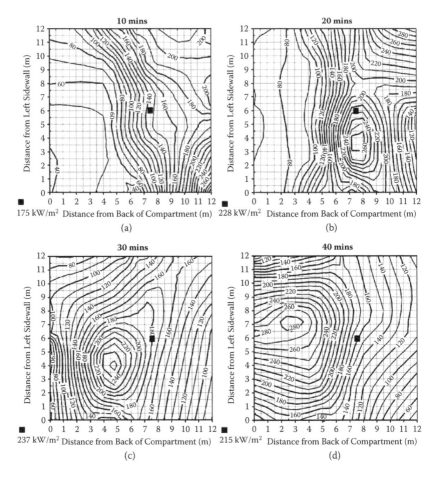

Figure 3.6 Heat flux maps for a 12 ′ 12 ′ 3 m high compartment just under the ceiling at various times after ignition. Bold numbers below the plots indicate readings from the billet at the position of the black square. Ventilation was provided from the right of. (After Welch, S., Jowsey, A., Deeny, S., Morgan, R., and Torero, J.L., BRE Large Compartment Fire Tests—Characterising Post-Flashover Fires for Model Validation, *Fire Safety Journal*, 42, pp. 548–567, 2007, by permission of Elsevier.)

3.4.1.1 Clifton's model

The earliest representation of travelling fires appears to be that of Clifton (1996), who developed a model for fires in large compartments in which the assumption of uniform burning cannot be applied. Clifton proposed that a fire in a large compartment could be described by dividing the compartment into a number of "design areas" and assuming that fully developed burning

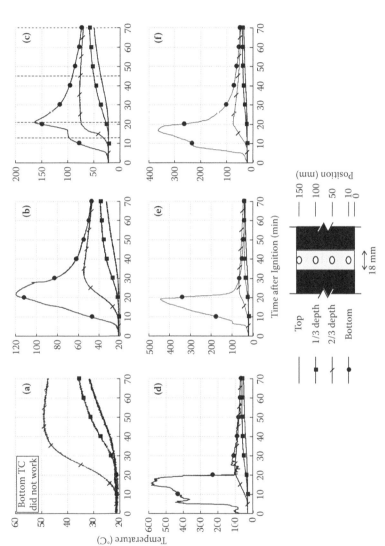

Figure 3.7 Temperature data from the Dalmarnock tests. In each case, temperatures were recorded at several depths (as indicated below the plots) with n a concrete slab heated by a compartment fire. (a)–(f) refer to different locations on the ceiling of the compartment, which had a plan area of approximately 3.5 ´ 4.5 m. TC = thermocouple.

occurs only in certain design areas at any time before moving to adjacent areas. The gas temperature-time relationship within a design area is assumed to follow a natural fire curve; a modified version of the then-current ENV Eurocode (pre-Eurocode) parametric fire was proposed as suitable by Clifton. The size of a design area is recommended to be the upper-bound area for which the natural fire being used is applicable, which is 100 m² in the case of the ENV Eurocode. Ventilation to a design area is taken to be available, either directly from external boundaries that have suffered glazing failure, or via burnt-out areas of the fire compartment. Based on these assumptions and a specified fire load, a time during which fully developed burning occurs in each design area is obtained, with 20 min found to be typical. Several further phenomena are discussed by Clifton and incorporated in the model with varying degrees of sophistication. These include "preheating" of design areas, cooling of design areas after burnout, time-varying external ventilation conditions, and application of the model to fire compartments with a variety of shapes. An example of the model is shown in conceptual terms in Figure 3.8.

Despite its pioneering nature and many merits, Clifton's model does not appear to have been widely used in structural fire engineering. This is perhaps due to the rather involved procedure necessary to implement it, limited availability of the original document describing the method and lack of robust experimental validation.

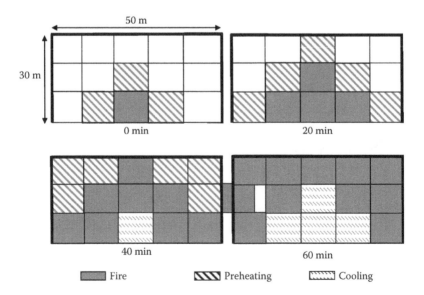

Figure 3.8 Conceptual illustration of Clifton's (1996) model of travelling fires applied to a compartment with ventilation on the lower edge.

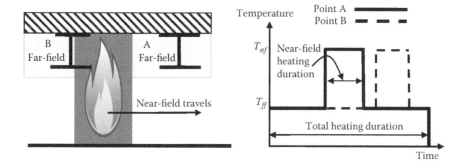

Figure 3.9 Conceptual illustration of Rein's travelling fire model. Left: The distribution of near- and far-field regions within a large compartment. Right: The variation of gas temperature over time at points A and B on the compartment ceiling.

3.4.1.2 Rein's model

More recently, Rein et al. (2007c) and Stern-Gottfried et al. (2009) developed an alternative method for modelling travelling fires in large compartments within building structures. They suggest that, due to localised burning, the gas temperatures to which the ceiling of a fire compartment is exposed consist of "near-field" temperatures, resulting from direct impingement of flames on the structure, together with "far-field" temperatures due to a layer of hot gases. Figure 3.9 shows this model of travelling fires in conceptual terms.

The near field only affects a percentage of the ceiling area at any time but travels within a compartment until the entire fuel load has been consumed. The size of the near field in a real fire would be governed by the available ventilation. However, since ventilation conditions in fire compartments are usually difficult to predict, the size of the near field is an input to the model that must be varied parametrically by the designer to determine the most structurally deleterious fire. The near-field temperature T_{nf} to be used in the model is dependent on flame temperature, so it is somewhat dependent on the type of fuel being consumed. For an office fire, 1200–1300°C (Drysdale 1998) is typical. The time when near-field temperatures are experienced by each point in a compartment depends on the fuel load density and the heat release rate per unit area. Based on the assumption that the fuel load is evenly distributed across the compartment, for a typical office fire Stern-Gottfried et al. (2009) give this time as 19 min, which corresponds well with the suggested time of 20 min for fully developed burning in each design area in Clifton's method.

Far-field temperatures are highly dependent on the compartment considered (Figure 3.10). The temperature of the gases that accumulate under a ceiling as the result of a fire plume decays as distance from the plume

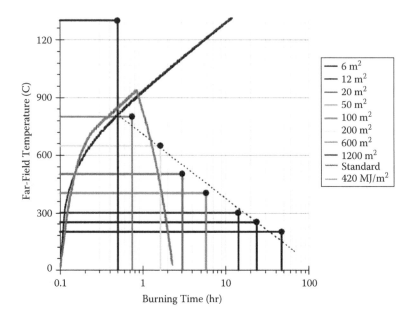

Figure 3.10 Predictions of far-field temperatures for different sizes of travelling fire in a 1250 m² compartment using a Rein near-field/far-field approach. Each curve represents the far-field temperature and total fire duration for an assumption about the size of near-field burning. The Standard Fire curve and Eurocode parametric curve predictions for a fire load 420 MJ/m² are shown for comparison. (After Rein, G., et al., *The Dalmarnock Fire Tests: Experiments and Modelling: Chapter 3: A Priori Modelling of Fire Test One*, School of Engineering and Electronics, University of Edinburgh, UK, 2007b. With permission.)

increases. Based on large-scale tests, Alpert (1972) suggested the following relationship between peak gas temperature T and radial distance from the centre of a fire plume r:

$$T - T_\infty = \frac{5.38}{H}\left(\frac{\dot{Q}}{r}\right)^{2/3}$$

(3.13)

where $\dot{Q}$ is the heat release rate in the fire plume, H is the height of the compartment and T_∞ is the temperature distant from the fire plume. Rein et al. suggest the heat release rate per unit area is around 500 kW/m² for a typical office fire. To produce a manageably simple description of compartment temperatures for use in design, Stern-Gottfried et al. (2009) simplified

Alpert's relationship by using a fourth-power average to give a single characteristic far-field temperature $\bar{T}_{ff}$ for a compartment.

$$\bar{T}_{ff} = \frac{\left[\int_{r_{nf}}^{r_{ff}} T^4 \, dr\right]^{0.25}}{\left(r_{ff} - r_{nf}\right)^{0.25}} \tag{3.14}$$

where r_{ff} is the radius of the far field (taken as the distance to the compartment boundary), and r_{nf} is the radius of the near field, as determined by the size of fire being considered. A fourth-power average was used in Equation (3.14) due to the fourth-power dependence of radiative heat transfer on absolute temperature. Since convective heat transfer depends just on temperature difference, the equation is thus a conservative approximation of true compartment temperatures. Substituting Equation (3.13) in Equation (3.14) and rearranging gives

$$(r_{ff} - r_{nf})\bar{T}_{ff}^4 = \int_{r_{nf}}^{r_{ff}} \left[\frac{5.38}{H}\left(\frac{\dot{Q}}{r}\right)^{\frac{2}{3}} + T_\infty\right]^4 dr$$

On integration, this gives

$$(r_{ff} - r_{nf})\bar{T}_{ff}^4 = \frac{3}{5}\left[\frac{5.38}{H}\dot{Q}^{\frac{2}{3}}\right]^4 \left(-r_{ff}^{-5/3} + r_{nf}^{-5/3}\right) + 4T_\infty\left[\frac{5.38}{H}\dot{Q}^{\frac{2}{3}}\right]^3 \left(-r_{ff}^{-1} + r_{nf}^{-1}\right)$$

$$+18T_\infty^2\left[\frac{5.38}{H}\dot{Q}^{\frac{2}{3}}\right]^2 \left(-r_{ff}^{-1/3} + r_{nf}^{-1/3}\right) + 12T_\infty^3\left[\frac{5.38}{H}\dot{Q}^{\frac{2}{3}}\right]\left(r_{ff}^{1/3} - r_{nf}^{1/3}\right) + T_\infty^4 \tag{3.15}$$

which may be used to evaluate $\bar{T}_{ff}$, with all temperatures in Kelvin.

The path that the near field follows within a compartment is not described in Rein's method and will in reality depend on the ventilation conditions, point of ignition and arrangement of fuel in the compartment. There is some experimental evidence that a fire will travel away from a ventilation source further into a compartment (Cooke 1998). Numerical analyses by Röben (2010) found that, in composite steel-concrete framed buildings, structural behaviour depends only weakly on the path taken by a linearly travelling fire. Law (2010) obtained similar results for travelling fires taking various paths in a concrete-framed structure, but more work is needed to generalize these results. Similarly, compartment temperatures during cooling are not directly addressed by Rein. Röben studied cases for which it was assumed that once the entire compartment has burned out, the gas temperature returned to ambient uniformly over the compartment.

Example 3.2

Determine the nature of the Rein fires for an office compartment 27 × 22.5 × 3 m high if the fuel load q is 570 MJ/m².
  Calculations are as follows:

  Assume: Heat release rate per unit area $\dot{Q}'' = 500$ kW/m²
  Fire size = 25% of total area
  Near-field temperature = 1200°C
  Far-field length r_{ff} (worst case) =27/2 = 13.5 m
  Burning time in a given area = $q/\dot{q}'' =$570 × 10⁶/500 × 10³ = 19 min
  Total duration of fire = 19/0.25 = 76 min

  For a circular fire,

  $$r_{nf} = [(27 \times 22.5 \times 0.25)/\pi]^{0.5} = 7 \text{ m}$$

so the far-field temperature will be given according to Equation (3.8) as 727°C if T_∞ is taken as 20°C.

Thus, in this case the near field will consist of 152 m² (= 27 × 22.5 × 0.25) of high (1200 °C) temperatures, while temperatures elsewhere are 727°C. Each part of the compartment will experience near-field temperatures for a period of 19 min. Rein's method does not specify the path the high-temperature region will follow within the compartment, and this must be specified by the analyst; as discussed, there is evidence that a path travelling away from the area of ventilation may be the most realistic. Rein's method aims to determine a "family" of fires with different sizes of near-field burning; similar calculations to those provided here produce the results in Table 3.3 for different assumptions about the near-field size.

Table 3.3 Possible fire design scenarios for a horizontal travelling fire

Percentage area burning	Total heat release rate, MW	Near-field temperature, °C	Far-field temperature, °C	Total duration of fire, min
5	15	1200	324	380
10	30	1200	462	190
25	76	1200	727	76
50	152	1200	1021	38
100 (uniform fire)	304	1200	1200	19

3.4.2 Vertically travelling fires

Although structures are typically designed with vertical compartmentation to prevent fire spread between floors, there are examples of accidental interfloor fire spread occurring and of multiple-floor fires. Such fires can occur as the result of malicious actions, as in the World Trade Center attacks; due to fire occurring when compartmentation has been removed for refurbishment, as occurred in the Windsor Tower fire (Intemac 2005); or simply due to a failure of compartmentation, such as during the fire that occurred in the School of Architecture at the University of Delft (Zannoni 2008). Given these events, and since the primary purpose of structural fire design is life safety, some consideration of interfloor fire spread may be desirable, particularly in multistorey structures, for which evacuation times are long and consequently structural stability is required for an extended period.

It appears that the only studies to date on vertical fire spread between floors have attempted to estimate the interfloor ignition times in the Windsor Tower fire. Fletcher et al. (2006) report on two studies that estimated the interfloor ignition times in this fire to be between 6 and 20 min. The large difference in these estimates probably reflects the difficulties in observing real fire behaviour in a multistorey building, and in any case interfloor spread times will probably vary widely according to specific building details. However, it is notable that even the largest estimate is shorter than a typical compartment fire. Therefore, it is reasonable to assume that if interfloor fire spread occurs, simultaneous fires on adjacent floors are likely. Röben et al. (2010) considered the implications of such fires for composite steel-concrete structures and found that varying the assumed interfloor spread rate affected the structural response, but they were unable to conclude whether a rapid or slow rate was structurally more serious.

3.5 COMPUTER MODELS OF COMPARTMENT FIRES

A variety of computer models of compartment fires is available. Software exists that can represent the very wide range of physical phenomena known to affect fire behaviour, including compartment geometry, heat release rates of burning fuel, complex ventilation conditions, turbulent gas flow, soot production and many others. Using such software is time consuming, and in many cases the complexity of the output makes it difficult to use in a design situation. For these reasons, computer models of fire behaviour are currently little used in structural design work except for very unusual structural layouts.

Pettersson's model of a compartment fire is the simplest example of a class of compartment fire models known collectively as "zone models", some of which are available as computer programs. These all represent compartment temperatures by considering energy, mass and momentum conservation with various levels of sophistication. "Single-zone" models, such as Pettersson's, assume that all the gases within a compartment are at an equal temperature, whereas "two-zone" models divide the gases into an upper hot zone and a lower cooler zone. The more sophisticated zone models allow for factors, such as compartment boundaries with varying thermal properties and multiple compartment openings, to be included in analyses. Models that account for the interaction between fires in more than one compartment are also available. All zone models require numerical solution and, due to the inherent assumptions, are only valid for a small range of enclosure geometries. A useful source of information on the various models available is the work of DiNenno (2002).

The CFD models of fire growth and behaviour have been available for some years. CFD modelling is a numerical approach to representing fluids that divides a fluid domain into small volumes and considers conservation of mass, energy, and so on within each volume. If the greater resolution of fire behaviour available from CFD models is required, great care must be taken when obtaining and interpreting predictions as the output can be influenced hugely by even minor differences in input data. Obtaining the full range of input data needed in a sufficiently accurate and precise manner will be impractical for most structural engineering problems. The implications of not having the correct input data were highlighted in a study in which Rein et al. (2007a) compared the blind predictions from nine different analysts using CFD models of a fire in a very well-defined compartment. The predictions varied widely. Rein et al. concluded that at present CFD predictions of fire growth are not sufficiently reliable to be used in engineering design unless directly supported by experimental validation. However, they also note that, if the use of CFD models is restricted to predicting gas temperatures for a given fire heat release rate, good predictions can be made. The use of CFD in this way is likely to be advantageous for structures with complex compartment geometries, for which the use of zone models cannot be justified.

Chapter 4

Heat transfer

4.1 INTRODUCTION

Heat transfer forms the link between the high-temperature environment to which a structure in fire is exposed. The theory of heat transfer is well established, and methods of heat transfer analysis, either analytically or numerically, are well developed. However, the choice of input data, mainly thermal boundary conditions and material thermal properties, to be used in heat transfer analysis for fire engineering of structures requires clarification. This is the main focus of this chapter. After presenting a brief introduction to the basics of heat transfer and its modelling, this chapter demonstrates the sensitivity of heat transfer results to variations in input data for heat transfer analysis, leading to some recommendations on how to select appropriate input data for heat transfer analysis.

In materials that decompose, contain water, or both, mass transfer is also involved during the heat transfer process. In specialized situations such as predicting concrete spalling, timber burning, composite material decomposition or intumescent coating expansion, combined heat and mass transfer analysis will be necessary. Fortunately, for most structures exposed to fire, it is sufficient to perform heat transfer analysis only, but modifications may have to be made in some material thermal properties to approximate the effects of mass transfer. How this may be done is discussed further in Chapter 5 on material properties.

The three basic mechanisms of heat transfer are conduction, convection and radiation. In structural fire engineering, heat transfer analysis is usually about solving the heat conduction problem using convection and thermal radiation as the thermal boundary condition.

Except in the simplest cases, it is not possible to find analytical solutions to a heat transfer problem. Therefore, in fire safety applications problems of heat transfer are usually solved either experimentally or numerically. Experiments are expensive to run, and their results can only be applied to the specific situations that have been tested. Numerical analysis of heat transfer can be more general and is becoming more widely used. An understanding of the three heat transfer modes is essential for the selection of appropriate factors for carrying out numerical heat transfer analyses.

4.2 BASICS OF HEAT TRANSFER

4.2.1 Fourier's law of heat conduction

The basic equation for one-dimensional steady-state heat conduction is Fourier's law of heat conduction. It is expressed as

$$\dot{Q} = -\lambda \frac{dT}{dx}$$

(4.1)

where, referring to Figure 4.1, dT is the temperature difference across an infinitesimal thickness dx. $\dot{Q}$ is the rate of heat transfer (heat flux) across the material thickness. The negative sign in Equation (4.1) indicates that heat flows from the higher temperature to the lower temperature.

The coefficient λ is the thermal conductivity of the material. Detailed information on thermal conductivity of materials at high temperatures is presented in Chapter 5.

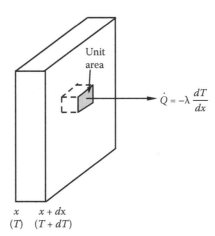

Figure 4.1 Heat conduction in one dimension. (From Wang Y.C., *Steel and Composite Structures, Behaviour and Design for Fire Safety*, p. 171, Spon Press, London, a Taylor & Francis publication, 2002. With permission.)

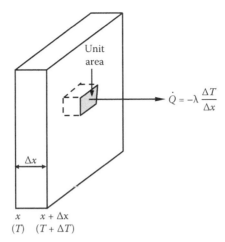

Figure 4.2 Heat transfer through finite thickness with constant thermal conductivity. (From Wang Y.C., *Steel and Composite Structures, Behaviour and Design for Fire Safety*, p. 172, Spon Press, London, a Taylor & Francis publication, 2002. With permission.)

4.2.2 One-dimensional steady-state heat conduction in a composite element

For small heat conduction thickness, Equation (4.1) may be replaced by its finite difference equivalent:

$$\dot{Q} = -\lambda \frac{T_2 - T_1}{\Delta x} \quad \text{or} \quad T_1 - T_2 = \dot{Q}\frac{\Delta x}{\lambda} \tag{4.2}$$

where (refer to Figure 4.2) T_1 and T_2 are temperatures at the two sides of a material and Δx is the material thickness. The term $\Delta x/\lambda$ expresses the thermal resistance of the material.

The term $\Delta x/\lambda$ is referred to as thermal resistance and is frequently used in derivation of heat transfer equations. To illustrate the usefulness of this quantity, consider a construction element with a number of layers of different materials, shown in Figure 4.3, under steady-state heat conduction. Steady-state heat conduction means that the quantities involved do not change with time. According to the principle of energy conservation, with the heat flow across each layer of material being the same, using Equation (4.2) the following equation may be derived:

$$\dot{Q} = -\lambda_{12}\frac{T_2 - T_1}{\Delta x_{12}} = -\lambda_{23}\frac{T_3 - T_2}{\Delta x_{23}} = \ldots -\lambda_{i,i+1}\frac{T_{i+1} - T_i}{\Delta x_{i,i+1}}\ldots = -\lambda_{n,n+1}\frac{T_{n+1} - T_n}{\Delta x_{n,n+1}}$$

$$\tag{4.3}$$

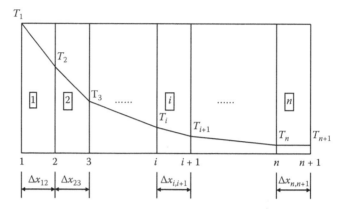

Figure 4.3 Steady-state conduction in a multilayer composite element. (From Wang Y.C., *Steel and Composite Structures, Behaviour and Design for Fire Safety*, p. 173, Spon Press, London, a Taylor & Francis publication, 2002. With permission.)

After rearrangement of Equation (4.3), the following equations are obtained:

$$T_1 - T_2 = \dot{Q}\frac{\Delta x_{12}}{\lambda_{12}} = \dot{Q}R_{12}$$

$$T_2 - T_3 = \dot{Q}\frac{\Delta x_{23}}{\lambda_{23}} = \dot{Q}R_{23}$$

...   (4.4)

$$T_i - T_{i+1} = \dot{Q}\frac{\Delta x_{i,i+1}}{\lambda_{i,i+1}} = \dot{Q}R_{i,i+1}$$

...

$$T_n - T_{n+1} = \dot{Q}\frac{\Delta x_{n,n+1}}{\lambda_{n,n+1}} = \dot{Q}R_{n,n+1}$$

where for the ith layer

$\lambda_{i,i+1}$ is the thermal conductivity;
$R_{i,i+1} = \Delta x_{i,i+1}/\lambda_{i,i+1}$ is the thermal resistance;
T_i and T_{i+1} are temperatures on the two sides, and
$\Delta x_{i,i+1}$ is the thickness.

Summing Equations (4.4) gives

$$T_1 - T_{n,n+1} = \dot{Q}\sum_{i=1}^{n} R_{i,i+1}$$
     (4.5)

Therefore, if temperatures on the two exterior sides of the construction (T_1 and T_{n+1}) are known, Equation (4.5) may be used to obtain the heat flux $\dot{Q}$. This value is then substituted back into Equation (4.4) to obtain temperatures at different interior locations of the construction element (i.e. T_2, T_3, ... , T_n).

4.2.3 Thermal boundary conditions

In structural fire safety applications, the exterior temperatures of a construction element are unknown variables. Instead, the surfaces of the element are in contact with fluids of known temperatures. For example, one side of the element may be in contact with the fire and the other side with ambient-temperature air. To determine the temperature distribution in the construction element, these fluid temperatures are used as boundary conditions.

Refer to the example in Figure 4.4. When applying thermal boundary conditions, it is often assumed that the heat exchange between the fluid and the element surface is related to the temperature difference at the interface. Therefore, the following holds:

On the fire side:

$$\dot{Q} = h_{fi}(T_{fi} - T_1) \tag{4.6}$$

On the ambient-temperature air side:

$$\dot{Q} = h_a(T_{n+1} - T_a) \tag{4.7}$$

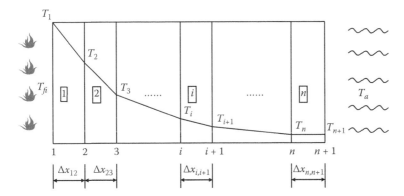

Figure 4.4 Boundary conditions for one-dimensional heat conduction. (From Wang Y.C., *Steel and Composite Structures, Behaviour and Design for Fire Safety*, p. 174, Spon Press, London, a Taylor & Francis publication, 2002. With permission.)

where T_{fi} and T_a are the fire and air temperatures, respectively.

Quantities h_{fi} and h_a are the overall heat exchange coefficients on the fire and air side, respectively. Methods to obtain these two values are discussed in Sections 4.3 and 4.4.

Equations (4.6) and (4.7) can be rewritten as

$$(T_{fi} - T_1) = \dot{Q}/h_{fi} = \dot{Q}\,R_f$$

$$(T_{n+1} - T_a) = \dot{Q}/h_a = \dot{Q}\,R_a \tag{4.8}$$

where R_f and R_a may be regarded as the thermal resistance of the fire and ambient-temperature air layers, respectively.

Combining Equations (4.8) and (4.5) gives

$$T_f - T_a = R\dot{Q}, \quad \text{with} \quad R = R_f + \sum_{i=1}^{n} R_{i,i+1} + R_a \tag{4.9}$$

where R is the overall thermal resistance including the fire layer, the construction element and the ambient-temperature air layer.

4.2.4 Transient heat transfer

For structures in fire, the heat transfer process is invariably transient, with temperatures changing with time. The simplest case of transient heat transfer is the change in temperature T with time of a mass m with a net heat gain of $\dot{Q}$. Without considering mass change, the equation is

$$\dot{Q} = m \times c \times \frac{dT}{dt} \tag{4.10}$$

where c is specific heat of the material, and t is the time.

Detailed information on specific heat of common structural and fire protection materials can be found in Chapter 5.

4.3 CONVECTIVE HEAT TRANSFER COEFFICIENTS

An important part of performing heat transfer analysis for fire engineering of structures is to determine appropriate heat transfer coefficients at the solid-fluid (fire or ambient temperature air) interface. Since both convective and radiant heat transfer may be present, the overall heat transfer coefficient contains two parts: the convective part and the radiant part. Convective heat transfer applies only when the fluid is in contact with the solid surface. Radiant heat transfer will always occur, whether or not the fluid is in contact with the solid surface.

Heat convection is a complex subject, and this section only gives, without derivation, the convective heat transfer coefficients that are necessary for modelling temperature distributions in structures in fire.

4.3.1 Types of flow

Fluid movement passing a solid surface is either forced convection or natural convection. In the context of structural fire engineering, convective heat transfer is usually treated as natural convection because there is no external force driving the fluid. Within each category, there are two types: The flow is either laminar or turbulent. If fluid movement is in a continuous path without mixing with adjacent paths, it is called laminar or streamline flow. On the other hand, if eddy motion of small fluid elements occurs, this produces fluctuations in their flow velocities, both in the direction of the surface and perpendicular to it, and introduces fluid mixing.

4.3.2 Dimensionless numbers

Studying heat transfer usually involves the use of a number of dimensionless quantities. These dimensionless quantities help extend the range of applicability of the limited number of small-scale experimental studies. The convective heat transfer coefficient (hereafter simplified as h_c) is related to the Nusselt (Nu) number using the following equation:

$$Nu = \frac{h_c L}{k}, \text{ giving } h_c = \frac{Nu.k}{L} \tag{4.11}$$

where L is the characteristic length of the solid surface, and k is the thermal conductivity of the fluid. The thermal conductivity of air at different temperatures may be approximated using the following equation (Smith 1981):

$$\lambda_{air} = \lambda_{air,0} \left(\frac{T}{T_0} \right)^{0.8} \tag{4.12}$$

in which T is gas temperature in Kelvin, T_0 is the ambient temperature in Kelvin, and $\lambda_{air,0}$ is the thermal conductivity of air at ambient temperature T_0, $\lambda_{air,0} = 0.0246$ W/mK.

Table 4.1 Convective heat transfer coefficients for natural convection

Surface configuration	Flow type	Condition	Characteristic dimension	B	m
Vertical plate (or cylinder) with height L (e.g. a wall)	Laminar	$10^4 < Ra < 10^9$	L	0.59	1/4
	Turbulent	$Ra > 10^9$	L	0.13	1/3
Horizontal plate with area A and perimeter p (e.g. ceiling)	Laminar	$10^5 < Ra < 10^7$	L for square	0.54	1/4
	Turbulent	$Ra > 10^7$	$(L + W)/2$ for rectangle	0.14	1/3
			$0.9D$ for circular disk		
			A (area)/p (perimeter) for others		

Source: Adapted from Drysdale, D. (1998), *An Introduction to Fire Dynamics*, 2nd ed., Wiley, New York.

As explained in Section 4.3.1, natural convection may be assumed for structural fire engineering purposes. Under natural convection, heat exchange between the fluid and the solid surface depends not only on the fluid properties but also on how the surface is located in relation to the fluid (i.e. whether the surface is perpendicular or parallel to the fluid and whether the surface is above or below the fluid).

The general equation for the Nusselt number for natural convection is

$$Nu = B.Ra^m \tag{4.13}$$

where B and m are numbers whose values are given in Table 4.1 for convective heat transfer to horizontal and vertical plates.

Ra is the Raleigh number and is the product of the Grashof (Gr) number and the Prandtl number (Pr), giving

$$Ra = Gr.Pr \tag{4.14}$$

The Grashof number is defined as

$$Gr = \frac{gL^3\beta\Delta T}{v^2} \tag{4.15}$$

where g is the gravity acceleration, β is the coefficient of thermal expansion of the fluid and ΔT is the temperature difference between the fluid and the solid surface; v ($=\mu/\rho$) is the relative viscosity of the fluid (air), and μ and ρ are the viscosity and density of the fluid, respectively.

According to the ideal gas law, the coefficient of thermal expansion of air at different temperatures is

$$\beta = \frac{1}{T} \tag{4.16}$$

where T is the absolute temperature of air.

The Prandtl number is defined as

$$Pr = \frac{\mu C_{air}}{\lambda_{air}} \tag{4.17}$$

where λ_{air} is the thermal conductivity, and C_{air} is the specific heat of air [approximately $C_{air} = 1$ kJ/(kg.K)]. The value of Pr is very close to 0.7.

4.3.3 Approximate values of convective heat transfer coefficients for fire safety

Due to the temperature-dependent nature of the equations in the previous section and the fact that the required surface temperature necessary for the calculations is unknown, precise calculation of the convective heat transfer coefficient can be a lengthy process. Fortunately, in most cases of heat transfer analysis for structural fire engineering, radiation is the dominant mode of heat transfer, and temperature calculations will not be very sensitive even to very large variations in the convective heat transfer coefficient. Simplification is possible.

Using realistic dimensions for L in Equation (4.15) to calculate the value of Ra, according to the condition in Table 4.1, convection is most likely to be turbulent. Therefore, $m = 1/3$ from Table 4.1. The convective heat transfer coefficient is

$$h_c = B \times \lambda_{air} \times \left[\left(\frac{g * Pr}{v^2} \right) \right]^{1/3} \left(\frac{\Delta T}{T} \right)^{1/3} \tag{4.18}$$

Substituting values of B (≈ 0.14), g ($= 9.81$ m/s^2), and Pr (≈ 0.7) into Equation (4.18) gives

$$h_c = 0.266 k_{air} \times v^{-2/3} \left(\frac{\Delta T}{T} \right)^{1/3} \tag{4.19}$$

Since $\left(\frac{\Delta T}{T} \right)_{max} \approx 1$, Equation (4.19) gives maximum values of h_c of around 10 W/(m^2.K). These values are considerably lower than the recommended values in Eurocode EN 1991-1-2 (Committee of European Normalization [CEN] 1991–2002) for structural fire design, being 25 W/(m².K) under the nominal standard fire exposure condition, 50 W/(m².K) under the nominal hydrocarbon fire condition, and 35 W/(m².K) under the parametric fire condition. The higher values recommended in the Eurocode account for some aspects of forced convection. However, in postflashover fires, the construction temperatures are not very sensitive to large variations in h_c. On the ambient-temperature air side, the Eurocode recommends a constant value of 4 W/(m².K). This is a lower-bound value calculated using Equation (4.19). This Eurocode recommended value is to ensure that the convective heat loss from the air side is at the lower bound to give the highest possible temperature in the structure as a conservative estimate.

4.4 RADIANT HEAT TRANSFER COEFFICIENT

When radiant thermal energy passes a medium, any object within the path can absorb, reflect and transmit the incident thermal radiation. Use absorptivity α, reflectivity ρ, and transmissivity τ to represent the fractions of incident thermal radiation that a body absorbs, reflects and transmits, respectively, giving

$$\alpha + \rho + \tau = 1 \tag{4.20}$$

In general, the three factors in Equation (4.20) are functions of the temperature, the electromagnetic wavelength and the surface properties of the incident body. Simplifications are usually made for structural fire engineering calculations. An extreme case is when all the incident thermal radiation is absorbed by the body ($\alpha = 1$). Such an ideal body is called a black body.

4.4.1 Total power of black-body thermal radiation

The black-body thermal radiation is of fundamental importance to radiant heat transfer. It has many special properties, but the most important one is that it is a perfect emitter. This means that no other body can emit more thermal radiation per unit surface area than a black body at the same temperature. The total amount of thermal radiation E_b emitted by a black-body surface is a function of its temperature only and is given by the Stefan-Boltzmann law:

$$E_b = \sigma T^4 \tag{4.21}$$

where σ is the Stefan-Boltzmann constant, which is equal to 5.67×10^{-8} W/(m²K⁴); and T is the absolute temperature in Kelvin.

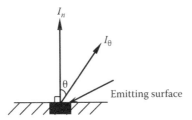

Figure 4.5 Directional intensity of radiant heat. (From Wang Y.C., *Steel and Composite Structures, Behaviour and Design for Fire Safety*, p. 180, Spon Press, London, a Taylor & Francis publication, 2002. With permission.)

4.4.2 Intensity of directional thermal radiation

Equation (4.20) gives the total thermal radiation of a unit area of black-body surface (the emitter), but thermal radiation is not uniformly distributed in space. The directional dependence is expressed by the Lambert law:

$$I_\theta = I_n \cos\theta \qquad (4.22)$$

Referring to Figure 4.5, I_n is the intensity of thermal radiation in the normal direction to the emitting surface. I_θ is the intensity of thermal radiation in the direction at an angle θ to the normal direction of the emitting surface.

The intensity of thermal radiation is defined as the radiant heat flux per unit area of the emitting surface per unit subtended solid angle. The intensity of directional thermal radiation can be derived to give

$$I_n = \frac{E_b}{\pi} \qquad (4.23)$$

4.4.3 Exchange of thermal radiation between black-body surfaces

Figure 4.6 shows two black-body surfaces A_1 and dA_2, with A_1 the emitting surface. It is required to find the incident thermal radiation on dA_2. In Figure 4.6, dA_1 is a small area on A_1 and points 1 and 2 are at the centres of dA_1 and dA_2. θ_1 is the angle between the normal to dA_1 and line 1–2, and θ_2 is the angle between the normal to dA_2 and line 1–2.

If the total thermal radiation per unit surface area of A_1 is E_{b1}, the intensity of incident thermal radiation in direction 1–2 is $(E_{b1}/\pi)\cos\theta_1$. The effective area of dA_2 normal to direction 1–2 is $dA_2.\cos\theta_2$, giving a subtended

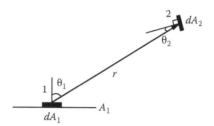

Figure 4.6 Radiant heat exchange between a finite and an infinitesimal area. (From Wang Y.C., *Steel and Composite Structures, Behaviour and Design for Fire Safety*, p. 181, Spon Press, London, a Taylor & Francis publication, 2002. With permission.)

solid angle of dA_2 on the centre of dA_1 of $dA_2.\cos \theta_2/r^2$. Therefore, the thermal radiation from dA_1 incident on dA_2 is

$$d\dot{Q}_{dA_1 \to dA_2} = E_{b1} \frac{\cos \theta_1 \cos \theta_2}{\pi r^2} dA_1 dA_2 \qquad (4.24)$$

and the total thermal radiation from A_1 incident on dA_2 is

$$\dot{Q}_{A_1 \to dA_2} = \int_{A_1} E_{b1} \frac{\cos \theta_1 \cos \theta_2}{\pi r^2} dA_1\, dA_2 = \Phi E_{b1} dA_2 \qquad (4.25)$$

4.4.4 Configuration (view) factor Φ

On the right-hand side of Equation (4.25), $E_{b1}dA_2$ is the maximum incident thermal radiation power on dA_2, and this occurs when dA_2 is completely surrounded by A_1. In most cases, the incident thermal radiation on dA_2 is much less. The factor Φ is used to represent the fraction of thermal radiation from A_1 incident on dA_2. This factor is often referred to as the configuration or view factor because it only depends on the spatial configuration between A_1 and dA_2. The configuration factor will not be greater than 1.0.

The configuration factor is an important value. However, the integration contained in Equation (4.25) can be laborious to carry out. Since it depends only on the spatial arrangement of a surface (A_1) and a view point (dA_2), values of the configuration factor for many situations have already been calculated and are presented in a number of textbooks.

Table 4.2 Configuration factors for two common geometrical arrangements

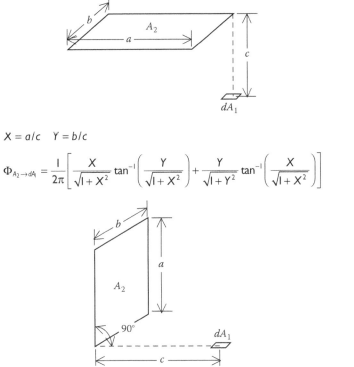

$X = a/c \quad Y = b/c$

$$\Phi_{A_2 \to dA_1} = \frac{1}{2\pi}\left[\frac{X}{\sqrt{1+X^2}} \tan^{-1}\left(\frac{Y}{\sqrt{1+X^2}} \right) + \frac{Y}{\sqrt{1+Y^2}} \tan^{-1}\left(\frac{X}{\sqrt{1+X^2}} \right) \right]$$

$X = a/b \quad Y = c/b \quad A = 1/\sqrt{X^2 + Y^2}$

$$\Phi_{A_2 \to dA_1} = \frac{1}{2\pi}\left[\tan^{-1}\left(\frac{1}{Y} \right) - AY \tan^{-1} A \right]$$

Source: Adapted from DiNenno, P. (2002), *The SFPE Handbook of Fire Protection Engineering*, 3rd ed., Society of Fire Protection Engineers, Bethesda, MD, USA.

For example, Table 4.2 is extracted from the Society of Fire Protection Engineering (SFPE) handbook (SFPE 2002) for two common geometric arrangements.

The configuration factor is additive, so that the configuration factor for a complex surface may be obtained by dividing the surface into a number of simple zones whose view factors can be obtained from readily available prepared tables. The total configuration factor is simply the summation of all the configuration factors of the subdivided areas.

4.4.5 Exchange area

Equation (4.25) gives the incident thermal radiation from a finite area to a point. In realistic cases, it is often required to calculate the incident thermal radiation from one finite area to another finite area. For two finite areas A_1 and A_2, Equation (4.25) now becomes

$$\dot{Q}_{A_1 \to A_2} = \iint_{A_1, A_2} E_{b1} \frac{\cos\theta_1 \cos\theta_2}{\pi r^2} dA_1 \, dA_2 = F_{1 \to 2} A_1 E_{b1} \tag{4.26}$$

The factor $F_{1 \to 2}$ is the "integrated configuration factor". It represents the fraction of radiant heat flux emitted by surface A_1 and incident on surface A_2. Since $A_1 E_{b1}$ is the total thermal radiation from surface A_1, this is the maximum that can be incident on A_2. Therefore, $F_{1 \to 2}$ cannot be greater than 1. Similar to the configuration factor, values of the integrated configuration factor for many arrangements of two surfaces have been calculated and are available from textbooks, for example, the *SFPE Handbook of Fire Protection Engineering* (DiNenno 2002).

4.4.6 Radiant heat transfer of grey-body surfaces

No real material emits and absorbs radiation according to laws of the black body. In general, an additional term is necessary to define the radiant energy of an emitting surface. This is the emissivity ε. This term is defined as the ratio of the total energy emitted by a surface to that of a black-body surface at the same temperature. Thus, the total radiant energy emitted by a general surface is

$$E = \varepsilon \sigma T^4 \tag{4.27}$$

In general, the emissivity of a surface depends on the wavelength of radiant energy, the temperature of the surface and the angle of radiation. However, if the emissivity is independent of these factors, the radiant surface is called a grey-body surface. For simplicity, grey-body radiation is adopted in fire engineering calculations. According to Kirchhoff's law, the emissivity of a surface is equal to its absorptivity. Therefore, if there is no transmission of radiant energy through a grey-body surface, the reflectivity of the surface is $1 - \varepsilon$, according to Equation (4.20).

4.4.7 Network method for radiant heat transfer between grey-body surfaces

With the introduction of emissivity and reflectivity, the heat exchange between grey-body surfaces is complicated, and analytical equations can

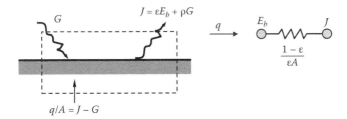

Figure 4.7 Network representation of a grey-body surface.

be obtained only for very simple cases. To solve the problem, a network approach may be used.

First, consider one isolated surface as shown in Figure 4.7. It has a total incident radiation G per unit area and total radiosity J per unit area. The total radiosity J is made up of two parts: emitted radiation εE_b and reflected radiation ρG. Thus, the net radiant heat flux per unit area is $q/A = J - G$. Eliminate the quantity G to express q in terms of J as follows:

$$J = \varepsilon E_b + \rho G = \varepsilon E_b + (1-\varepsilon)G \Rightarrow G = \frac{J - \varepsilon E_b}{1-\varepsilon} \tag{4.28}$$

$$\frac{q}{A} = J - G = \frac{\varepsilon}{1-\varepsilon}(E_b - J)$$

$$or \quad q = \frac{E_b - J}{(1-\varepsilon)/\varepsilon A} \tag{4.29}$$

This may be represented by a branch of the equivalent electric network model shown on the right in Figure 4.7. In the electrical network, E_b and J are the voltages of the two nodes; $(1 - \varepsilon)/\varepsilon A$ is the electrical resistance and q is the electric current.

Now, consider radiant heat exchange between two grey-body surfaces shown in Figure 4.8. Applying Equation (4.26) but replacing E_{b1} and E_{b2} by J_1 and J_2 for the total radiosity gives

$$q_{1-2} = J_1 A_1 F_{12}$$

$$q_{2-1} = J_2 A_2 F_{21}$$

$$q_{1-2|net} = J_1 A_1 F_{12} - J_2 A_2 F_{21} = (J_1 - J_2)A_1 F_{12} \tag{4.30}$$

$$q_{1-2|net} = \frac{J_1 - J_2}{1/A_1 F_{12}}$$

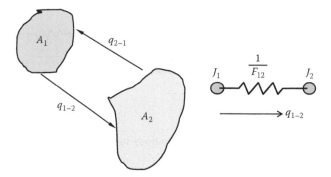

Figure 4.8 Network representation of radiant heat transfer between two grey-body surfaces.

This can again be represented by a branch of an equivalent electric network with voltages J_1 and J_2 at two nodes and electric resistance $1/(A_1F_{12})$, as shown on the right in Figure 4.8.

Thus, an entire network can be built for radiant heat transfer between a group of grey-body surfaces. It is important to understand that the network should include all participating radiating surfaces. For example, Figure 4.9 shows the complete network for radiant heat transfer between three surfaces that see each other but nothing else.

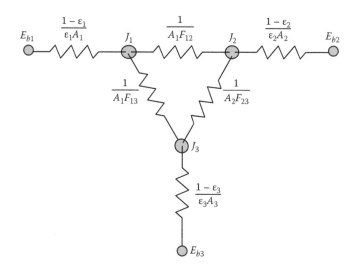

Figure 4.9 Network representation of three mutual radiating grey-body surfaces.

Accordingly, three electric balance equations can be written for the three nodes J_1, J_2 and J_3 as follows:

$$Node \quad 1: \frac{E_{b1} - J_1}{(1 - \varepsilon_1)/\varepsilon_1 A_1} + \frac{J_2 - J_1}{1/A_1 F_{12}} + \frac{J_3 - J_1}{1/A_1 F_{13}} = 0$$

$$Node \quad 2: \frac{E_{b2} - J_2}{(1 - \varepsilon_2)/\varepsilon_2 A_2} + \frac{J_1 - J_2}{1/A_1 F_{12}} + \frac{J_3 - J_2}{1/A_2 F_{23}} = 0 \qquad (4.31)$$

$$Node \quad 3: \frac{E_{b3} - J_3}{(1 - \varepsilon_3)/\varepsilon_3 A_3} + \frac{J_1 - J_3}{1/A_1 F_{13}} + \frac{J_2 - J_3}{1/A_2 F_{23}} = 0$$

With these three equations, the three unknowns (J_1, J_2, J_3) can be found. From this, the radiant heat exchanges between the different surfaces (q_{12}, q_{13}, q_{23}) can easily be calculated by making use of Equation (4.30).

The network method may be extended to radiant heat exchange between multiple grey-body surfaces that see each other but nothing else. Part of the network is sketched in Figure 4.10.

For the network at node i, the following equation may be written:

$$\frac{\varepsilon_i}{1 - \varepsilon_i}(E_{bi} - J_i) + \sum_j F_{ij}(J_j - J_i) = 0 \qquad (4.32)$$

All the equations can then be assembled to give the following matrix form:

$$[A][J] = [C] \qquad (4.33)$$

After solving for J for all the nodes, the radiant heat exchange between any two surfaces can be obtained by using Equation (4.30).

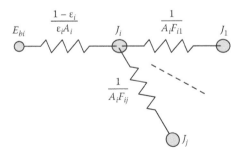

Figure 4.10 Partial network representation of a multiple grey-body surface radiant heat transfer system.

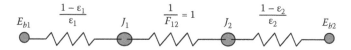

Figure 4.11 Network representation of radiant heat transfer between unit area of two parallel and infinitely large grey-body surfaces.

Example 4.1

As an example to demonstrate application of the network method, consider radiant heat transfer between two parallel grey-body surfaces that are infinitely large compared to the distances between them, so that the configuration factor between the two surfaces is 1. This applies to radiant heat transfer involving a structural surface that is in contact with fire. The equivalent network for unit areas of the two surfaces is shown in Figure 4.11. In this case, J_1 and J_2 are the total radiosity per unit surface 1 and surface 2.

For node 1:

$$\frac{E_{b1} - J_1}{(1 - \varepsilon_1)/\varepsilon_1} = J_1 - J_2 \tag{4.34}$$

For node 2:

$$\frac{J_2 - E_{b2}}{(1 - \varepsilon_2)/\varepsilon_2} = J_1 - J_2 \tag{4.35}$$

Therefore

$$J_1 - J_2 = \frac{E_{b1} - E_{b2}}{1/\varepsilon_1 + 1/\varepsilon_2 - 1} = \varepsilon_r \sigma \left(T_1^4 - T_2^4 \right) \tag{4.36}$$

In Equation (4.36), ε_r is often referred to as the resultant emissivity and

$$\varepsilon_r = \frac{1}{1/\varepsilon_1 + 1/\varepsilon_2 - 1} = \frac{\varepsilon_1 \varepsilon_2}{\varepsilon_1 + \varepsilon_2 - \varepsilon_1 \varepsilon_2} \tag{4.37}$$

Following Equation (4.6), the radiant heat transfer coefficient between the two surfaces in this example is obtained as

$$h_r = \varepsilon_r \sigma \left(T_2^2 + T_1^2 \right) \left(T_2 + T_1 \right) \tag{4.38}$$

4.5 SOME SIMPLIFIED SOLUTIONS FOR HEAT TRANSFER

To obtain temperature distributions in a construction element exposed to fire attack, numerical procedures are generally necessary. However, for the two common cases of unprotected and protected steelwork exposed to fire attack, simple analytical solutions have been derived to enable their temperatures to be calculated quickly.

These simple analytical solutions have been derived by using the "lumped-mass method," in which the entire steel mass has the same temperature. The validity of this assumption depends on the rate of heat transfer within the material, which depends on its thermal conductivity and its thickness. Considering a plate totally immersed in air that experiences a sudden rise in temperature, detailed theoretical consideration (Carslaw and Jaeger 1959) suggests that, provided the Biot number is less than 0.1, the plate may be assumed to have a uniform temperature distribution and the lumped-mass method may be applied. This requirement is satisfied for steel plate thicknesses up to about 100 mm.

4.5.1 Temperatures of unprotected steelwork in fire

Figure 4.12 shows the cross section of a steel element subject to fire attack on all sides. Assuming that the steel temperature is T_s and the fire temperature is T_{fi}, the heat balance equation may be written as

$$V\rho C_s \frac{dT_s}{dt} = h(T_{fi} - T)A_s \tag{4.39}$$

where V and A_s are the volume and the exposed surface area of the steel element, respectively; ρ is the density, and C_s is the specific heat of steel.

The left-hand side of Equation (4.39) is the heat required to increase the steel temperature during an infinitesimal period of time, and the right-hand side is the heat input from fire to the steel element. Using a step-by-step

Figure 4.12 An unprotected steel section in fire. (From Wang Y.C., *Steel and Composite Structures, Behaviour and Design for Fire Safety*, p. 189, Spon Press, London, a Taylor & Francis publication, 2002. With permission.)

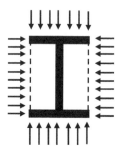

Figure 4.13 Assumed radiant heat transfer boundary of a section with concave surfaces.

approach and assuming that the time increment is small ($\Delta t < 5$ s), the steel temperature increase ΔT_s may be calculated using

$$\Delta T_s = \frac{h}{\rho C_s} \frac{A_s}{V} (T_{fi} - T_s) \Delta t \qquad (4.40)$$

The ratio A_s/V in Equation (4.40) is often referred to as the section factor of the steel element.

Equation (4.40) forms the basis of the temperature calculation equation in Eurocodes EN 1993-1-2 and EN 1994-1-2 (CEN 1993–2005b, 1994–2005b).

It should be pointed out that in the EN 1993-1-2 version of Eurocode 3 (CEN 1993–2005b), a correction factor is applied. The effect of this correction factor is to change the section factor A_s/V to that of the box enclosing the steel section, as shown in Figure 4.13. Nominally, this box represents the effective boundary of the steel section to the radiant heat flux and is introduced to allow for the assumption that radiant heat transfer to the concave surfaces is blocked by the shadow of the cross section. However, examination of the background documents (Franssen 2006; Wickstrom 2005) indicates that the principal reason for introducing this so-called shadow effect is due to the different values of resultant emissivity recommended in the two different versions of Eurocode 3 Part 1.2, being 0.5 in ENV 1993-1-2 (CEN 1993–1995) and 0.7 in EN 1993-1-2 (CEN 1993–2005b). Therefore, the real purpose of the shadow effect is to compensate for the overestimation of steel temperature in EN 1993-1-2 caused by using a higher value of resultant emissivity.

4.5.2 Temperatures of protected steelwork in fire

Figure 4.14 shows the cross section of a protected steel element under fire attack. Assume that the temperatures of the steel section and the fire are

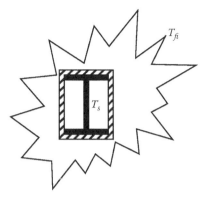

Figure 4.14 A protected steel section in fire. (From Wang Y.C., *Steel and Composite Structures, Behaviour and Design for Fire Safety*, p. 191, Spon Press, London, a Taylor & Francis publication, 2002. With permission.)

T_s and T_f, respectively. According to Equation (4.9), the heat transfer from fire through the fire protection to the steel section is

$$\dot{Q}_{con} = \frac{1}{1/h + d_p/k_p}(T_{fi} - T_s)A_s\Delta t \tag{4.41}$$

where t_p and k_p are the thickness and thermal conductivity of the fire protection material, respectively; h is the overall heat transfer coefficient of fire.

As a first approximation, assume that the fire protection temperature is the average of the fire temperature and the steel temperature, $T_p = \frac{1}{2}(T_{fi} + T_s)$. The total heat required to increase the temperatures of the steel and the fire protection is

$$\dot{Q}_{req} = C_s\rho_s V\Delta T_s + C_p\rho_p t_p A_p \frac{1}{2}(\Delta T_{fi} + \Delta T_s) \tag{4.42}$$

where A_p is the exposed surface area of the fire protection.

Assuming that the fire protection is thin, A_p may be taken as the surface area of the steel element. Equating Equation (4.42) to Equation (4.41) gives the increase in steel temperature as

$$\Delta T_s = \frac{(T_{fi} - T_s)A_p/V}{(1/h + t_p/k_p)C_s\rho_s\left(1 + \frac{1}{2}\phi\right)}\Delta t - \frac{1}{\left(\frac{2}{\phi} + 1\right)}\Delta T_{fi} \tag{4.43}$$

where

$$\phi = \frac{C_p \rho_p}{C_s \rho_s} t_p \frac{A_p}{V}$$

The term $1/h$ in Equation (4.43) is the thermal resistance of fire. If the fire protection is thick and has low thermal conductivity, the effect of including $1/h$ will be small and may be ignored. However, when using fire protection materials with high thermal conductivity (such as sprayed concrete), the influence of the thermal resistance of fire can be significant (Wong and Ghojel 2003).

When deriving Equation (4.43), a number of assumptions have been made. More detailed theoretical considerations by Wickstrom (1982, 1985) suggest that Equation (4.44) may be used to give more accurate predictions of the temperature in a lightly protected steel element:

$$\Delta T_s = \frac{(T_{fi} - T_s)A_p / V}{(t_p / k_p)C_s \rho_s \left(1 + \frac{1}{3}\phi\right)} \Delta t - (e^{\phi/10} - 1)\Delta T_{fi} \tag{4.44}$$

As explained, the thermal resistance of the fire should be included by replacing t_p / k_p by $1/h + t_p / k_p$.

The time increment should not be too large. When using Equation (4.44), the time increment Δt should not exceed 30 s.

Because of the second term in Equation (4.44), it is possible that, at the early stage of increasing fire temperature, the steel temperature may decrease. In this case, the steel temperature increase should be taken as zero.

4.5.3 Section factors

Equations (4.40) and (4.44) clearly indicate that the temperature rise in a steel element is directly related to the section factor of the steel element. Consider a unit length of a steel element where the end effects are ignored; the section factor may alternatively be expressed as H_p/A. Here, H_p is the fire-exposed perimeter length of the steel cross section, and A is the cross-sectional area of the steel element.

Sometimes, for example in the case of unprotected steelwork, it can be more useful to obtain different temperatures in different parts of the steel cross section. In this case, the so-called element factor may be used. For example, element factors for different parts of an unprotected universal beam/column section are given in Figure 4.15.

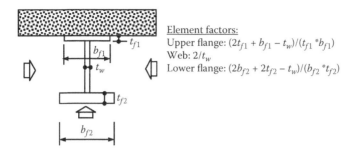

Element factors:
Upper flange: $(2t_{f1} + b_{f1} - t_w)/(t_{f1}*b_{f1})$
Web: $2/t_w$
Lower flange: $(2b_{f2} + 2t_{f2} - t_w)/(b_{f2}*t_{f2})$

Figure 4.15 Element factors for an unprotected steel section. (From Wang Y.C., *Steel and Composite Structures, Behaviour and Design for Fire Safety*, p. 193, Spon Press, London, a Taylor & Francis publication, 2002. With permission.)

Research studies by Ding and Wang (2009) and Dai et al. (2009) on steel connections indicate that the simplified calculation methods can be extended to calculating connection component temperatures, provided that the appropriate section factors for the components are used. Further information is given in Chapter 8.

4.6 IMPORTANCE OF USING APPROPRIATE THERMAL PROPERTIES OF MATERIALS

To use Equations (4.40) and (4.44), it is necessary to have available information on the thermal properties (thermal conductivity k, density ρ and specific heat C) of steel, concrete and insulation materials. Chapter 5 presents detailed material thermal properties.

For protected steelwork, it is particularly important to have reliable information on the thermal properties of the fire protection material. It is not appropriate to assume that their thermal properties are constant as at their ambient-temperature values.

To demonstrate the importance of using temperature-dependent thermal properties, Figures 4.16 and 4.17 compare steel temperatures calculated using Equation (4.44), without including the thermal resistance of the fire, for rock fibre fire protection and vermiculite fire protection, respectively. The section factor was 150 m^{-1}. For simplicity, the specific heat of steel was 650 J/kg.K. The calculations were performed for 60 min of the standard fire exposure. For rock fibre fire protection, the density was 160 kg/m^3, and the specific heat value was 900 J/kg.K. For vermiculite fire protection, the density was 600 kg/m^3, and the specific heat value was 900 J/kg.K. The rock fibre thickness was 10 mm, and the vermiculite thickness was 15 mm. The temperature-dependent thermal conductivities of rock fibre and vermiculite were calculated using

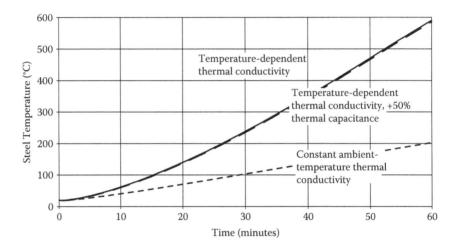

Figure 4.16 Comparison of steel temperatures using rock fibre fire protection.

the equations in Sections 5.3.3.1 and 5.3.3.4. The ambient-temperature thermal conductivity of rock fibre and vermiculite was 0.0257 W/(m.K) and 0.1638 W/(m.K), respectively.

For vermiculite (Figure 4.17), the dependence of its thermal conductivity on temperature is not very strong. Therefore, the difference between the calculated steel temperatures using temperature-dependent and temperature-independent thermal conductivities was relatively small.

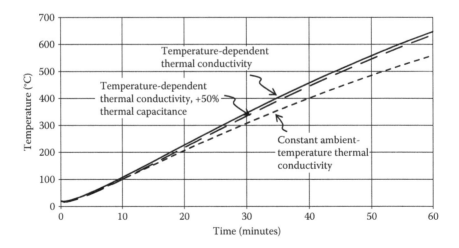

Figure 4.17 Comparison of steel temperatures using vermiculite fire protection.

Nevertheless, the difference in steel temperature can still be substantial. For example, at 60 min of fire exposure, the steel temperature calculated using the temperature-dependent thermal conductivities was 647°C, whilst that using the temperature-independent thermal conductivity was 560°C, a difference of 87°C.

If the thermal conductivity of the fire protection material is strongly temperature dependent, such as rock fibre, using a temperature-independent value will result in grossly erroneous results, as shown in Figure 4.16. At 60 min of fire exposure, using the temperature-dependent thermal conductivity gave a steel temperature of 590°C, but using the temperature-independent thermal conductivities gave a steel temperature of only 203°C.

In contrast, the influence of thermal capacitance (density multiplied by specific heat) is minor. For example, by increasing the thermal capacitance of rock fibre by 50%, the calculated steel temperature at 60 min of fire exposure was reduced by 5°C from 590°C to 585°C. For vermiculite fire protection, due to its higher density, changing the thermal capacitance had a more noticeable effect. However, the calculated steel temperature was still not very sensitive to large changes in thermal capacitance. For example, by increasing the thermal capacitance of vermiculite by 50%, the calculated steel temperature at 60 min of fire exposure was reduced by only 11°C, from 647°C to 636°C.

As a conclusion, it is important to obtain accurate temperature-dependent thermal conductivity values of fire protection materials in performance-based structural fire engineering calculation.

4.7 EFFECTS OF THERMAL BOUNDARY CONDITIONS

In Section 4.2.3, it was explained that there is a large difference between the values of convective heat transfer coefficient based on convection theory and the simplified values recommended in Eurocode EN 1991-1-2 (CEN 1991–2002). There is also uncertainty in the value of the resultant emissivity to be used in calculating the radiant heat transfer coefficient. Unprotected steel temperatures will be most sensitive to changes in the convective and radiant heat transfer coefficients. Figure 4.18 gives indicative variations in unprotected steel temperatures using different convective heat transfer coefficients [10, 25 and 50W/(m².K)] and resultant emissivities (0.3, 0.5, 0.7). For illustration purposes, the standard fire exposure condition was used, and the steel section factor was 100 m⁻¹.

There are significant differences in the calculated steel temperatures. However, a distinction should be made between the effects of changing the convective heat transfer coefficient and those of changing the resultant emissivity. Comparing the steel temperatures obtained using the same emissivity (0.5) but different convective heat transfer coefficients, the

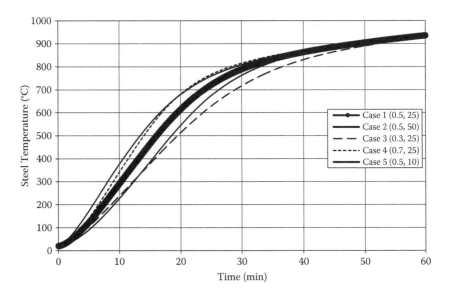

Figure 4.18 Variations in unprotected steel temperature for different convective heat transfer coefficients and emissivity values.

maximum differences in the calculated steel temperatures occurred when the rate of steel temperature rise was fast (around 15 min in Figure 4.18). Therefore, the difference in time at which the same steel temperature was reached was relatively insignificant. This contrasts with the steel temperatures calculated using the same convective heat transfer coefficient (25 W/m².K) but different resultant emissivity values (0.3, 0.5, 0.7). Substantial differences in the steel temperatures were maintained even when the rate of increase in steel temperature was low (around 30 min in Figure 4.18). This would give a large difference in the time when the same steel temperature is attained.

This result suggests that, whilst the calculated steel temperatures are affected by both the convective heat transfer coefficient and the resultant emissivity used, it is more important to obtain accurate values of resultant emissivity.

4.8 BRIEF INTRODUCTION TO NUMERICAL ANALYSIS OF HEAT TRANSFER

It is generally not possible to obtain analytical solutions to general problems of heat transfer, and numerical methods are necessary. To help the reader understand heat transfer software packages, this section gives some basic equations of three-dimensional transient-state heat transfer.

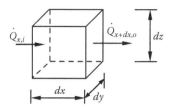

Figure 4.19 Heat conduction in a three-dimensional control volume. (From Wang Y.C., *Steel and Composite Structures, Behaviour and Design for Fire Safety*, p. 198, Spon Press, London, a Taylor & Francis publication, 2002. With permission.)

Refer to Figure 4.19 for a heat-conducting element $dx.dy.dz$ with its edges parallel to the Cartesian coordinates x, y and z. Assume no internal heat generation.

Consider heat transfer in the x direction. The total heat inflow to the element is

$$\dot{Q}_{x,i} = -k_x \frac{\partial T}{\partial x} dydz \tag{4.45}$$

The total heat outflow from the element is

$$\dot{Q}_{x+dx,o} = -k_{x+dx} \frac{\partial \left(T + \frac{\partial T}{\partial x} dx \right)}{\partial x} dydz = -k_{x+dx} \left(\frac{\partial T}{\partial x} + \frac{\partial^2 T}{\partial x^2} dx \right) dydz \tag{4.46}$$

Assume that the thermal conductivity of the element is constant within an infinitesimal space, so that $k_x = k_{x+dx}$, the net heat inflow into the control volume in the x direction is

$$\Delta \dot{Q}_x = \dot{Q}_{x,i} - \dot{Q}_{x+dx,o} = k_x \frac{\partial^2 T}{\partial x^2} dxdydz = k_x \frac{\partial^2 T}{\partial x^2} dV \tag{4.47}$$

Similarly, net heat inflows into the control volume in y and z directions are

$$\Delta \dot{Q}_y = k_y \frac{\partial^2 T}{\partial y^2} dV \tag{4.48}$$

and

$$\Delta \dot{Q}_z = k_z \frac{\partial^2 T}{\partial z^2} dV \tag{4.49}$$

According to the principle of energy conservation, the total heat accumulated in the element equals the heat necessary to increase the element temperature, giving

$$k_x \frac{\partial^2 T}{\partial x^2} + k_y \frac{\partial^2 T}{\partial y^2} + k_z \frac{\partial^2 T}{\partial z^2} = \rho C \frac{\partial T}{\partial t} \tag{4.50}$$

The problem of numerical heat transfer analysis is to find numerical solutions for temperatures to Equation (4.50).

The theoretical basis of numerical heat transfer is well established, and many standard textbooks are available for detailed study. Many numerical heat transfer packages have also been developed by researchers and commercial finite element organizations. A key issue in obtaining credible numerical heat transfer simulation results is to apply appropriate boundary conditions and to use appropriate thermal properties of materials.

4.9 CONCLUDING REMARKS

This chapter has presented an introduction to heat transfer for the purpose of carrying out performance-based structural fire engineering analysis. The theory of heat transfer is mature, and various tools are available. However, this cannot guarantee correctness of the heat transfer analysis results. To obtain credible results, it is important that appropriate boundary conditions are applied and reliable thermal properties of materials are used. In particular, the thermal conductivity of fire protection materials will have the most significant influence on heat transfer analysis results. In general, it is not appropriate to assume that the thermal conductivity of a fire protection material is temperature independent and then use its ambient-temperature value. Temperature dependency of material thermal conductivity should be reliably quantified.

Chapter 5

Material properties

5.1 INTRODUCTION

Accurate and appropriate information on the ways in which materials of construction are affected by high surrounding temperatures is clearly an important prerequisite to performance-based fire engineering design of structures. It allows the designer either to model the structural behaviour under different scenarios of fire development and protection or to strike a balance between specifying materials with inherently favourable fire-resistant properties with the need for economical construction. The relevant material properties are usually grouped into two categories; thermal properties, which control the temperatures achieved by elements of structure, and temperature-dependent mechanical properties, which control the structural response.

The degree of depth to which thermomechanical properties need to be specified depends on the nature of any particular case. For example, in the majority of structural fire engineering cases the relatively short duration of a fire allows mechanical properties to be considered as time independent, but in certain cases if the structure stays at high temperatures for long periods of time dependence may need to be considered explicitly.

Buildings constructed by any method and using any combination of materials are subjected to fires from time to time, and at least the safety of their occupants needs to be considered. In the case of large and multistorey buildings where there is a significant design input by structural engineers, the range of framing materials can be limited mainly to steel and reinforced concrete. However, where insulating (fire protection) materials are used to slow the heating rates of the main structure, their properties are also essential information for structural fire engineering design calculations.

A considerable body of research knowledge exists on material behaviour at elevated temperatures, and this is well reported, both in the original research papers and in specialised textbooks. For performance-based structural fire engineering, it is necessary to have an adequate subset of this knowledge, and for most purposes the structural Eurocodes, which were launched in final form between 2000 and 2005, provide excellent representations of the thermal and structural constitutive relationships of the main materials of construction based on recent research. In this chapter, there is no attempt to present an all-embracing review of research findings; its emphasis is on providing usable models of the relevant material properties that can be introduced into structural simulations at elevated temperatures. The Eurocode characteristics are used whenever they provide a realistic model of the relevant properties, with amplification as necessary.

5.2 STRUCTURAL MATERIALS

5.2.1 Relevant thermal properties

The mechanisms and equations of heat transfer from a fire, which create temperature distributions in elements of the surrounding structure, have been covered in Chapter 4. The material thermal properties whose meaning needs to be understood in carrying out practical structural fire engineering design can be summarised as emissivity, thermal conductivity, specific heat capacity and thermal expansion.

5.2.1.1 Emissivity

Emissivity is essentially a property of the surface of a material rather than an intrinsic property of the material itself. It represents the relative ability of its surface to radiate its absorbed heat, expressed as the ratio of heat radiated by the particular material surface to the heat radiated by a perfect "black body" at the same temperature. The ideal black body would have an emissivity $\varepsilon = 1.0$, so that any real material surface has an emissivity that is a positive number below 1.0. In physics, the property absorptivity is also defined, which is the ratio of incident heat absorbed by the particular material surface to the heat absorbed by a perfect black body, again with the ideal black-body surface having absorptivity $\alpha = 1.0$. In reality, both emissivity and absorptivity values depend on the radiation wavelength, but for the very restricted band of infrared wavelengths concerned in the heating of structure by fires, this is not significant. According to Kirchhoff's law of radiation, $\alpha = \varepsilon$, and in the case of heat transfer, this can be assumed correct.

It is logical that the colour and texture of a material surface control its emissivity to a large extent; duller and darker surfaces tend to have emissivities closer to 1.0 than more polished and lighter surfaces, which may be thought of as more reflective. If one considers the structure surfaces during a building fire in these terms, it is clear that ascribing a single value of emissivity to a material is subject to considerable uncertainty because

Table 5.1 Typical emissivity values for materials at 300 K

Surface material	Emissivity ε	Surface material	Emissivity ε
Aluminium (heavily oxidized)	0.2–0.31	Iron (polished)	0.14–0.38
Aluminium (highly polished)	0.039–0.057	Iron (plate, rusted red)	0.61
Aluminium (rough)	0.07	Iron (dark grey surface)	0.31
Asbestos board	0.96	Iron (rough ingot)	0.87–0.95
Asphalt	0.93	Masonry (plastered)	0.93
Brick (red, rough)	0.93	Mild steel	0.20–0.32
Brick (fireclay)	0.75	Plaster	0.98
Cast iron (newly turned)	0.44	Plaster (rough)	0.91
Cast iron (turned and heated)	0.60–0.70	Plastics	0.91
Chromium (polished)	0.058	Sand	0.76
Concrete	0.85	Steel (oxidized)	0.79
Concrete (rough)	0.94	Steel (polished)	0.07
Concrete tiles	0.63	Stainless steel (weathered)	0.85
Glass (smooth)	0.92–0.94	Stainless steel (polished)	0.075
Glass (Pyrex)	0.85–0.95	Steel (galvanized, old)	0.88
Granite	0.45	Steel (galvanized, new)	0.23
Gypsum	0.85	Wrought iron	0.94

Source: Data from http://www.engineeringtoolbox.com/emissivity-coefficients-d_447.html.

of the effects of different fire fuel types and ventilation conditions on the colouration and texture of surfaces. A sample of values at 300 K is given in Table 5.1.

It should be borne in mind that, in fire engineering codes of practice, such as the Eurocodes EN 1993-1-2 (Committee of European Normalization [CEN] 1993–2005b) and EN 1994-1-2 (CEN 1994–2005b), the emissivity values given for materials are based on the assumption of heating in a Standard Fire test furnace, rather than the conditions of an actual building fire. For structural and stainless steels, single values of 0.7 and 0.4 are used, respectively. For concrete surface, EN 1994-1-2 prescribes a single value of 0.7.

5.2.1.2 Thermal conductivity

The thermal conductivity of a material is the coefficient that dictates the rate at which heat is conducted through the material. It relates the rate of energy conduction per second (in watts) per square metre at a point to the temperature gradient (in degrees centigrade or Kelvin per metre) at this point. Over a short distance, across which the thermal gradient can be considered constant, within a material it may be used to calculate the energy (joules) conducted per unit area over a reasonably brief time period.

Table 5.2 Typical thermal conductivity values for materials

Material at 25°C unless otherwise stated		Conductivity (W/m.K)		
Air (gas)		0.024		
Aluminium		250		
Brick	Dense	1.31		
	Brickwork	0.69		
	Fireclay brick (500°C)	1.4		
Concrete	Lightweight	0.42		
	Normal weight	1.7		
	Cement mortar	1.73		
Glass	Window	0.96		
Insulation materials	General	0.035-0.16		
	Asbestos board	0.744		
	Glass wool	0.04		
	Gypsum plaster	0.48		
	Gypsum or plasterboard	0.17		
	Plastics (foamed insulation materials)	0.03		
	Plaster, metal lath	0.47		
	Rock wool insulation	0.045		
	Vermiculite	0.058		
Iron		25°C	125°C	225°C
	Pure	80	68	60
	Cast		55	
	Wrought		59	
Steels		25°C	125°C	225°C
	Carbon	54	51	47
	Stainless	16	17	19
	Carbon 1%		43	
	Chrome nickel (18% Cr, 8% Ni)		16.3	
Rock	Solid	2-7		
	Sandstone	1.7		

Source: Data from http://www.engineeringtoolbox.com/thermal-conductivity-d_429.html.

Typical ambient temperature values of construction materials are given in Table 5.2. If materials are not defined precisely, ranges are given.

5.2.1.2.1 Thermal conductivity of structural steels

It can be seen from Table 5.2 that the thermal conductivity of steels changes with temperature. A simplified representation of the change of thermal

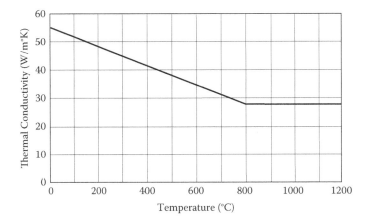

Figure 5.1 EN 1993-1-2 representations of the variation of thermal conductivity of steel with temperature (CEN 1993–2005b).

conductivity of carbon steel with temperature, defined in EN 1993-1-2, is shown in Figure 5.1; this applies to steel reinforcement as well as structural sections. For use with simple design calculations, the constant value of 45 W/m.K was suggested in the prestandard version of the code, and this should be a conservative assumption above 300°C where the full relationship is not required.

The variation is expressed in EN 1993-1-2 as

$$\theta_a = 20°C \text{ to } \theta_a = 800°C: \qquad \lambda_a = 54 - 3.33 \times 10^{-2}\,\theta_a \quad W/m.K$$

$$\theta_a = 800°C \text{ to } \theta_a = 1200°C: \quad \lambda_a = 27.3 \quad W/m.K \tag{5.1}$$

where θ_a is the steel temperature in degrees centigrade.

5.2.1.2.2 Thermal conductivity of concrete

The thermal conductivity λ_c of concrete logically depends on its chemistry, the thermal conductivity of each of its individual constituents and their proportions, and the moisture content, the type of aggregate and the cement type. If concrete is dry, the aggregate type has the most significant influence on its conductivity. However, as the moisture content of the concrete increases, its thermal conductivity increases. Hence there may be considerable deviation of thermal conductivity values from any prescribed value in particular cases.

EN 1994-1-2 provides curves specifying the upper and lower limits of thermal conductivity with temperature for normal-weight concretes

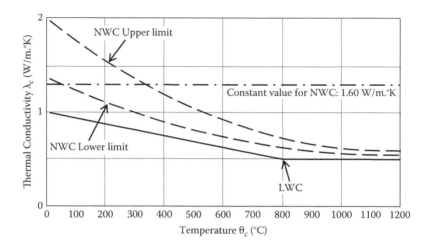

Figure 5.2 EN 1994-1-2 representations of the thermal conductivity of normal-weight concrete (NWC) and lightweight concrete (LWC) as functions of temperature (CEN 1994–2005b).

(Figure 5.2) based on large numbers of tests on concrete in composite structures. In the U.K. National Annex, only the upper-limit curve is permitted, as is recommended in the generic Eurocode. In simple calculation models for normal-weight concrete, a constant value of thermal conductivity of 1.6 W/m.K may be used. A single bilinear variation is provided for lightweight concrete.

The Eurocode curves for the limiting values of thermal conductivity λ_c of normal-weight concrete are given over the whole 20–1200°C range by

(Upper limit) $\lambda_c = 2 - 0.2451 \times 10^{-2}\,\theta_c + 0.0107 \times 10^{-4}\,\theta_c^2$ W/m.K

(Lower limit) $\lambda_c = 1.36 - 0.136 \times 10^{-2}\,\theta_c + 0.0057 \times 10^{-4}\,\theta_c^2$ W/m.K

$$(5.2)$$

where θ_c is the concrete temperature in degrees centigrade.

The Eurocode variation of thermal conductivity for lightweight concrete is given by

$\theta_c = 20°C$ to $\theta_c = 800°C$: $\lambda_c = 1 - \theta_c/1600$ W/m.K

$\theta_c = 800°C$ to $\theta_c = 1200°C$: $\lambda_c = 0.5$ W/m.K (5.3)

Table 5.3 Typical specific heat capacity values for materials

Material		Specific heat capacity, c_p (J/kg.K)
Aluminium	At 0°C	0.87
Brick	Hard	1.0
	Common	0.9
	Firebrick	1.05
Concrete	Lightweight	0.96
	Stone (normal weight)	0.75
Glass	Window	0.84
Insulation materials	Asbestos board	0.84
	Glass wool	0.67
	Plaster, light	1.0
	Plaster, sand	0.9
	Rockwool insulation	0.84
	Vermiculite	0.84
Iron	Pure 20°C	0.46
Rock	Solid	0.84
	Granite	0.79
	Sandstone	0.92

Source: Data from http://www.engineeringtoolbox.com/specific-heat-solids-d_154.html.

5.2.1.3 Specific heat capacity

The specific heat capacity of a material is the amount of energy (in joules) that it needs to gain to raise the temperature of unit mass (1 kg) of the material by 1°C. This varies to some extent with temperature for some materials, so the values in Table 5.3 for some materials used in construction should be taken only as nominal values, generally referring to the ambient range of temperatures.

5.2.1.3.1 Specific heat capacities of structural steels

The change of specific heat capacity of steel with temperature, in the form given in EN 1993-1-2, is shown in Figure 5.3 and can be seen to change gradually over the majority of the usable range. However, its value undergoes a dramatic change, apparently becoming extremely high within the range 700–800°C. The apparent sharp rise to an "infinite" value at about 735°C is actually a pragmatic way of taking account of the heat input needed to allow the endothermic process of the crystal structure phase change of carbon steel from body-centred to face-centred structures to take place. If heat transfer calculation to steelwork is to be

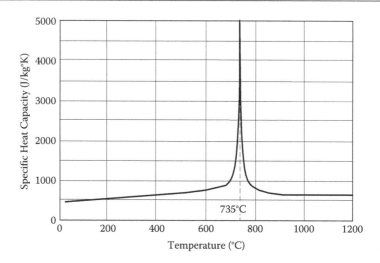

Figure 5.3 EN 1993-1-2 representation of the variation of the specific heat capacity of carbon steels with temperature (CEN 1993–2005b).

performed analytically or numerically, then this apparent "spike" in the specific heat capacity allows this to be done using normal heat conduction calculations.

Eurocode EN 1993-1-2 expresses this characteristic for all carbon steels as a sequence of equations:

$$\theta_a = 20°C \text{ to } \theta_a = 600°C: \quad c_a = 425 + 7.73 \times 10^{-1}\,\theta_a - 1.69 \times 10^{-3}\,\theta_a^2 + 2.22 \times 10^{-6}\,\theta_a^3 \text{ J/kg.K}$$

$$\theta_a = 600°C \text{ to } \theta_a = 735°C: \quad c_a = 666 + 13002/(738 - \theta_a) \text{ J/kg.K}$$

$$\theta_a = 735°C \text{ to } \theta_a = 900°C: \quad c_a = 545 + 17820/(\theta_a - 731) \text{ J/kg.K}$$

$$\theta_a = 900°C \text{ to } \theta_a = 1200°C: \quad c_a = 650 \text{ J/kg.K} \tag{5.4}$$

where θ_a is the steel temperature in °C.

5.2.1.3.2 Specific heat capacities of stainless steels

The change of specific heat capacity of all stainless steels with temperature, in the form given in EN 1993-1-2, is shown in Figure 5.4. It can be seen to change gradually over the entire usable range, in contrast to carbon steels.

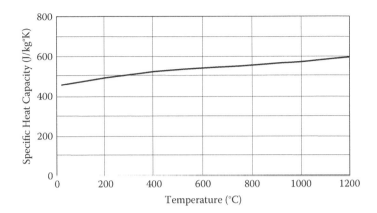

Figure 5.4 EN 1993-1-2 representation of the variation of the specific heat capacity of all stainless steels with temperature (CEN 1993–2005b).

EN 1993-1-2 expresses this characteristic for all stainless steels as a sequence of equations:

$$\theta_a = 20°C \text{ to } \theta_a = 1200°C: \quad c_a = 450 + 0.28\ \theta_a - 2.91 \times 10^{-4}\ \theta_a^2 \\ + 1.34 \times 10^{-7}\ \theta_a^3 \text{ J/kg.K} \tag{5.5}$$

where θ_a is the steel temperature in degrees centigrade.

5.2.1.3.3 Specific heat capacities of concrete

The specific heat of concrete c_c is also influenced by aggregate type, mixture rate and moisture content. Aggregate type is significant particularly in the case of concrete with calcareous aggregate, for which the specific heat increases suddenly because of chemical changes at a temperature of about 800°C. The moisture content is significant at temperatures up to 200°C because the specific heat of wet concrete is twice that for dry concrete. Clearly, the heat capacity of the free water has to be added to that of the dry concrete; in addition, the latent heat input needed to convert free water to pressurised steam within the concrete's pore structure has to be taken into account. Since the steam takes time to percolate out of the concrete, especially where it is some distance from the free surface, the temperature range over which the apparently increased specific heat capacity applies can extend between 100°C and about 200°C, and it is impossible to define this range precisely.

Because of the great variation possible in the constituents and water content of concrete, in EN 1994-1-2 the specific heat capacity of normal-weight concrete (Figure 5.5) is treated fairly simply. The value increases

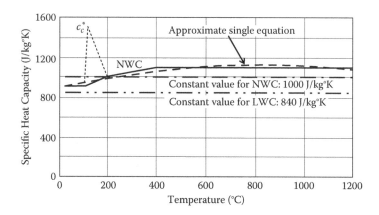

Figure 5.5 Specific heat capacity of normal-weight concrete (NWC) and lightweight concrete (LWC) as a function of temperature (CEN 1994–2005b).

gently between 100°C and 400°C, with constant values either side of this range. Between 100°C and 200°C, a spike can be added if the effect of the loss of free water as steam needs to be taken into account. If simplified models need a value for specific heat capacity, then a constant value of 1000 J/kg.K is permitted.

For lightweight concrete, once again there is considerable potential variation in the aggregates that can be used and in the processes by which the concrete is made, so a single conservative value of 840 J/kg.K is used.

For normal-weight concrete, the EN 1994-1-2 equations representing the variation shown in Figure 5.5 are

$$\theta_c = 20°C \text{ to } \theta_c = 100°C: \quad c_c = 900 \text{ J/kg.K}$$

$$\theta_c = 100°C \text{ to } \theta_c = 200°C: \quad c_c = 800 + \theta_c \text{ J/kg.K}$$

$$\theta_c = 200°C \text{ to } \theta_c = 400°C: \quad c_c = 900 + 0.5 \, \theta_c \text{ J/kg.K}$$

$$\theta_c = 400°C \text{ to } \theta_c = 1200°C: \quad c_c = 1100 \text{ J/kg.K} \tag{5.6}$$

An alternative approximate version that uses a single polynomial equation, shown as a broken line in Figure 5.5, is also provided. This may be advantageous to some software developers:

$$\theta_c = 20°C \text{ to } \theta_c = 1200°C: \quad c_c = 890 + 0.562 \, \theta_c - 3.4 \times 10^{-4} \, \theta_c^2 \text{ J/kg.K} \tag{5.7}$$

In simple calculation models, the constant value 1000 J/kg.K may be used.

Table 5.4 Variation of NWC specific
heat peak due to vaporising
free-water content

Water content, %	c_c^* , J/kg. K
3	2020
10	5600

If the spike between 100°C and 200°C needs to be used, the temperature at which the peak occurs cannot be specified, but its value c_c^* is given for two particular values of moisture content in Table 5.4. A 3% water content would be considered fairly normal inside buildings, and 10% probably indicates exposed elements in a wet climate.

5.2.1.4 Thermal expansion

In most simple fire engineering calculations, the thermal expansion of materials is neglected, but for steel members that support a concrete slab, on the upper flange the differential thermal expansion caused by shielding of the top flange and the heat-sink function of the concrete slab cause a "thermal bowing" towards the fire in the lower range of temperatures. When more advanced calculation models are used, it is also necessary to recognise that thermal expansion of the structural elements in the fire compartment is resisted by the cool structure outside this zone, and that this causes behaviour that is considerably different from that experienced by similar members in unrestrained furnace tests.

It is therefore necessary to appreciate the way in which thermal expansion of steel and concrete varies with temperature.

5.2.1.4.1 Thermal expansion of carbon steels

The Eurocodes EN 1993/EN 1994-1-2 curves of thermal strain of all carbon steels, which are representative of the steels used routinely for different grades of structural steel sections, is shown as the continuous curve in Figure 5.6. A gradually increasing gradient is halted in the range 750°C to 860°C, during which the major phase change is taking place, and resumes at the end of this range. The linearised version of this, which is only defined in EN 1994-1-2, is shown as the dashed mean line.

These characteristics are represented by curve-fitting equations in both Eurocodes:

$\theta_a = 20°C$ to $\theta_a = 750°C$: $\Delta l/l = 1.2 \times 10^{-5}\theta_a + 0.4 \times 10^{-8}\theta_a^2 - 2.416 \times 10^{-4}$

$\theta_a = 750°C$ to $\theta_a = 860°C$: $\Delta l/l = 1.1 \times 10^{-2}$

$\theta_a = 860°C$ to $\theta_a = 1200°C$: $\Delta l/l = 2 \times 10^{-5}\theta_a - 6.2 \times 10^{-3}\theta_a$ (5.8)

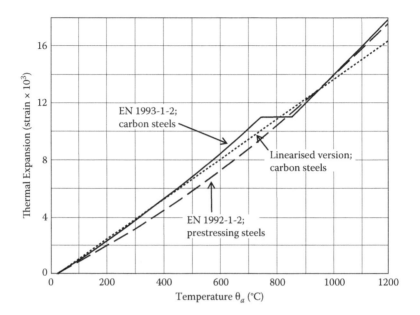

Figure 5.6 Variation of EN 1993/EN1994-1-2 thermal expansion of carbon steel with temperature (CEN 1993–2005b, 1994–2005b).

The linearised version from EN 1994-1-2 is

$$\Delta l/l = 1.4 \times 10^{-5}\, \theta_a$$

5.2.1.4.2 Thermal expansion of prestressing steel

A single expression is given in EN 1992-1-2 for thermal strain of the high-strength steels used as prestressing tendon:

$$\theta_p = 20°\text{C to } \theta_p = 1200°\text{C}: \Delta l/l = -2.016 \times 10^{-4} + 10^{-5}\, \theta_p + 0.4 \times 10^{-8}\, \theta_p^2$$

$$(5.9)$$

This is also shown for comparison in Figure 5.6.

5.2.1.4.3 Thermal expansion of austenitic stainless steels

Eurocode EN 1993-1-2 provides a curve of thermal strain for austenitic stainless steels, the most common type that may be used for load-bearing structural members. These have no phase changes comparable to that which affects carbon steels, so the expansion shown in Figure 5.7 is monotonic

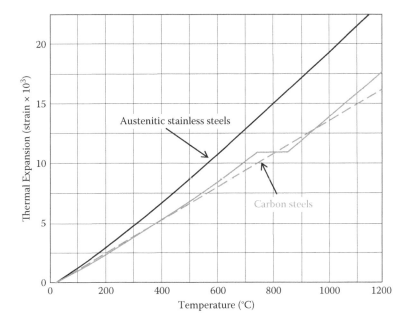

Figure 5.7 Variation of Eurocode EN 1993/EN 1994-1-2 thermal expansion of austenitic stainless steel with temperature (CEN 1993–2005b, 1994–2005b).

with temperature. It can be seen that austenitic stainless steels expand due to temperature increase by an order of 50% more than carbon steels.

The EN 1993-1-2 curve-fitting equation is

$$\theta_a = 20°C \text{ to } \theta_a = 1200°C: \quad \Delta l/l = (16 + 4.79 \times 10^{-3}\,\theta_a - 1.243 \times 10^{-6}\,\theta_a^2)$$
$$\times (\theta_a - 20) \times 10^{-6} \tag{5.10}$$

5.2.1.4.4 Thermal expansion of concrete

Thermal expansion of concrete is mainly influenced by the characteristics of its aggregate type because this is its major constituent by volume. Aggregate may vary from place to place, depending on local circumstances, but in normal weight concrete tends to be either siliceous (including gravels and crushed granite) or calcareous (crushed limestones). Lightweight aggregates are generally by-products of industrial processes, such as sintered blast furnace slag or pulverised fuel ash and expanded shale. By and large, thermal strains change in a regular manner, but where abrupt changes happen, these tend to be caused by chemical or crystal structure processes.

Thermal strains of siliceous aggregates are mainly influenced, in the normal range of structure temperatures in fire, by the phase transition from the α- to

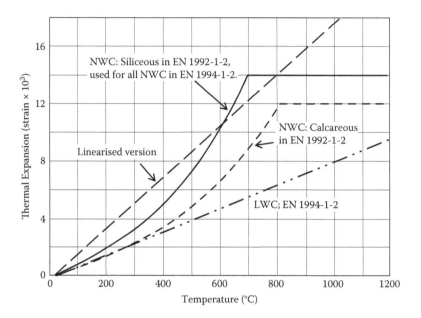

Figure 5.8 Variation of Eurocodes EN 1992-1-2 and EN 1994-1-2 total thermal expansion of concretes with temperature (CEN 1992–2005b, 1994–2005b).

β-crystal structure of quartz, which occurs at 573°C and is accompanied by a volume expansion of 0.45%. In the case of calcareous aggregates, chemical changes to the structure of limestone, mainly concerned with the breakdown from calcium carbonate to calcium oxide that occurs at an accelerating rate from about 550°C, are the most significant influences; industrial production of quicklime takes place at temperatures in the range 900–1000°C.

The thermal strain of concrete (Figure 5.8) is given for both siliceous and calcareous normal-weight concretes in EN 1992-1-2 (CEN 1992–2005b); in EN 1994-1-2, only the higher expansion of siliceous concrete is taken as a single curve representing all normal-weight concretes. It is not obvious that this represents a conservative approach for composite structures because of the effects of differential expansion of steel sections and concrete composite slabs on lateral deflections, whereas in concrete construction the net expansion might be considered the more significant effect. The curve-fitting equations given in the Eurocodes are as follows:

Siliceous-aggregate normal-weight concrete in EN 1992-1-2 or all normal-weight concrete in EN 1994-1-2:

$\theta_c = 20°C$ to $\theta_c = 700°C$: $\Delta l/l = -1.8 \times 10^{-4} + 9 \times 10^{-6}\theta_c + 2.3 \times 10^{-11}\theta_c^3$

$\theta_c = 700°C$ to $\theta_c = 1200°C$: $\Delta l/l = 1.4 \times 10^{-2}$ (5.11)

Calcareous-aggregate normal-weight concrete in EN 1992-1-2:

$\theta_c = 20°C$ to $\theta_c = 805°C$: $\Delta l/l = -1.2 \times 10^{-4} + 6 \times 10^{-6}\,\theta_c + 1.4 \times 10^{-11}\,\theta_c^3$

$\theta_c = 805°C$ to $\theta_c = 1200°C$: $\Delta l/l = 1.2 \times 10^{-2}$ $\qquad$ (5.12)

In EN 1994-1-2, a linearised version of thermal expansion is given for use with simplified calculation methods:

$\theta_c = 20°C$ to $\theta_c = 1200°C$: $\Delta l/l = 1.8 \times 10^{-5}\,(\theta_a - 20)$ $\qquad$ (5.13)

For lightweight concrete, no information is given in EN 1992-1-2, but in EN 1994-11-2 a simple linearised relationship is given for all types:

$\theta_c = 20°C$ to $\theta_c = 1200°C$: $\Delta l/l = 8 \times 10^{-6}\,(\theta_a - 20)$ $\qquad$ (5.14)

5.2.1.4.5 Thermal expansion coefficients

In time-based fire engineering software applications, it may be useful to calculate free thermal strain in incremental rather than absolute terms, as the time into a fire scenario increases. The coefficients of thermal expansion given as the temperature derivatives of the Eurocode expressions for absolute thermal strain are shown in Figure 5.9.

5.2.2 Mechanical properties

Most construction materials suffer a progressive loss of both stiffness and strength with increase of temperature. This occurs at temperatures considerably below the materials' melting points. From the structural viewpoint, the materials science background to this behaviour is less relevant than obtaining accurate representations of the stress-strain-temperature properties in a form that is appropriate for the structural analysis being conducted, and this is the emphasis here.

5.2.2.1 Rate dependency of mechanical properties

It is conventional in most structural engineering design analysis to consider structural behaviour as quasi-static, although it is obvious that there is always a dynamic aspect to all structural behaviour as loads change. When loads change slowly and a structure embodies enough structural damping and internal friction to make vibration behaviour of minor importance, the quasi-static assumption is reasonable in that inertial forces are not significant in contributing to the strength of the structure. However, material mechanical properties themselves may be time or strain-rate dependent.

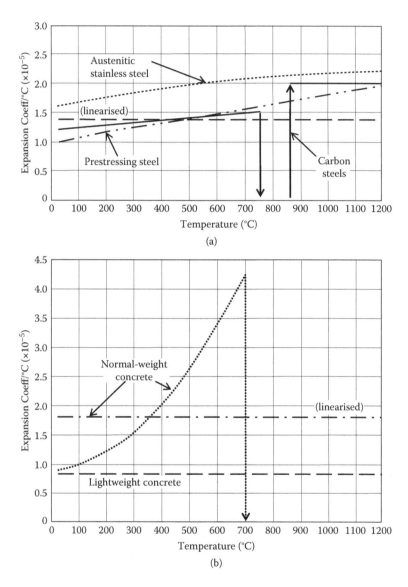

Figure 5.9 Variation of EN 1993-1-2/EN 1994-1-2 thermal expansion coefficients of (a) steel and (b) concrete with temperature (CEN 1993–2005b, 1994–2005b).

The most important phenomena that link stress to strain rate in many materials are generally grouped together under the general title of creep. This refers to the tendency of materials under stresses below yield to continue to strain after an initial stress has been established. Creep is usually viewed by engineers in terms of the S-shaped curve illustrated by Figure 5.10 for

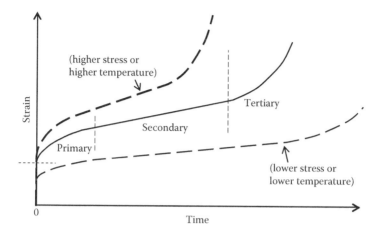

Figure 5.10 Qualitative illustration of creep straining of a material with time. The temperature and stress level are assumed to be constant.

a material at a fixed temperature and a constant stress level and is classified qualitatively into the stages primary, secondary and tertiary, which are divided by changes of curvature. In the primary stage, which starts as the stress is applied, there is initially a rapid strain rate that diminishes progressively to a roughly constant, but very low, rate that defines the secondary stage. The tertiary stage is generally only experienced when the material is in the vicinity of its yield strength; it either may never occur or may only happen after extremely long loading times. In Figure 5.10, the stages are shown for the middle curve, which is for a constant stress level. The upper and lower curves represent the behaviour of the same material at higher and lower stress levels, or temperatures, respectively.

5.2.2.2 Transient and isothermal materials testing methods

Temperatures in an accidental fire in a building vary with time, and clearly the generation of temperatures in the structural materials within and surrounding the fire is dependent on the nature of the fire temperature curve to which they are subjected. Even in standard testing situations, when elements of structure are subjected to one of the standard time-temperature curves (commonly known as the Standard Fire curve of the International Organisation for Standardisation [ISO] 834-1, 1975; EN 1991-1-2 [CEN 1991–2002] or BS 476-22 of the British Standards Institution [BSI] 1987), time is an important parameter, together with the heat transfer properties of the sections and their materials, in determining the temperatures generated. Therefore, for any material for which creep can occur under stress and elevated temperature, it is necessary to find a way of taking account of the effect of creep on the structural behaviour in fire.

In feasible numerical modelling for structural fire engineering design, it is generally impractical to include strain rate dependency in the material constitutive relationships used. A practical way of compensating for time dependency of material stress-strain characteristics at high temperatures is to conduct materials testing, which takes extended periods of time compared with normal testing procedures. The objective is to achieve strains and strain rates that cover a reasonable part of the normal response range of structural members in building fires. This is easier to achieve if the fire atmosphere temperature regime is assumed to be the ISO 834 Standard Fire, effectively representing standardised isolated member tests, rather than the models of natural fires used in performance-based design. A notable example of extended-period testing, which is generally referred to as transient testing, was reported by Kirby and Preston (1988) on structural carbon steels of the contemporary U.K. grades 43A and 50B, with nominal yield strengths of 275 and 355 N/mm², respectively. In their transient tests, tensile specimens were held at different constant stress levels as their temperatures were controlled to increase at different constant rates, and their total strains were recorded as temperatures increased. Mechanical strains were then obtained by removing thermal expansions from the total values. Tests were conducted at heating rates of 2.5, 5, 10 and 20°C/min. Specimen results can be seen, as curves of strain against temperature for grade 50B (S355) steel heated at 10°C/min, in Figure 5.11.

By re-interpolating from these eight curves at constant stress, a family of stress-strain curves can be generated for different temperature levels, which are specific to this particular heating rate; similar families of curves can be generated for each of the heating rates at which tests are conducted. Another form of presentation of the results is to plot curves of the degeneration of "strength" with temperature for all the strain rates. Since high-temperature stress-strain curves are curvilinear, the strengths used are proof stresses at particular strain values. Two examples taken from Kirby and Preston's (1988) results are shown in Figures 5.12a and 5.12b for 0.2% and 1.0% proof stresses, respectively. The figures show the ranges of these proof stresses taken from a large number of tests on carbon steel specimens at constant temperature ("steady-state" tests), together with curves representing the minimum values taken from the transient tests at different heating rates. It can be seen that the minimum 0.2% proof stress from transient tests lies at least 10% below the minimum of the steady-state range between 400°C and 800°C, which is the temperatures range reported. However, for the 1.0% proof stress the minimum from transient tests is almost indistinguishable from the lower end of the steady-state range. Since 2% is a high working strain for steels, at which an individual stress-strain curve at any temperature has effectively

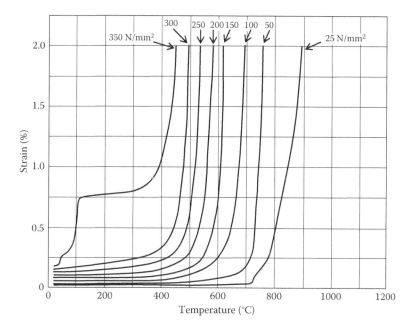

Figure 5.11   Results of transient tensile tests (Kirby & Preston 1988) on grade 53B steel at different constant stress levels, under 10°C/min heating rate. (From Kirby, B.R., and Preston, R.R., High Temperature Properties of Hot-Rolled Structural Steels for Use in Fire Engineering Studies, *Fire Safety Journal*, 13(1), pp. 27–37, 1988. With permission from Elsevier.)

reached a plateau (although at lower temperatures additional strength can be mobilised through work hardening at much larger strains), whereas at 0.1% the curve is still rising significantly, this shows the effect of time dependency as simply adding creep strain to the short-term mechanical strain at any given stress level.

5.2.2.3 Eurocode EN 1993-1-2 stress-strain curves for structural carbon steels

For structural fire engineering design purposes, the degeneration of strength with steel temperature is much more significant than the associated strain level (or in other words the degeneration of the elastic modulus). The approach taken in EN 1993-1-2 is therefore to devise a format for high-temperature stress-strain curves that is systematic in retaining a linear-elastic range, has a good representation of the curvilinear range, and (most important) takes a conservative approach to the effect of creep on yield strength without specifying heating rates. Hence EN 1993-1-2 specifies stress-strain curves that vary only with temperature and are based on

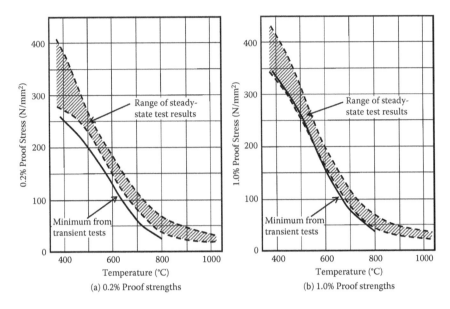

Figure 5.12 Degeneration of 0.2% and 1.0% proof strengths from transient and isothermal tensile tests (Kirby & Preston 1988) on grade 53B steel; range of results
from steady-state tests compared with the minimum values from transient
tests. (From Kirby, B.R., and Preston, R.R., High Temperature Properties of
Hot-Rolled Structural Steels for Use in Fire Engineering Studies, *Fire Safety
Journal*, 13(1), pp. 27–37, 1988. With permission from Elsevier.)

the lower bound of transient test results. Since strain hardening is of no
significance above 400°C, there is a general form of stress-strain curve
that ignores the possibility of strain hardening, which can be applied at all
temperatures. For temperatures below 400°C, a nominal ultimate strength
above yield can be used as an enhancement of the general curve. The curves
are constructed as shown in Figure 5.13:

- A linear range runs from zero stress to the limit of proportionality $f_{p,\theta}$.
  The gradient of this part forms an elastic (Young's) modulus $E_{a,\theta}$, which
  is degraded as the steel temperature rises.
- An ellipse is constructed to be tangential, both to the final point of the
  initial line (at $f_{p,\theta}$, $\varepsilon_{p,\theta}$) and to the horizontal at the yield point where
  the strain is 2% and the stress is the degraded yield strength $f_{y,\theta}$.
- In the general model, a stress plateau at stress $f_{y,\theta}$ runs from the yield
  point to the strain $\varepsilon_{t,\theta}$, which is set at 15%, at which the stress begins
  to reduce.

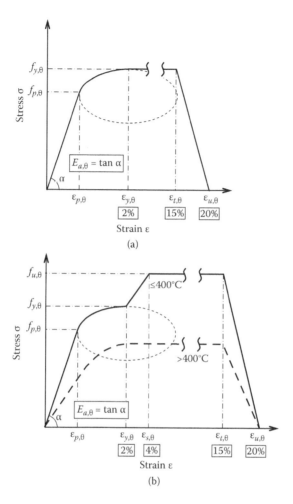

Figure 5.13 Eurocode 3 (CEN 1993–2005b) models for uniaxial stress-strain characteristics of carbon steels with temperature q. (a) General model for all temperatures; (b) alternative model with strain hardening below 400°C.

- For temperatures below 400°C, the alternative model represents
strain hardening with a straight line from the yield point to an ulti-
mate stress of 1.25 $f_{y,\theta}$ at a strain $\varepsilon_{s,\theta}$ of 4%. A horizontal plateau then
runs to $\varepsilon_{t,\theta}$, again at 15%.
- The final, negative-gradient linear, phase then represents the final
fracture process. The stress reaches zero at a strain $\varepsilon_{u,\theta}$ of 20%, representing fracture.

The EN 1993-1-2 equations, which can be used in spreadsheet software to reproduce the general model stress-strain curves, are

$\varepsilon = 0$ to $\varepsilon = \varepsilon_{p,\theta}:$ $\qquad$ $\sigma = \varepsilon E_{a,\theta}$ $\qquad\qquad$ [So $f_{p,\theta} = \varepsilon_{p,\theta} E_{a,\theta}$]

$\varepsilon = \varepsilon_{p,\theta,\theta}$ to $\varepsilon = \varepsilon_{y,\theta} = 0.02:$ $\quad$ $\sigma = f_{p,\theta} - c + \frac{b}{a}\sqrt{a^2 - (0.02 - \varepsilon)^2}$

$\varepsilon = \varepsilon_{y,\theta}$ to $\varepsilon = \varepsilon_{t,\theta} = 0.15:$ $\quad$ $\sigma = f_{y,\theta}$ $\qquad\qquad\qquad\qquad$ (5.15)

$\varepsilon = \varepsilon_{t,\theta}$ to $\varepsilon = \varepsilon_{u,\theta} = 0.20:$ $\quad$ $\sigma = f_{y,\theta}[1 - (\varepsilon - 0.15)/0.05]$

The terms a, b and c in these equations are calculated using

$$c = \frac{(f_{y,\theta} - f_{p,\theta})^2}{(\varepsilon_{y,\theta} - \varepsilon_{p,\theta})E_{a,\theta} - 2(f_{y,\theta} - f_{p,\theta})} \qquad (5.16)$$

This can then be substituted into

$$a = \sqrt{(0.02 - \varepsilon_{p,\theta})(0.02 - \varepsilon_{p,\theta} + c/E_{a,\theta})} \qquad (5.17)$$

$$b = \sqrt{c(0.02 - \varepsilon_{p,\theta})E_{a,\theta} + c^2} \qquad (5.18)$$

To construct stress-strain curves from these equations at any steel temperature θ, it is necessary to use data giving the unknown coordinates of the elastic modulus, the limit of proportionality and the yield strength as functions of temperature. Carbon steels can be specified at different strength grades, with the most common being S235, S275 and S355; in terms of stress, their proportional degradations with temperature are extremely close to one another. Since all the common grades are to be covered by this common set of equations, these are most conveniently defined in terms of reduction factors for $E_{a,\theta}$, $f_{p,\theta}$ and $f_{y,\theta}$. The reduction factors for $E_{a,\theta}$ and $f_{y,\theta}$ for structural carbon steels are shown graphically in piecewise-linear form in Figure 5.14. All three are tabulated in Table 5.5. Although melting does not happen until well over 1500°C, only 23% of the ambient-temperature strength remains at 700°C. At 800°C, this has reduced to 11% and at 900°C to 6%.

A set of EN 1993-1-2 stress-strain curves for S275 steel at temperatures from 20°C to 800°C is shown in Figure 5.15.

5.2.2.4 Reduction factors for other carbon steel components

Although the carbon steels used in structural engineering have similar chemical compositions, some types of components achieve higher

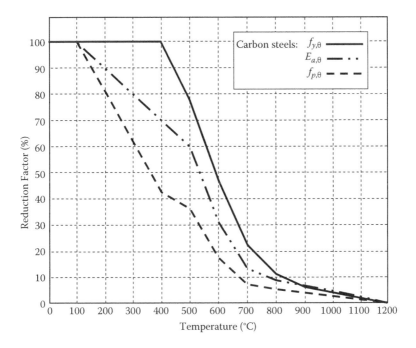

Figure 5.14 EN 1993-1-2 strength reduction factors for structural carbon steel at high temperatures (CEN 1993–2005b).

Table 5.5 Reduction factors for hot-rolled carbon steels, including reinforcing bars, at temperature θ, relative to the value of f_y or E_a at 20°C.

Steel temperature θ (°C)	Yield strength $k_{y,\theta} = f_{y,\theta}/f_y$	Proportional limit $k_{p,\theta} = f_{p,\theta}/f_y$	Elastic modulus $k_{E,\theta} = f_{a,\theta}/f_a$
20	1.000	1.000	1.000
100	1.000	1.000	1.000
200	1.000	0.807	0.900
300	1.000	0.613	0.800
400	1.000	0.420	0.700
500	0.780	0.360	0.600
600	0.470	0.180	0.310
700	0.230	0.075	0.130
800	0.110	0.050	0.090
900	0.060	0.0375	0.0675
1000	0.040	0.0250	0.0450
1100	0.020	0.0125	0.0225
1200	0.000	0.0000	0.0000

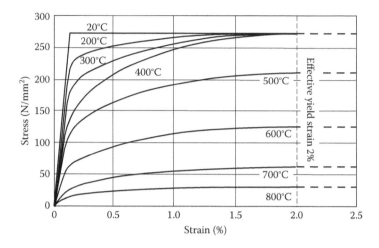

Figure 5.15 Stress-strain curves of hot-rolled carbon steels with temperature for S275 steel (EN 1993-1-2 curves; CEN 1993–2005b).

ambient-temperature strengths (at the expense of some ductility) than their parent steels:

- by heat treatment (quenching and tempering), which applies to bolts
- by cold-working, which applies to some drawn reinforcing bars.

In both cases, heating causes higher loss of strength than that for normal carbon steel components, largely because the ambient-temperature strengths have been greatly increased from that of the original steel.

For welds, very rapid cooling from the molten state during the welding process causes an unusual distribution of steel crystal structures across the welded region, as well as considerable residual stresses. When the weld is heated in fire to temperatures that are significant in terms of the carbon steel phase diagram, the times for which these temperatures are retained are slow enough to allow the necessary crystal structure changes to take place.

The EN 1994-1-2 reduction factors for the elastic modulus and yield strength of cold-worked reinforcement, whose standard grade is S500, are shown graphically in Figure 5.16 and are tabulated in Table 5.6. It is unlikely that reinforcing bars or mesh will reach very high temperatures in a fire, given the insulation provided by the concrete if normal cover specifications are maintained. The very low ductility of S500 steel (it is only guaranteed at 5%) may be of more significance where high mesh strains occur in slabs.

For bolts and welds, the elastic modulus is not considered significant in EN 1993-1-2, so only the strength reduction factors, which apply to both tension and shear, are given in Figure 5.16 and Table 5.6. As whole-structure modelling

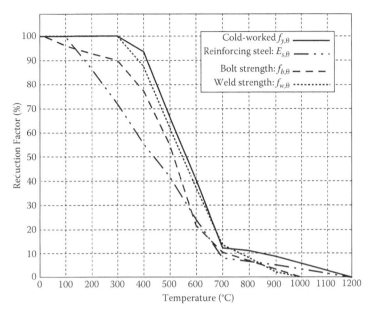

Figure 5.16 EN 1993/EN 1994-1-2 reduction factors for strength and modulus of cold-worked reinforcement and strengths of bolts and welds at high temperatures (CEN 1993–2005b, 1994–2005b).

Table 5.6 Reduction factors at temperature θ for cold-worked reinforcing bars, steel bolts and welds

| Steel temperature θ (°C) | Cold-worked reinforcement | | | Bolts | Welds |
	Yield strength $k_{y,\theta}$	Proportional limit $k_{p,\theta}$	Elastic modulus $k_{E,\theta}$	Strength $k_{b,\theta}$	Strength $k_{w,\theta}$
20	1.000	1.000	1.000	1.000	1.000
100	1.000	0.960	1.000	0.968	1.000
150	—	—	—	0.952	1.000
200	1.000	0.920	0.870	0.935	1.000
300	1.000	0.810	0.720	0.903	1.000
400	0.940	0.630	0.560	0.775	0.876
500	0.670	0.440	0.400	0.550	0.627
600	0.400	0.260	0.240	0.220	0.378
700	0.120	0.080	0.080	0.100	0.130
800	0.110	0.060	0.060	0.067	0.074
900	0.080	0.050	0.050	0.033	0.018
1000	0.050	0.030	0.030	0.000	0.000
1100	0.030	0.020	0.020	—	—
1200	0	0	0	—	—

for advanced performance-based design develops, however, it is likely that a model for the elastic modulus of bolts in joints will need to be developed.

5.2.2.5 An alternative form of stress-strain curve for numerical analysis: the Ramberg-Osgood equation

For the purposes of practical design that is subject to scrutiny by building control authorities, there is a clear incentive for designers to use material constitutive properties that have the stamp of approval conferred by a current code of practice; therefore, the Eurocode definitions of high-temperature steel properties outlined previously are very important in performance-based design modelling. They do, however, have the disadvantage of being discontinuous in terms of the rate of change of the gradient (the tangent modulus) at the points that define the limit of proportionality and yield, even though the values of stress and tangent modulus are continuous. In numerical analysis, it may be more convenient to express the stress-strain curves as continuous functions, and the most flexible way of doing this is to use Ramberg-Osgood (Ramberg and Osgood 1943) equations. These take the general form

$$\varepsilon_\theta = \left(\frac{\sigma_\theta}{A_\theta}\right) + 0.01\left(\frac{\sigma_\theta}{B_\theta}\right)^{N_\theta} \tag{5.19}$$

where ε_θ and σ_θ represent strain and stress, respectively, at temperature θ; and A_θ, B_θ and N_θ are temperature-dependent coefficients. These coefficients can be varied to change the shape of a curve; basically, A_θ represents the initial elastic modulus and B_θ the yield strength.

The index N_θ enables the curvature to be adjusted. Figure 5.17 shows the results of a curve fit by El-Rimawi (1989) in plotting the variation of

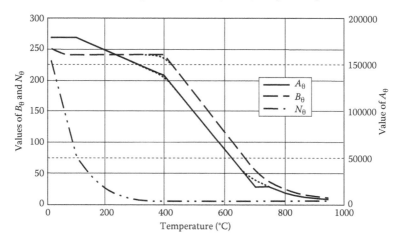

Figure 5.17 Typical variation of coefficients for the Ramberg-Osgood equation for steel with temperature.

the coefficients with temperature. – The solid lines represent his original equations:

$\theta = 20°C$ to $\theta = 100°C$: $A_\theta = 180 \times 10^3$

$B_\theta = 0.00134\theta^2 - 0.26\theta + 254.67$

$N_\theta = 237 - 1.58\theta$

$\theta = 100°C$ to $\theta = 400°C$: $A_\theta = 194 \times 10^3 - 140\theta$

$B_\theta = 242$

$N_\theta = 15.3 \times 10^{-7} \times (400 - \theta)^{3.1} + 6$

$\theta = 400°C$ to $\theta = 700°C$: $A_\theta = (295.33 - 0.3933\theta) \times 10^3$

$B_\theta = 492.667 - 0.6266\theta$

$N_\theta = 6$

$\theta = 700°C$ to $\theta = 800°C$: $A_\theta = (30.5 - 0.015\theta) \times 10^3$

$B_\theta = 306 - 0.36\theta$

$N_\theta = 0.4\theta - 22$ (5.20)

The dotted curves at 400°C and 700°C represent a polynomial rounding of the sharp intersections of the expressions for A_θ and B_θ. It has been found that, although the stress-strain curves produced for any temperature by the original expressions are perfectly adequate in most cases, there may be some anomalies in analytical results for cases in which buckling is a major factor because of the sharp changes of slope at these temperatures, and that creating a gradual change eliminates these. The stress-strain curves shown in Figure 5.18, and the previous equations, assume that the steel's ambient-temperature yield strength is 250 N/mm², and that its elastic modulus is 180×10^3 N/mm². If it is necessary to convert the curves for steel with different strength and modulus, this can be done simply by factoring A_θ and B_θ by a_θ and b_θ, respectively, as follows:

$a_\theta = E_{20}/(180 \times 10^3)$

$b_\theta = f_{y,20}/250$ (5.21)

Ramberg-Osgood curves do not, by the nature of their equation, ever peak, so they must be used with care when very high strain levels are anticipated. It is also noticeable that the stress-strain curves plotted in Figure 5.18, particularly the curve at 300°C as rising above ambient-temperature yield at about 1.5% strain. This is in line with some tensile

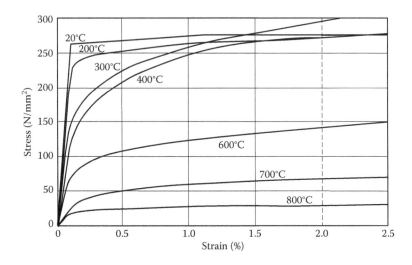

Figure 5.18 Stress-strain curves of hot-rolled carbon steels with temperature for S250 steel (Ramberg-Osgood curves).

test results, and it must be remembered the lower-temperature curves do experience strain hardening from about this level of strain, so it is not as surprising as it seems.

5.2.2.6 Biaxial properties

Structural steel is a ductile and homogeneous material, with properties that can be considered, at least within the range of strains relevant for most engineering purposes, as having the same properties in tension and compression. It is of course true that final failure mechanisms at very high strains differ considerably; necking, which marks the eventual postpeak path culminating in fracture, does not happen in compression. In numerical modelling applications, it is rarely necessary to perform three-dimensional stress analysis because structural components tend to be composed of plates. However, plane-stress conditions are common in the planes of webs and flanges, and where it is necessary to consider stress fields in plate elements, it is necessary at least to have a criterion that links the biaxial principal stresses to identify yield.

Two classical multiaxial failure criteria tend to be used for ductile materials, and there is little virtue in going any further for carbon steels. The Tresca (1864) criterion is based on the maximum absolute value of shear stress obtained from the three Mohr circles that describe the three-dimensional stress state at any point. This implies for plane stress that, where the two principal stresses are both compressive or both tensile, the greater of these principal stresses is compared with the yield stress. Where the principal stresses are of different signs, there is a linear relationship linking the uni-axial stresses:

$$(\sigma_1/f_y) - (\sigma_2/f_y) = 1, \text{ between } \sigma_1 = f_y, \sigma_2 = 0, \text{ and } \sigma_1 = 0, \sigma_2 = -f_y$$

$$(\sigma_2/f_y) - (\sigma_1/y) = 1, \text{ between } \sigma_1 = 0, \sigma_2 = f_y, \text{ and } \sigma_1 = -f_y, \sigma_2 = 0$$

$$(5.22)$$

These yield surfaces are shown in Figure 5.19a; the elevated-temperature yield surfaces are scaled with the yield strength reduction factors appropriate to their temperatures, from Table 5.5.

It is more usual for steel to use the energy-based von Mises (1913) yield criterion, which coincides with Tresca for uniaxial stress. As can be seen from Figure 5.19b, this criterion produces elliptical yield surfaces for plane stress:

$$(\sigma_1^2 - \sigma_1\sigma_2 + \sigma_2^2) = f_y^2 \qquad (5.23)$$

5.2.2.7 High-temperature mechanical properties of concrete

Concrete also loses its strength properties as its temperature increases, although a variety of parameters contribute to the relevant characteristics of any given concrete element in the structure.

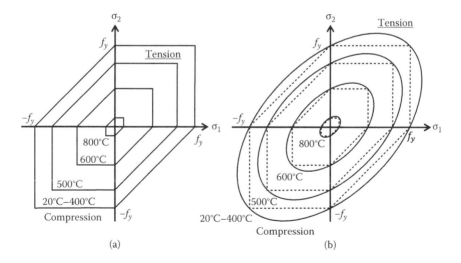

Figure 5.19 Failure surfaces of steel for plane stress: (a) Tresca; (b) von Mises (Tresca shown as dotted lines, circumscribed by von Mises).

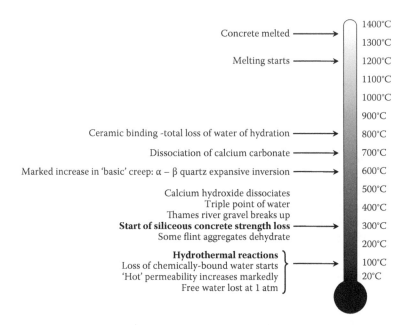

Figure 5.20 Chemical and physical changes to concretes as temperatures increase.

In thinking about the change of concrete properties with temperature, it is most useful to consider as a context the physical and chemical changes that happen to the main constituents of the concrete (the cement paste matrix and the aggregate) at various temperatures. A useful qualitative view, given by Khoury (2000), is reproduced as Figure 5.20. This shows clearly that even at temperatures as low as 100°C free water is lost from the matrix, increasing its permeability. Further chemical changes happen to calcium hydroxide and carbonate between 300°C and 800°C; siliceous aggregates also experience considerable changes in this temperature range.

The stress-strain curves at different temperatures for concrete have a significant difference in form from those for steel. Concrete is a brittle material whose strength in tension is hard to guarantee, and in normal structural design any tensile strength is ignored; this is the normal situation in Eurocode 4 design for fire, although in whole-structure analysis it is useful to make use of some of the tensile strength that undoubtedly exists. Some guidance for this exists in EN 1992-1-2.

5.2.2.8 EN 1992/EN 1994-1-2 stress-strain curves for concrete in compression

Eurocodes EN 1992-1-2 and EN 1994-1-2 present representations only of the compressive stress-strain behaviour of concretes. The curves are all curvilinear

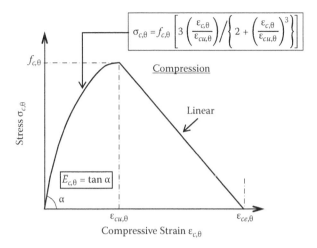

Figure 5.21 EN 1992/EN 1994-1-2 model for uniaxial compressive stress-strain characteristics of siliceous (conservative for calcareous) concretes with temperature q (CEN 1992–2005b, CEN 1994–2005b).

up to a maximum compressive strength, and this occurs at strains that progressively increase with temperature, up to 600°C, followed for each curve by
a linear descending branch. In EN 1994-1-2 design, the stress-strain curves
are based on concretes with siliceous aggregates, which are assumed to be
conservative in terms of strength at high temperatures compared with those
with calcareous aggregates. However, EN 1992-1-2 also provides information
for calcareous-aggregate concrete. The Eurocode curves for all concretes are
constructed as shown in Figure 5.21, in which compression is plotted as the
top-right quadrant; this will be the case in all the uniaxial stress-strain plots
for concrete. The initial curvilinear part is drawn in the form

$$\varepsilon_c, \theta = 0 \text{ to } \varepsilon_{c,\theta} = \varepsilon_{cu,\theta}: \quad \sigma_{c,\theta} = f_{c,\theta}\left[3\left(\frac{\varepsilon_{c,\theta}}{\varepsilon_{cu,\theta}}\right)\middle/\left\{2+\left(\frac{\varepsilon_{c,\theta}}{\varepsilon_{cu,\theta}}\right)^3\right\}\right] \quad (5.24)$$

Beyond the maximum compressive strength at a given temperature, a
descending part is defined as a straight line to the defined strain $\varepsilon_{c,\theta}$, at
which the stress is zero. The factors controlling the values of compressive
strength $f_{c,\theta}$ for both normal-weight and lightweight concretes, the corre-
sponding strain $\varepsilon_{cu,\theta}$ and the final strain at zero stress $\varepsilon_{s,\theta}$ are defined by the
data given in Table 5.7.

The reduction factors for strength are shown graphically in Figure 5.22.
Lightweight concretes are treated as if they all degrade similarly with temperature, independent of the aggregate type used. A set of normalised EN

Table 5.7 EN 1992/EN 1994-1-2 reduction factors at temperatures θ for normal-weight and lightweight concretes and the strains at ultimate and zero stresses

Concrete temperature θ (°C)	Compressive strength (NWC) EN 1992-1-2 and EN 1994-1-2 Siliceous $k_{c,\theta} = f_{c,\theta}/f_c$	Compressive strength (NWC) EN 1992-1-2 Calcareous $k_{c,\theta} = f_{c,\theta}/f_c$	Compressive strength (LWC) $k_{c,\theta} = f_{c,\theta}/f_c$	Strain at ultimate stress $\varepsilon_{cu,\theta}$	Strain at zero stress $\varepsilon_{ce,\theta}$
20	1	1	1	0.0025	0.0200
100	1	1	1	0.0040	0.0220
200	0.95	0.97	1	0.0055	0.0250
300	0.85	0.91	1	0.0070	0.0270
400	0.75	0.85	0.88	0.0100	0.0300
500	0.60	0.74	0.76	0.0150	0.0325
600	0.45	0.60	0.64	0.0250	0.0350
700	0.30	0.43	0.52	0.0250	0.0375
800	0.15	0.27	0.40	0.0250	0.0400
900	0.08	0.15	0.28	0.0250	0.0425
1000	0.04	0.06	0.16	0.0250	0.0450
1100	0.01	0.02	0.04	0.0250	0.0475
1200	0	0	0	—	—

1992/EN 1994-1-2 stress-strain curves for normal-weight concretes over a range of constant temperatures are shown in Figure 5.23.

Each of the normalised stress-strain curves for siliceous and calcareous-aggregate normal-weight concretes shown in Figure 5.23 covers only the behaviour at constant temperature. It is important to note that concrete, after cooling to ambient temperature, does not regain its initial compressive strength. The heating process damages the internal structure of the material permanently; there is even some increase in this internal damage during the cooling back to 20°C. The residual strength $f_{c,\theta,20°C}$ after cooling to 20°C from the maximum temperature θ_{max} depends on this maximum temperature. EN 1994-1-2 presents a model of this behaviour to be used with natural (parametric) fire heating:

- The residual strength $f_{c,\theta,20°C}$ after cooling from θ_{max} to 20°C is calculated according to Eurocode 4 Part 1–2 Annex C as

$$\theta_{max} = 20°C \text{ to } \theta_{max} = 100°C : \quad f_{c,\theta_{max},20°C} = k_{c,\theta_{max}}f_c$$

$$\theta_{max} = 100°C \text{ to } \theta_{max} = 300°C : \quad f_{c,\theta_{max},20°C} = [0.95 - 0.185(\theta_{max} - 100)/200]f_c$$

$$\theta_{max} > 300°C : \quad f_{c,\theta_{max},20°C} = 0.9k_{c,\theta_{max}}f_c \quad (5.25)$$

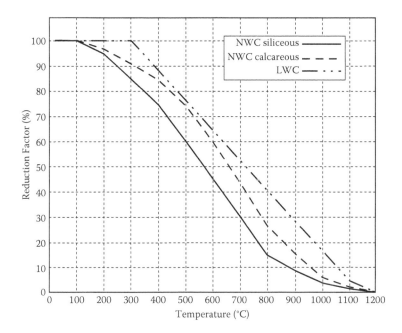

Figure 5.22 EN 1992/EN 1994-1-2 reduction factors for strengths of normal-weight concretes with siliceous and calcareous aggregates and lightweight concretes at high temperatures (CEN 1992–2005b, CEN 1994–2005b).

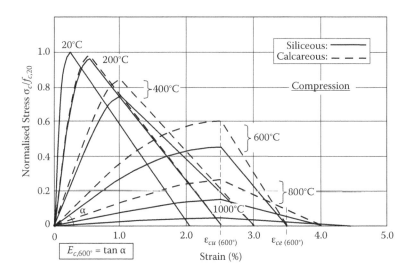

Figure 5.23 Normalised stress-strain curves of siliceous and calcareous normal-weight concretes with temperature (EN 1992-1-2 and EN 1994-1-2 curves).

The second of these equations is clearly incorrect because it does not give common values with the first and third at 100°C and 300°C, respectively. Since the intention is clearly to interpolate linearly between $k_{c,\theta_{max}}f_c$ and $0.9k_{c,\theta_{max}}f_c$, it is suggested that, until a clarification is available, the equation should be

$$\theta_{max} = 100°C \text{ to } \theta_{max} = 300°C : f_{c,\theta_{max},20°C} = \left[\frac{k_{c,100°C}(300 - \theta_{max}) + k_{c,300°C}(\theta_{max} - 100)}{200}\right]f_c$$

(5.26)

- The strain $\varepsilon_{cu},\theta_{max}$ at which this residual strength occurs after cooling from θ_{max} is identical to that at maximum stress $f_{cu,\theta_{max}}$ at the temperature θ_{max}. The slope of the descending branch is the same as that in the high-temperature curve at θ_{max}.
- The initial curvilinear rising part of the curve is calculated using the same equation as for the other concrete curves.
- Elevated-temperature stress-strain curves after cooling to a temperature θ from θ_{max} can be linearly interpolated between those at θ_{max} and 20°C.
- An example of residual strength calculation is given in Figure 5.24.

5.2.2.9 Tensile strength for advanced modelling

Although concrete's tensile strength is conventionally ignored in ambient-temperature design, it is well known that it generally has a tensile strength of a few percent of its compressive (cylinder or cube) strength. There is little

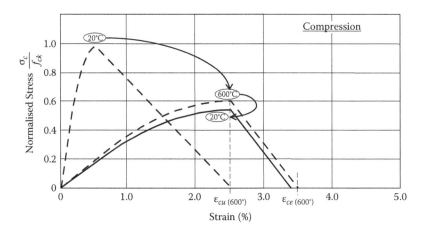

Figure 5.24 Stages in the generation of the residual stress-strain curve at 20°C for cal-careous-aggregate concrete after heating to 600°C.

reason to use this in simple fire resistance design of isolated structural elements such as beams or columns. It would clearly be unconservative to attempt to make use of concrete's tensile strength to gain a small advantage in assessing structural resistance, given the uncertainty that must exist about its exact value and the degree to which the concrete remains bonded to its reinforcement. However, when integrated large-deflection numerical modelling is being used to model the behaviour of a large subregion of a building, it is necessary to assess the integrity of separating elements, particularly of concrete floor slabs. This is particularly vital for the thin, lightly reinforced slabs used with downstand steel beams in composite flooring systems. To predict through-depth tensile cracks to check integrity, it is necessary to use a tensile stress-strain relationship that includes a rule for brittle cracking.

Although EN 1994-1-2 provides no guidance about the tensile strength of concrete, either at ambient or elevated temperatures, Table 3.1 of EN 1992-1-1 summarises relationships that relate ambient-temperature tensile strength, as a 5%–95% fractile range, to the concrete's characteristic compressive (cylinder) strength. The mean tensile strengths for the common range (C25/30 to C50/60) of concrete grades are 8% (6% to 11%) to 10% (7% to 13%) of the characteristic values. If tensile strengths are needed for elevated-temperature analysis in performance-based design, EN 1992-1-2 Annex E provides a simple reduction of the ambient-temperature tensile strength $f_{ck,t}$ as

$$f_{ck,t,\theta} = k_{c,t,\theta} f_{c,k,t} \tag{5.27}$$

in which

$$\theta = 20°C \text{ to } \theta = 100°C: \quad k_{c,t,\theta} = 1.0$$
$$\theta = 100°C \text{ to } \theta = 600°C: \quad k_{c,t,\theta} = 1.0 - (\theta - 100)/500 \tag{5.28}$$

The reduced tensile strength from 100°C to 600°C is clearly a very small proportion of the concrete's ambient-temperature characteristic compressive strength; at 400°C, it is on the order of 4% of f_{ck}, as illustrated in Figure 5.25. Although tensile failure is brittle in nature, it may be necessary for analytical purposes to give it a falling branch to zero stress; this may be of the same gradient as that of the falling branch of the compressive curve at the same temperature.

5.2.2.10 Biaxial failure surfaces for concrete

If it is required to predict either through-depth or layer-by-layer failure in slabs, then it is necessary to have a criterion for the failure by cracking or crushing of concrete under biaxial stress. It is rarely necessary in numerical

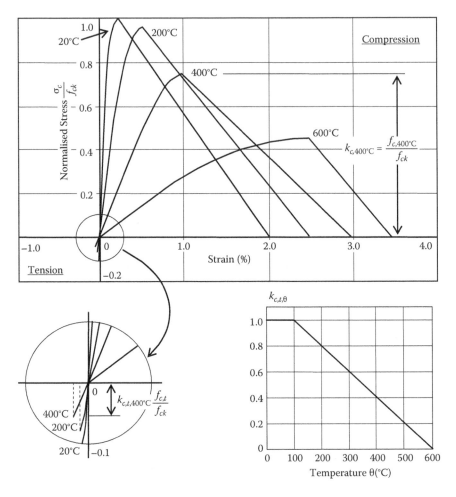

Figure 5.25 Uniaxial tensile strength of concrete at elevated temperatures according to EN 1992-1-2 (CEN 1992–2005b).

analysis of building structures to consider full three-dimensional stress fields, and for slabs in particular the through-depth stresses are effectively zero, so it is reasonable to assume plane-stress conditions at any planar level of the slab that is parallel to its midsurface.

There is no single mathematical relationship that links the combination of planar principal stresses that defines failure of concrete. As a generally brittle material, several adaptations of well-known yield criteria used for ductile materials have been defined as approximations to the brittle fracture interaction surfaces (or curves in plane-stress systems) observed in tests. Even at ambient temperature, the test data are probably inadequate

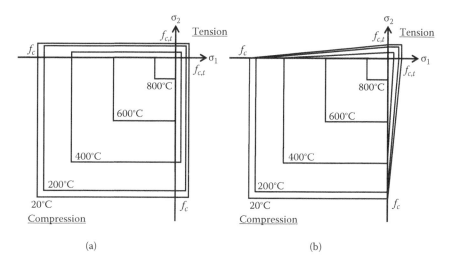

$$(a) \qquad\qquad\qquad\qquad (b)$$

Figure 5.26 Failure surfaces for plane stress: (a) independent uniaxial failure stresses; (b) the Mohr-Coulomb failure surface.

to properly define failure surfaces or the concrete parameters that control them, and at elevated temperatures virtually no test data exist. The most logical approach in these circumstances is to scale the failure surfaces with the reduction factors $k_{c,\theta}$ and $k_{c,t,\theta}$ for compression and tension so that they are compatible with the uniaxial peak stresses at any temperature.

The simplest possible biaxial failure criterion is to consider the principal stresses as independently capable of causing failure in either tension or compression. This seems to work well for cast iron, given equal strengths in tension and compression, but is less successful for concrete. If scaled to be compatible with uniaxial failure stresses, this criterion creates a family of square boundaries in plane stress, as shown in Figure 5.26a.

The Mohr-Coulomb (Coulomb 1776) criterion, which is actually more appropriate for frictional soil mechanics problems than for concrete, gives a more logical interaction for plane stress in the principal tension-compression quadrants, although this is simply a linear transition between the tension and compression failure stresses. For plane stress, this is illustrated in Figure 5.26b. In the tension-tension and compression-compression quadrants, the greater of the two principal stresses initiates failure. This criterion is given by the inner boundary of the six straight lines:

$$(\sigma_1/f_{c,t}) - (\sigma_2/f_c) = 1, \sigma_1 = f_{c,t}, \sigma_1 = -f_c$$
$$(\sigma_2/f_{c,t}) - (\sigma_1/f_c) = 1, \sigma_2 = f_{c,t}, \sigma_2 = -f_c$$

$$(5.29)$$

Observations make it clear that principal stresses in most materials interact to affect when fracture is experienced. However, no general criterion exists that can be generally used from ductile to brittle materials. Two notable attempts have been made to formulate general peak-stress failure criteria, as developments from the well-known energy-based von Mises yield criterion. These are due to Drucker and Prager (1952) and Bresler and Pister (1958).

The Drucker-Prager criterion attempts to cover failure of a range of materials including as concrete, under combinations of principal stresses, and can usually be implemented in numerical analysis programs. In its plane-stress form, this criterion is given by

$$2f_c f_{c,t} - (f_c - f_{c,t})(\sigma_1 + \sigma_2) - (f_c + f_{c,t})\sqrt{\sigma_1^2 - \sigma_1\sigma_2 + \sigma_2^2} = 0 \qquad (5.30)$$

where σ_1 and σ_2 are the principal stresses, and f_c and $f_{c,t}$ are the absolute values of the uniaxial compressive and tensile strengths of the concrete, respectively. For uniaxial stress, this fits the test values of f_c and $f_{c,t}$.

This produces a curved failure surface aligned at 45° to the principal stress axes, which is compatible with the uniaxial failure stresses for a homogeneous material. The shape of the failure surface is controlled only by the values of f_c and $f_{c,t}$.

The Bresler-Pister criterion (1958) is capable of adjusting the shape of the failure surface to reflect uniaxial and biaxial failure test results. For plane stress, it takes the form

$$c_0 + c_1(\sigma_1 + \sigma_2) + c_2(\sigma_1 + \sigma_2)^2 - \frac{1}{\sqrt{3}}\sqrt{\sigma_1^2 - \sigma_1\sigma_2 + \sigma_2^2} = 0 \qquad (5.31)$$

The constants c_0, c_1, and c_2 in this expression are defined using the strengths f_c and $f_{c,t}$ in uniaxial tension and compression, together with the strength $f_{c,b}$ in equal biaxial compression. This last term allows the shape of the failure surface to be made more consistent with test data. The constants are then defined as

$$c_1 = \left(\frac{f_{c,t} - f_c}{\sqrt{3}(f_{c,t} + f_c)}\right)\left(\frac{4f_{c,b}^2 - f_{c,b}(f_{c,t} + f_c) + f_{c,t}f_c}{4f_{c,b}^2 + 2f_{c,b}(f_{c,t} - f_c) - f_{c,t}f_c}\right)$$

$$c_2 = \left(\frac{1}{\sqrt{3}(f_{c,t} + f_c)}\right)\left(\frac{f_{c,b}(3f_{c,t} - f_c) - 2f_{c,t}f_c}{4f_{c,b}^2 + 2f_{c,b}(f_{c,t} - f_c) - f_{c,t}f_c}\right)$$

$$c_0 = \frac{f_c}{\sqrt{3}} + c_1 f_c - c_2 f_c^2 \qquad (5.32)$$

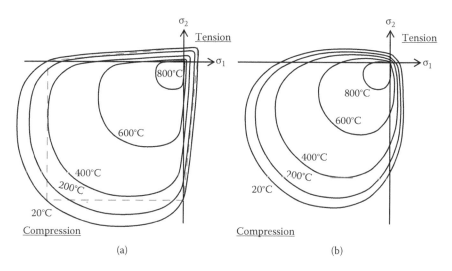

Figure 5.27 Failure surfaces for plane stress: (a) Drucker-Prager (from Drucker, D.C., and Prager, W., Soil Mechanics and Plastic Analysis or Limit Design, *Quarterly of Applied Mathematics,* 10, pp. 157–165, 1952); (b) Bresler-Pister (from Bresler, B., and Pister, K.S., Strength of Concrete under Combined Stresses, *Journal of American Concrete Institute,* September, pp. 321–345, 1958).

Examples of the Drucker-Prager and Bresler-Pister failure surfaces for biaxial plane stress at ambient and elevated temperatures are shown in Figures 5.27a and 5.27b, respectively.

Experimental evidence at ambient temperature (Kupfer et al. 1969) seems to support failure surfaces of shapes similar to the Bresler-Pister surface, but with a tension-tension quadrant similar to that of Mohr-Coulomb, with little diminution of uniaxial tensile strength by combination with a second principal stress.

5.2.2.11 Load-induced transient strain of concrete

The transient and steady-state (or isothermal) philosophies of high-temperature mechanical testing of structural materials have been discussed in Section 5.2.2.2. For concrete, the choice between these methods can lead to quite radically different results. It has been found since the 1970s that there is a "beneficial" influence of compressive loading, which "compacts" the concrete during heating and inhibits the development of cracks (Khoury et al. 1985). Conventional stress-strain curves, conducted at constant temperature, show a significant degradation of the strength and modulus of elasticity and a considerable increase in the ultimate strain between curves at different successive temperatures, whilst there is sufficient experimental

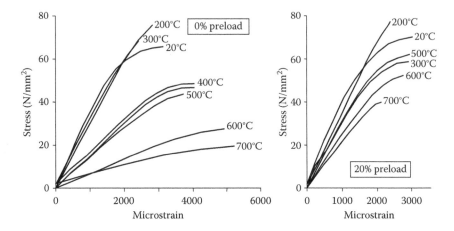

Figure 5.28 Influence of load history on the stress-strain relationships of concrete at various temperatures. (From Khoury, G.A., Effect of Fire on Concrete and Concrete Structures, *Progress in Structural Engineering and Materials*, 2(4), pp. 429–447, 2000.)

evidence to indicate that the temperature effect is much less marked if concrete is heated having previously been loaded, as is demonstrated in Figure 5.28. This remarkable effect of load history is very important, and the correct representation of elevated-temperature mechanical behaviour should be incorporated in structural modelling, at least where concrete structure is likely to be critical. However, the phenomenon is not adequately covered in codes and standards.

In accidental fire conditions, structural materials usually experience transient heating, so transient tests obviously simulate the actual situation of compressed concrete members in fire rather better than steady-state tests. Transient tests show an important property unique to concrete. The difference between the total stain measured in a transient test and the thermal expansion is much larger than the sum of the instantaneous stress-related strain (obtained from loaded steady-state tests) and the creep strain. This difference is illustrated in Figure 5.29, which shows temperature-strain plots for concrete with and without preload, heated for the first time.

This additional strain component (excluding the instantaneous stress-related strain and the creep strain) is defined as load-induced transient strain (LITS), on which work has been done by Anderberg and Thelandersson (1976), Khoury et al. (1985; Khoury 2000), and Schneider and Horvath (2003). It is of very large magnitude, of comparable order to the thermal expansion, and is non-recoverable. LITS appears to be effectively time independent and is a function of temperature, stress and stress history at

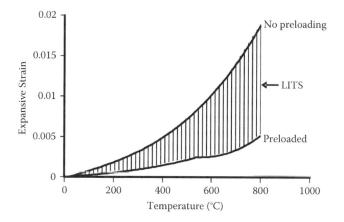

Figure 5.29 LITS, as the difference between total strain of preloaded and non-preloaded concrete, heated for the first time.

temperatures above 100°C. From the microstructural point of view, it may result from dehydration (loss of pore water and chemically bonded water) and chemical decomposition. From the phenomenological point of view, it may be due to the preloading, the transient heating, or the interaction of both, but this is still in need of further experimental investigation.

The division into components of total strain measurements of concrete at high temperatures is still subject to some disagreement among investigators. It is generally agreed that the total strain comprises four parts: thermal strain, instantaneous stress-related strain, creep strain and transient strain. Khoury et al. (1985) introduced a classification in which LITS is actually composed of three components: instantaneous stress-related strain, creep strain and transient strain. They further classified these components by

- separating the elastic and plastic parts of the instantaneous stress-related strain, and
- combining the plastic strain with the creep strain and transient strain to form a component defined as transient creep.

Thus, Khoury et al. defined LITS as a combination of elastic strain and transient creep, from which the elastic strain can be obtained by unloading and measuring the strain recovery at the end of a test. They also suggest the use of a "master LITS" curve for Portland-cement-based concrete in general for temperatures up to 450°C, irrespective of the type of aggregate or cement blend used.

5.2.2.11.1 Anderberg model

The Anderberg model was derived on a phenomenological basis directly linked to data from tests on quartzite-aggregate concrete carried out by Anderberg and Thelandersson (1976) and is presented in a form that is easy to implement in computational structural analysis. The total strain of uniaxially compressed concrete under heating is defined as the sum of four components, each defined as compressive positive:

- Thermal strain ε_{th}, including shrinkage, measured from unloaded specimens under heating;
- Instantaneous stress-related strain ε_σ, derived from the stress-strain relationship determined from steady-state tests;
- Creep strain ε_{cr}, the time-dependent strain recorded under constant load at constant temperature;
- Transient strain ε_{tr}, accounting for the effect of loading-heating history and determined from transient tests under constant load.

These strain components, with compressive strain taken as positive, are expressed as

$$\varepsilon = \varepsilon_{th}(T) + \varepsilon_\sigma(\tilde{\sigma},\sigma,T) + \varepsilon_{cr}(\sigma,T,t) + \varepsilon_{tr}(\sigma,T) \tag{5.33}$$

where T is temperature, σ is stress, t is time and $\tilde{\sigma}$ is stress history.

5.2.2.11.2 Thermal strain ε_{th}

The thermal strain ε_{th} is a simple function of temperature, which is taken directly from measured thermal strain curves. The EN 1992-1-2 (2004) equation [Equation (5.11)], with its signs changed, for thermal strain of siliceous-aggregate concrete, gives a close representation of Anderberg's curve and is written as

For $20°C \leq T \leq 700°C$ $\varepsilon_{th} = -\left(-1.8 \times 10^{-4} + 9 \times 10^{-6}T + 2.3 \times 10^{-11}T^3\right)$

For $700°C \leq T \leq 1200°C$ $\varepsilon_{th} = -14 \times 10^{-3}$ $\tag{5.34}$

The Anderberg and EN 1992-1-2 curves are shown in Figure 5.30b.

5.2.2.11.3 Instantaneous stress-related strain ε_σ

The instantaneous stress-related strain ε_σ is the usual uniaxial stress-strain relationship, determined from steady-state tests at any given

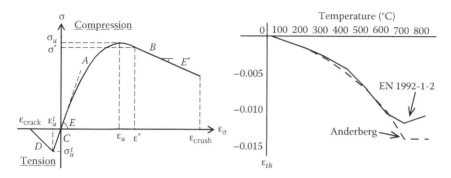

Figure 5.30 Anderberg and Thelandersson (1976) concrete strain component: (a) instantaneous stress-related strain; (b) thermal strain. (Based on Anderberg, Y., and Thelandersson, S., Stress and Deformation of Concrete at High Temperatures. *2 Experimental Investigation and Material Behaviour*, Bulletin 54, Lund Institute of Technology, Sweden, 1976.)

temperature. Anderberg's generalised description of the concrete stress-strain relationship is illustrated in Figure 5.30a. For compression, the curve consists of a parabolic branch followed by a linear descending branch until crushing occurs, whilst in the tension quadrant a linear increase up to an ultimate tensile stress is followed by a linear descending branch.

Anderberg's model for instantaneous stress-strain is defined as

A.   For $0 \leq \varepsilon_\sigma < \varepsilon^*$ $\sigma = \sigma_u \dfrac{\varepsilon_\sigma}{\varepsilon_u}\left(2 - \dfrac{\varepsilon_\sigma}{\varepsilon_u}\right)$

B.   For $\varepsilon^* \leq \varepsilon_\sigma < \varepsilon_{crush}$ $\sigma = E^*\varepsilon_\sigma + \sigma^*$

C.   For $0 \geq \varepsilon_\sigma > \varepsilon_u^t$ $\sigma = E\varepsilon_\sigma$ (5.35)

D. $\begin{cases} \text{For } \varepsilon_u^t \geq \varepsilon_\sigma > \varepsilon_{crack} & \sigma = \sigma_u^t \dfrac{\varepsilon_\sigma - \varepsilon_{crack}}{\varepsilon_u^t - \varepsilon_{crack}} \\[2ex] \text{For } \varepsilon_{crack} \geq \varepsilon_\sigma & \sigma = 0 \end{cases}$

in which

σ_u is the ultimate compressive stress;
ε_u is the strain at ultimate compressive stress;
$\varepsilon^* = \varepsilon_u(1 - E^*/E)$ is the strain at the transition between the parabolic branch and the linear descending branch in the compression quadrant;

$\sigma^* = \sigma_u (1 - E^*/E)^2$ is the corresponding compressive stress;

$E = 2\sigma_u / \varepsilon_u$ is the initial elastic modulus in the compression zone;

$E^* = -880 \text{MPa}$ is the slope of the descending branch in the compression zone;

ε_{crush} is the crushing strain;

σ_u^t is the ultimate tensile stress;

$\varepsilon_u^t = \frac{\sigma_u^t}{E}$ is the strain at the ultimate tensile stress;

$\varepsilon_{crack} = -0.04\%$ is the strain at which cracking starts; it is assumed to be invariant with temperature.

The stress-strain relationship is uniquely defined if the three parameters σ_u, ε_u and σ_u^t are known.

Due to the relatively modest effect of stress history on the high-temperature compressive strength σ_u, this is assumed to be a unique function of temperature, and the influence of stress history is neglected. The ratios of σ_u at various temperatures to the ambient-temperature compressive strength σ_{u0} are given in Table 5.8.

The strain ε_u at the ultimate compressive stress is significantly affected by stress history. It increases monotonically with increase of temperature for specimens that are unloaded during heating, whilst it is almost unaffected by temperature rise when specimens are loaded during heating. In Anderberg's model, this effect is taken into account by expressing the stress history in terms of the accumulated transient strain ε_{tr} and by relating ε_u both to temperature and to ε_{tr}. This is written as

$$\varepsilon_u (T, \tilde{\sigma}) = \max (\varepsilon_{u0}, \bar{\varepsilon}_u - \varepsilon_{tr}) \tag{5.36}$$

in which ε_{u0} is the strain at ultimate compressive stress at ambient temperature, and $\bar{\varepsilon}_u$ is the strain at ultimate compressive stress obtained from tests that are unloaded during the heating phase, simplified on the basis of test data as a linear function of temperature, and modelled as shown in Figure 5.31a.

Table 5.8 Ratio of compressive strength of concrete at high temperature σ_{uT} to the ambient-temperature strength σ_{u0}

T (C)	20	100	200	300	400	500	600	700	800
$\sigma_{uT} / \sigma_{u0}$	1	1	0.98	0.96	0.97	0.6	0.45	0.3	0.2

Source: Based on Anderberg, Y., and Thelandersson, S. (1976), *Stress and Deformation of Concrete at High Temperatures. 2 Experimental Investigation and Material Behaviour*, Bulletin 54, Lund Institute of Technology, Sweden.

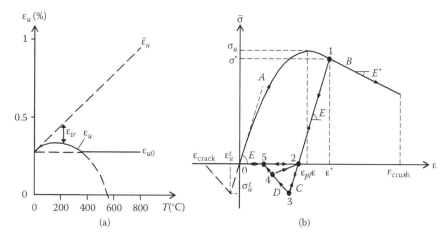

Figure 5.31 Anderberg and Thelandersson (1976): (a) mathematical models of the strains ε_u and $\bar{\varepsilon}_u$ at ultimate compressive stress; (b) unloading paths of the high-temperature stress-strain model. (Based on Anderberg, Y., and Thelandersson, S., Stress and Deformation of Concrete at High Temperatures. *2 Experimental Investigation and Material Behaviour*, Bulletin 54, Lund Institute of Technology, Sweden, 1976.)

The strain at ultimate stress ε_u is always lower than or equal to the strain at the ultimate compressive stress from tests that are unloaded during heating but is not less than that at ambient temperature.

The ultimate tensile stress σ_u^t is solely related to temperature. The ratios of σ_u^t to the ambient-temperature tensile strength σ_{u0}^t at various temperatures are given in Table 5.9.

The ambient-temperature tensile strength σ_{u0}^t is defined as

$$\sigma_{u0}^t = -0.28\left(0.8\sigma_{cube}\right)^{2/3} \tag{5.37}$$

where σ_{cube} is the cube strength at ambient temperature in megapascals.

Unloading is also taken into account in this model for both compression and tension, as illustrated in Fig. 5.30b. The arrows indicate the allowed directions of loading and unloading. Unloading from compression starts at an arbitrary point 1 and ends at point 2, at which the stress becomes tensile. The strain at point 2 is the permanent plastic strain $\varepsilon_{pl} = \varepsilon_\sigma - \frac{\sigma}{E}$.

Table 5.9 Ratio of tensile strength of concrete at high temperature σ_{uT}^t to the ambient-temperature tensile strength σ_{u0}^t

T (°C)	20	100	200	300	400	500	600	700	800
$\sigma_{uT}^t / \sigma_{u0}^t$	1	0.95	0.82	0.71	0.56	0.37	0.2	0.15	0.1

The loading and unloading paths from tension are illustrated as the sequence 2-3-4-2, which is valid if previous unloading has taken place in compression. Unloading from tension is assumed always to follow the route 4-2. Cracking is assumed to occur at point 5, at which the strain is $\varepsilon_{pl} + \varepsilon_{crack}$, and strain may then follow either route 5-2 or 5-0 and then return to compression. To account for an arbitrary strain history, such as unloading in compression before entering the tension zone, the stress-strain model can be rewritten in the generalized form

A.   For $0 \le \varepsilon_\sigma < \varepsilon^*$

$\sigma = \sigma_u \dfrac{\varepsilon_\sigma}{\varepsilon_u}\left(2 - \dfrac{\varepsilon_\sigma}{\varepsilon_u}\right)$

B.   For $\varepsilon^* \le \varepsilon_\sigma < \varepsilon_{crush}$

$\sigma = E^* \varepsilon_\sigma + \sigma^*$

C.   For unloading, $\varepsilon_\sigma \ge \varepsilon_{pl} + \varepsilon_u^t$

$\sigma = E\left(\varepsilon_\sigma - \varepsilon_{pl}\right)$

$$(5.38)$$

D. $\begin{cases} \text{For } \varepsilon_{pl} + \varepsilon_u^t > \varepsilon_\sigma \ge \varepsilon_{pl} + \varepsilon_{crack} \qquad \sigma = \sigma_u^t \dfrac{\varepsilon_\sigma - \varepsilon_{crack} - \varepsilon_{pl}}{\varepsilon_u^t - \varepsilon_{crack}} \\[2em] \text{For } \varepsilon_{pl} > \varepsilon_\sigma \text{ or } \varepsilon_{pl} + \varepsilon_{crack} > \varepsilon_\sigma \qquad \sigma = 0 \end{cases}$

where ε^* is the strain from which unloading takes place.

5.2.2.11.4 Creep strain ε_{cr}

The creep strain ε_{cr}, at constant temperature and constant stress, is modelled based on test data as

$$\varepsilon_{cr}\left(\sigma, T, t\right) = 5.3 \times 10^{-4} \frac{\sigma}{\sigma_u}\left(\frac{t}{180}\right)^{0.5} e^{3.04 \times 10^{-3}(T-20)}$$

$$(5.39)$$

In fire applications, the creep strain is often very small compared to the other three strain components due to the relatively short periods of accidental fires. Anderberg and Thelandersson (1976) indicate that it may be neglected in this case.

5.2.2.11.5 Transient strain ε_{tr}

The characteristics and importance of transient strain ε_{tr} have been described in Section 5.2.2.11. In the Anderberg and Thelandersson model, ε_{tr} is not evaluated from direct measurements in tests but by deducting the other three strain components ε_{th}, ε_σ and ε_{cr}, calculated using the models described previously, from the total strain measured in loaded transient tests. It is found that ε_{tr} has a reasonably linear relationship with stress and is a non-linear function of temperature, which may however be mapped approximately linearly on to ε_{th}. Therefore, it is formulated as

$$\varepsilon_{tr} = -k_{tr} \frac{\sigma}{\sigma_{u0}} \varepsilon_{th} \tag{5.40}$$

where k_{tr} is a dimensionless constant between 1.8 and 2.35 for quartzite-aggregate concrete.

This relationship does not agree perfectly with the test data at temperatures above 500°C. For this temperature range, the following formula should be used instead:

$$\Delta\varepsilon_{tr} = 0.1 \times 10^{-3} \Delta T \frac{\sigma}{\sigma_{u0}} \tag{5.41}$$

5.2.2.11.6 Khoury model

The Khoury model was originally developed on the basis of the phenomenon of strain decomposition by Khoury et al. (1985) based on experiments on different concretes at temperatures up to 600°C, without directly associating it with a stress-strain-temperature model.

The elastic (ε_{el}) and plastic (ε_{pl}) parts of the instantaneous stress-related strain ε_σ are separated, and the plastic strain ε_{pl} is combined with the creep strain ε_{cr} and the transient strain ε_{tr}, forming one strain component that is defined as "transient creep" $\varepsilon_{tr,cr}$:

$$\varepsilon_{tr,cr} = \varepsilon_{pl} + \varepsilon_{cr} + \varepsilon_{tr} \tag{5.42}$$

The total strain is then defined as the sum of the thermal strain ε_{th}, the elastic strain ε_{el} and the transient creep $\varepsilon_{tr,cr}$:

$$\varepsilon = \varepsilon_{th}(T, V_a) + \varepsilon_{el}(\sigma, T) + \varepsilon_{tr,cr}(\sigma, T, V_a) \tag{5.43}$$

Models of these three components are developed as functions of temperature T, stress σ and aggregate content V_a (as a percentage by volume) of concrete.

Another new concept is defined as LITS, which is the sum of the elastic strain ε_{el} and the transient creep $\varepsilon_{tr,cr}$. It is not difficult to deduce that LITS is actually the sum of the instantaneous stress-related strain ε_σ, the creep strain ε_{cr} and the transient strain ε_{tr} in the Anderberg model. Therefore, there is no fundamental difference between these two models in their ways of partitioning the strain components; the total strain is either

$$\varepsilon = \varepsilon_{th} + \varepsilon_{el} + \varepsilon_{tr,cr} = \varepsilon_{th} + LITS \quad \text{(Khoury)} \tag{5.44}$$

or

$$\varepsilon = \varepsilon_{th} + \varepsilon_{\sigma} + \varepsilon_{cr} + \varepsilon_{tr} \quad \text{(Anderberg)} \tag{5.45}$$

The elastic strain ε_{el} is defined as $\varepsilon_{el} = \frac{\sigma}{E_0}$, where E_0 is the initial elastic ("Young's") modulus at ambient temperature. A "master transient creep" curve is suggested for Portland-cement-based concretes in general for temperatures up to 450°C, irrespective of the type of aggregate or cement blend used. This master curve was developed based on experiments subject to the stress levels of 10%, 20% and 30% of the cold strength and on concrete of $V_a = 65\%$. Based on this master curve, $\varepsilon_{tr,cr}$ is assumed to be linearly related to stress and is normalised against the values obtained at $\sigma/\sigma_{u0} = 0.3$:

$$\varepsilon_{tr,cr}\left(T,\sigma\right) = \varepsilon_{tr,cr}\left(T,0.3\sigma_{u0}\right) \times \left(0.032 + 3.226\frac{\sigma}{\sigma_{u0}}\right) \tag{5.46}$$

in which $\varepsilon_{tr,cr}\left(T,0.3\sigma_{u0}\right)$ is a non-linear function of temperature, based on the master curve up to 590°C:

$$\varepsilon_{tr,cr}\left(T,0.3\sigma_{u0}\right) = -\left(A_0 + A_1 T + A_2 T^2 + A_3 T^3 + A_4 T^4\right) \times 10^{-6} \tag{5.47}$$

where $A_0 = 43$, $A_1 = -2.725$, $A_2 = -6.248 \times 10^{-2}$, $A_3 = 2.193 \times 10^{-4}$ and $A_4 = -2.769 \times 10^{-7}$.

However, the master curve does not give a good representation of Thames gravel concrete (or presumably all concretes with siliceous aggregates) above 400°C; therefore, for this type of concrete Equation (5.46) needs to be modified to

$$\varepsilon_{tr,cr}\left(T,0.3\sigma_{u0}\right) = -1.48 \times \left(B_0 + B_1 T + B_2 T^2 + B_3 T^3 + B_4 T^4 + B_5 T^5\right) \times 10^{-6} \tag{5.48}$$

where $B_0 = 1098.5$, $B_1 = -39.206$, $B_2 = 0.428$, $B_3 = -2.444 \times 10^{-3}$, $B_4 = 6.274 \times 10^{-6}$ and $B_5 = -5.948 \times 10^{-9}$.

The effect of aggregate content on transient creep is assumed to be linear; therefore, the following equation is given for $\varepsilon_{tr,cr}$ of concrete of an arbitrary V_a:

$$\varepsilon_{tr,cr}\left(T,\sigma,V_a\right)=\varepsilon_{tr,cr}\left(T,\sigma,65\%\right)\times\left(3.05-3.15\frac{V_a}{100}\right) \tag{5.49}$$

Although transient creep includes the time-dependent creep strain, since this is always negligible in short-term heating scenarios such as building fires, the transient creep is generally considered time independent. An update of the strain decomposition of this model was given by Khoury (2006), but no further development of the mathematical models of the strain components was given, and the original model is still valid for use in structural analysis.

5.2.2.11.7 Schneider model

The Schneider model was first published by Schneider in 1986, followed by periodic revisions. The most recent version (Schneider et al. 2003) is described here. The total strain is divided into thermal strain ε_{th} and mechanical strain ε_m:

$$\varepsilon=\varepsilon_{th}+\varepsilon_m \tag{5.50}$$

The mechanical strain ε_m is actually the LITS in the Khoury model and is the sum of the instantaneous stress-related strain ε_σ, the creep strain ε_{cr} and the transient strain ε_{tr} in the Anderberg model. The strain decompositions of these three models fundamentally agree with each other.

The thermal strain ε_{th} is assumed to be a nonlinear function of temperature:

$$\varepsilon_{th}=-\left(6.6\times10^{-11}T^3-1.7\times10^{-8}T^2+9\times10^{-6}T+1.369\times10^{-4}\right) \quad (T \leq 650°C)$$

$$\varepsilon_{th}=-\left(2\times10^{-8}T^2-3.94\times10^{-5}T+0.0342\right) \quad\quad (T >650°C)$$

$$\tag{5.51}$$

The mechanical strain ε_m is related to stress, temperature, type of concrete and moisture content and is written as

$$\varepsilon_m=\frac{\sigma}{E}\left(1+\varphi\right) \tag{5.52}$$

where σ is stress, E is the high-temperature Young's modulus and φ is a function related to temperature, type of concrete and moisture content.

The high-temperature Young's modulus E is a function of temperature and load history. The load history is expressed as a load factor α:

$$\alpha = \frac{\sigma_{history}}{\sigma_{u0}} \qquad (5.53)$$

The term $\sigma_{history}$ is not clearly defined in the literature. In the tests on which this model is based, the applied load is usually constant, so that $\sigma_{history}$ should be the constant applied stress. However, the stress history of a structural member will be much more complicated than this, so clarification is needed.

High-temperature values of Young's modulus E at $\alpha = 0$, $\alpha = 0.1$ and $\alpha = 0.3$ are given as functions of temperature:

For $20°C \le T \le 320°C$:

$$E(T, \alpha = 0) = E_0(1 - 3 \times 10^{-3}(T - 20) + 4.085 \times 10^{-6}(T - 20)^2)$$

$$E(T, \alpha = 0.1) = E_0(3 \times 10^{-6}T^2 - 2.311 \times 10^{-3}T + 1.04505) \qquad (5.54)$$

$$E(T, \alpha = 0.3) = E_0(2 \times 10^{-6}T^2 - 1.57 \times 10^{-3}T + 1.0306)$$

For $320°C \le T \le 450°C$:

$$E(T, \alpha = 0) = E_0 \left(\begin{array}{l} 0.89406 - 3.2445 \times 10^{-3}T + 1.0081 \times 10^{-5}T^2 \\ -1.2801 \times 10^{-8}T^3 \end{array} \right)$$

$$\qquad (5.55)$$

$$E(T, \alpha = 0.1) = E_0(-5.1 \times 10^{-6}T^2 + 2.73 \times 10^{-3}T + 0.262)$$

$$E(T, \alpha = 0.3) = E_0(-4 \times 10^{-6}T^2 + 2.5 \times 10^{-3}T + 0.342)$$

For $450°C \le T \le 600°C$:

$$E(T, \alpha = 0) = E_0(0.45 \times e^{-7.52 \times 10^{-3}(T-400)})$$

$$E(T, \alpha = 0.1) = E_0(-6 \times 10^{-8}T^3 + 1 \times 10^{-4}T^2 - 0.056T + 10.874) \qquad (5.56)$$

$$E(T, \alpha = 0.3) = E_0(1.767 \times e^{-0.0022T})$$

For $T > 600°C$:

$$E(T, \alpha = 0) = E_0 \left(e^{-4 \times 10^{-3}(T-20)} \right)$$

$$E(T, \alpha = 0.1) = E_0 (0.82554 \times e^{-0.0016T}) \qquad (5.57)$$

$$E(T, \alpha = 0.3) = E_0 (0.86 \times e^{-0.001T})$$

For $\alpha \geq 0.3$, E is found to be reasonably constant, so the equations for $E(T, \alpha = 0.3)$ are used for calculating $E(T, \alpha > 0.3)$. No guidance is given about interpolating E when α is between 0, 0.1 and 0.3.

The constant φ is formulated as

$$\varphi = C_1 \tanh(\gamma_w (T - 20)) + C_2 \tanh(\gamma_0 (T - T_g)) + C_3 \qquad (5.58)$$

where C_1, C_2, C_3, γ_0 and T_g are given in Table 5.10, and γ_w is a function of the moisture content w of the concrete as a percentage by weight, written as

$$\gamma_w = \min(0.3 \times 10^{-3} w^{0.5} + 2.2 \times 10^{-3}, 2.8 \times 10^{-3}) \qquad (5.59)$$

5.3 FIRE PROTECTION MATERIALS

5.3.1 General

To enable performance-based fire engineering calculations to be carried out, it is necessary to have reliable and accurate data of temperature-dependent thermal properties of the materials used. As presented in Section 5.2 of this chapter, comprehensive information on the thermal properties of structural load-bearing materials, such as steel and concrete, is available. However, it is much more difficult to obtain information on the thermal properties

Table 5.10 Parameters for calculating mechanical strain ε_m

	Quartzite concrete	Limestone concrete	Lightweight concrete
C_1	2.5	2.5	2.5
C_2	0.7	1.4	3
C_3	0.7	1.4	2.9
$\gamma_0(°C^{-1})$	7.5×10^{-3}	7.5×10^{-3}	7.5×10^{-3}
T_g (°C)	800	700	600

of fire protection materials. Partly this is due to the specific nature of the numerous types of fire protection material available, which have complicated and variable chemical reactions at high temperatures. Also in assessing fire protection materials, these data have not been requested because the existing approach, based on a standard fire resistance test, only concerns ensuring that the fire protection system passes the standard fire resistance test for a prescribed period of time. A further important factor contributing to this lack of information is that most fire protection materials are proprietary systems from different manufacturers, and commercial sensitivity prevents the publication of this type of information.

Performance-based fire engineering deals with not only the standard fire exposure but also natural fires and not only fixed limiting structural temperature but also variable temperatures. It is essential that the thermal property data should reflect the effects of changing material temperatures and fire exposures. The ambient temperature data, which are often supplied by fire protection material manufacturers, are not adequate for performance-based fire engineering purpose.

This section presents some elevated temperature thermal property data for a number of common generic fire protection materials. Among these fire protection materials, gypsum-based products and intumescent coatings are the most popular. This section presents more detailed information on these two types of fire protection material. In particular, intumescent coatings are a reactive material, and their thermal properties are not only temperature dependent but also fire exposure dependent. Research studies are starting to be carried out to develop reliable methods to quantify the thermal properties of intumescent coatings under different fire exposure conditions. This section presents some preliminary results.

5.3.2 Theoretical considerations

For heat transfer analysis, the required thermal properties include density, specific heat and thermal conductivity. Although these properties can be experimentally measured, it is necessary to have a theoretical framework to guide the experiments. Also, as demonstrated in the next section, a theoretical model enables a gap in experimental data to be filled.

5.3.2.1 Density

Many fire protection materials experience changes in density due to dehydration, decomposition of organic compounds or decarbonation during exposure to fire. Density is mass divided by volume. Since both mass and volume of fire protection material may change at high temperatures, both values should be quantified. For quantification of mass change, thermogravimetric analysis (TGA) may be used. TGA tracks the reduction of mass at increasing temperature. Volume change may be measured using a dilatometer.

Fortunately, most fire protection materials are lightweight. Therefore, the heat stored in fire protection material is low compared to that conducted through the material and that stored in the protected structural material. If accurate information is not available, a constant density may be used. This value is often given in the fire protection manufacturer's literature.

5.3.2.2 Specific heat

When a fire protection material increases in temperature, external energy is required. The required external energy is reduced if the fire protection material generates heat through exothermic reaction or increased if the fire protection material undergoes endothermic reaction or if water is evaporated. The amount of energy required to raise the temperature of unit mass material by 1°C is defined as the specific heat of the material. In simplistic heat transfer analysis, an equivalent specific heat may be used to represent the combined effects.

The equivalent specific heat is obtained by adding to the base value additional energy that is required due to endothermic reaction or evaporation of water or by deducting from the base value energy that is released during exothermic reaction. The base value is usually temperature dependent. However, the range of this change is usually small. Therefore, if the precise variation is not available, a constant value may be used.

To allow for heat generated/consumed during exothermic/endothermic chemical reactions and heat consumed during water evaporation, a simple method is to distribute the energy involved through the temperature duration of the chemical reactions/water evaporation. The precise distribution may be difficult to quantify, but since the degree of accuracy required is not high, a triangular distribution may be conveniently used. For example, Figure 5.32 shows the specific heat of a material with a base value of 1000 J/(kg.K) and 5% moisture by weight. There is no chemical reaction. The latent heat of evaporation of water is 2260 kJ/kg. Water evaporation is assumed to occur between 90°C and 150°C. The average additional specific heat is 2260000*0.05/(150 − 90) = 1883 J/(kg.K), assuming peak water evaporation occurs at 120°C. Then, the peak value of specific heat at 120°C is 1000 + 2*1883 = 4766 J/(kg.K).

5.3.2.3 Thermal conductivity

Among the three thermal properties of a fire protection material, thermal conductivity is the most important in affecting the temperature of the protected structure because of the lightweight nature of the fire protection material. Therefore, it is most important that this value is determined

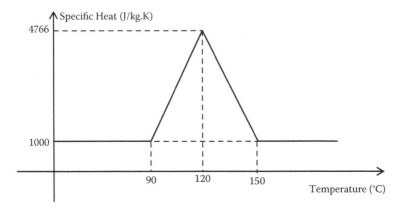

Figure 5.32 Effect of water evaporation on specific heat.

accurately. Manufacturers of fire protection materials usually list the ambient temperature thermal conductivity values in their literature. But, a constant thermal conductivity should not be used. As discussed further in this section, the thermal conductivity of fire protection material increases with temperature. Therefore, using a constant ambient temperature value will underestimate the protected structural temperature, leading to unsafe results. The temperature dependence of the thermal conductivity of fire protection materials should be quantified.

Methods are available to measure the thermal conductivity of material at high temperatures, for example, using the guarded hot plate method (American Society for Testing and Materials [ASTM] 1997). However, such test methods can be expensive, particularly if the test temperature is high. A theoretical model allows limited measurements to be extrapolated.

Common fire protection materials include intumescent coatings, gypsum plater, calcium silicate, mineral wool, rock wool, rock fibre, vermiculite and cementitious spray. A common feature of these materials is that they are porous. The thermal conductivity of a porous material with uniform distribution of pores may be calculated using Equation (5.60) (Russell 1935):

$$\lambda^* = \lambda_s \frac{\lambda_g \varepsilon^{\frac{2}{3}} + \left(1 - \varepsilon^{\frac{2}{3}}\right)\lambda_s}{\lambda_g \left(\varepsilon^{\frac{2}{3}} - \varepsilon\right) + \left(1 - \varepsilon^{\frac{2}{3}} + \varepsilon\right)\lambda_s} = \lambda_s \frac{\beta\varepsilon^{\frac{2}{3}} + \left(1 - \varepsilon^{\frac{2}{3}}\right)}{\beta\left(\varepsilon^{\frac{2}{3}} - \varepsilon\right) + \left(1 - \varepsilon^{\frac{2}{3}} + \varepsilon\right)} \quad (5.60)$$

where λ_s is the thermal conductivity of the solid, λ_g is the thermal conductivity of the gas in the pores, ε is the porosity, and $\beta = \lambda_g/\lambda_s$.

The thermal conductivity of the gas is an equivalent value that measures the overall heat transfer within the pores. When calculating this value, it is important to include the effect of radiant heat transfer within the pores. The effect of convection in the pores may be neglected if the pore size is small (less than 5 mm), which is usually the case. Therefore, the equivalent thermal conductivity value of gas may be calculated as

$$\lambda_g = \lambda_{g,cond} + \lambda_{g,rad} \tag{5.61}$$

where $\lambda_{g,cond}$ is the pure thermal conductivity of air due to heat conduction, and $\lambda_{g,rad}$ is the contribution to thermal conductivity of the gas due to radiant heat transfer within the pores.

The pure thermal conductivity of air at different temperatures may be calculated using the following equation (Smith 1981):

$$\lambda_{g,cond} = \lambda_{g0} \left(\frac{T}{T_0} \right)^{0.8} \tag{5.62}$$

in which T is gas temperature, T_0 is ambient temperature, and λ_{g0} (= 0.0246 W/m.K) is the thermal conductivity of air at ambient temperature T_0.

The radiant contribution to thermal conductivity of air at elevated temperature may be calculated using Equation (5.63) (Loeb 1954):

$$\lambda_{g,rad} = 4GdE\sigma T^3 \tag{5.63}$$

where E is the emissivity of the fire protection material, G is the ratio of the average width of the pore divided by the maximum width of the pore, and both quantities should be determined in the direction of heat transfer. G is 1 for laminar pores and cylindrical pores with axes parallel to the heat flow direction; G is $\pi/4$ for cylindrical pores with axes perpendicular to the heat flow direction; for spherical pores, $G = 2/3$. Here d is the dimension of the pore in the direction of thermal gradient, and σ is the Stefan-Boltzmann constant (5.67×10^{-8} W/m^2.K^4).

In porous fire protection materials, the pores may be different shapes and different sizes. For simplification, an equivalent uniform spherical pore size may be used.

5.3.3 Thermal properties of common fire protection materials

The equations in the preceding section may be used to provide sufficiently accurate temperature-dependent thermal conductivity values if detailed information such as porosity, distribution of pore shapes and sizes is available. However, for the majority of fire protection materials, such data are not readily available. The usefulness of these equations is to derive a simplistic expression of temperature dependency of the thermal conductivity of porous fire protection materials to enable extending the range of application of a limited amount of experimental data.

Consider Equation (5.60). The porosity of the fire protection material is usually high, approaching unity. The thermal conductivity of air within the pores (even including the effects of radiant heat transfer) is much lower than that of the solid. Therefore, the first term in the denominator of Equation (5.60) is negligible compared to the second term, and this enables the following relationship to be obtained:

$$\lambda^* = C_1 + C_2 \lambda_g \tag{5.64}$$

where C_1 and C_2 are constants.

Substituting Equations (5.62) and (5.63) into Equation (5.64) and recognising the weak dependence of $\lambda_{g,cond}$ on temperature, then the temperature-dependent thermal conductivity of a porous fire protection material may be expressed as

$$\lambda^* = \lambda_0^* + CT^3 \tag{5.65}$$

where λ_0^* may be considered as the thermal conductivity of the fire protection material at zero temperature, and T is the absolute temperature in kelvin. At ambient temperature, the thermal conductivity of air is very low compared to that of the solid. Therefore, λ_0^* approaches the thermal conductivity of the solid.

In the following, Equation (5.65) is used to fit expressions of temperature-dependent thermal conductivity for a number of common fire protection materials.

In all expressions that follow, the unit of temperature is kelvin and that of thermal conductivity is W/(k·m).

5.3.3.1 Rock fibre

The density of rock fibre is 155–180 kg/m³.

The base value of specific heat is about 900 J/(kg.K). An exothermic reaction happens at a temperature between 330°C and 600°C. Triangular distribution of heat generation may be assumed with a value of –1600 J/(kg.K) at 465°C, which gives the specific heat value at 465°C of –700J/(kg.K). This is shown in Figure 5.33.

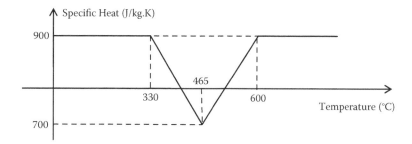

Figure 5.33 Specific heat of rock fibre. (Based on Benichou, N., and Sultan M.A., Thermal Properties of Components of Lightweight Wood-Framed Assemblies at Elevated Temperatures, *Fire and Materials*, 25, pp. 165–179, 2005.)

The thermal conductivity data reported by Benichou and Sultan (2005) may be used to fit the following expression:

$$\lambda_{rock\ fibre} = 0.022 + 0.1475\left(\frac{T}{1000}\right)^3 \tag{5.66}$$

where T is in Kelvin.

5.3.3.2 Mineral wool

The density of mineral wool is about 165 kg/m³.

A constant specific heat value of 840 J/(kg.K) may be used.

Based on the thermal conductivity data provided by the Engineering ToolBox Web site (http://www.engineeringtoolbox.com; accessed 7 June 2010), the relationship between thermal conductivity and temperature has been determined as follows:

$$\lambda_{mineral\ wool} = 0.03 + 0.2438\left(\frac{T}{1000}\right)^3 \tag{5.67}$$

Excellent correlation with the data provided by another Web site (http://www.insulationandlagging.co.uk; accessed 7 June 2010) gives confidence in the results.

5.3.3.3 Calcium silicate

Fire protection materials based on calcium silicate (CSH) can have a wide range of density, with the values reported in the literature being 200–1400 kg/m³.

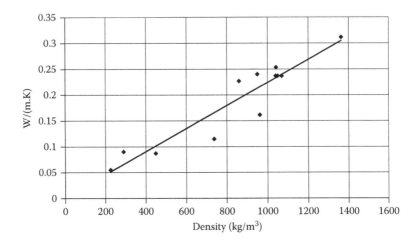

Figure 5.34 Relationship between initial thermal conductivity and density for calcium silicate.

The specific heat of calcium silicate is not changed at different densities. The base value is similar to that of gypsum, and a value of 900 J/(kg.K) may be used.

Changing the density of calcium silicate greatly affects its thermal conductivity. The following elevated temperature-thermal conductivity relationships were derived based on data from Arimil CFS (http://www.armilcfs.com, accessed 8 June 2010) for densities of 224–1362 kg/m³; from Christy Refractories (http://www.christyco.com/pdf/crc/Christy_Cal-Sil_Data_Sheet.pdf, accessed 8 June 2010) for densities of 860–1070 kg/m³; and from Do et al. (2007) for densities of 288 kg/m³ and 449 kg/m³.

Using Equation (5.65), the values of λ_0^* and C were derived for each reported density based on available elevated temperature-thermal conductivity values. Figure 5.34 shows the changes in λ_0^* as a function of temperature. The linear trend line in Figure 5.34 passes through the origin. This is conceptually correct because when the density of the material approaches zero (no solid), the equivalent thermal conductivity of the solid term should also approach zero. The trend line gives the following relationship:

$$\lambda_0^* = 0.23 \frac{\rho}{1000} \tag{5.68}$$

where density ρ is in kg/m³.

Koronthalyova and Matiasovsky (2003) studied the thermal conductivity of calcium silicate at ambient temperature for densities 200, 240 and

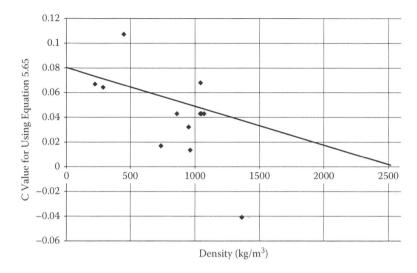

Figure 5.35 Relationship between *C* value and density for calcium silicate.

280 kg/m³. Using Equation (5.68), thermal conductivity values of 0.046, 0.0552 and 0.0644 W/(m.K) can be calculated. These values are very close to the reported thermal conductivity values of dry calcium silicate by Koronthalyova and Matiasovsky (2003).

The relationship between the *C* value of Equation (5.65) and temperature is shown in Figure 5.35. This figure shows that the *C* value can be highly variable at similar densities. Therefore, it may not be entirely appropriate to use density as the only factor determining the thermal conductivity of calcium silicate. However, because data are lacking and the purpose of this exercise is to find a reasonably accurate estimate of the thermal conductivity of calcium silicate, this is considered acceptable. In more refined fire engineering calculation of structures, more detailed information may have to be sought from the material's manufacturer.

Based on the trend line shown in Figure 5.35, the variation of the *C* value as a function of density may be calculated as follows:

$$C = 0.08 \times \frac{(2540 - \rho)}{2540} \tag{5.69}$$

where density ρ is in kg/m³.

The value 2540 kg/m³ represents the solid density of calcium silicate (Do et al 2007). Solid calcium silicate has zero porosity, so there is no contribution from radiant heat transfer to the thermal conductivity. For solid calcium silicate, it is conceptually correct to assume $C = 0$.

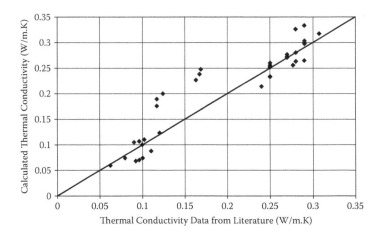

Figure 5.36 Overall comparison for thermal conductivity values of calcium silicate from different sources and from calculations using Equations (5.65), (5.68) and (5.69).

Substituting Equations (5.68) and (5.69) into Equation (5.65), the thermal conductivity of calcium silicate at elevated temperatures for different densities can be calculated. Figure 5.36 compares the calculated values and the reported thermal conductivity values from the aforementioned three different sources. Considering that the data were based on three different sources and there is very little detailed information on the composition and internal structure of the materials, the agreement between the source data and calculated values using the three simple equations is remarkably good.

5.3.3.4 Vermiculite

Vermiculite products can be gypsum based or cement based. There are limited data of elevated temperature properties. The expressions here are based on data from Armil CFS (http://www.armilcfs.com; accessed 8 June 2010) for densities of 375–600 kg/m³.

A base value of 900 J/(kg.K) may be used for specific heat.

For thermal conductivity, the same process as reported in the preceding section for calcium silicate was carried out, and the following relationship have been derived.

$$\lambda_0^* = 0.27 \frac{\rho}{1000} \tag{5.70}$$

$$C = 0.18 \times \frac{(1000 - \rho)}{1000} \tag{5.71}$$

where density ρ is in kilograms per cubic metre.

The value of 1000 kg/m³ in Equation (5.71) for calculating the value of C may be considered the density of solid vermiculite.

According to the Promat literature (http://www.promat-spray.com, accessed 10 June 2010), CAFCO300 (gypsum-based vermiculite sprayed fire protection) has a density of 315 kg/m³ and ambient temperature thermal conductivity value of 0.078 W/(m.K); CAFCO Mandolite CP2 (Portland cement-based vermiculite sprayed fire protection) has a density of 390 kg/m³ and ambient temperature thermal conductivity value of 0.095 W/(m.K). Equation (5.71) calculates values of 0.085 W/(m.K) and 0.105 W/(m.K) for 315 and 390 kg/m³ densities, respectively. These values correlate well with the values reported by Promat.

Suvorov and Skurikhin (2002, 2003) reported ambient temperature thermal conductivity values for vermiculite at densities between 400 and 1000 kg/m³. Figure 5.37 compares the literature data (points) and calculated results (line) using Equation (5.71) for different densities. The agreement is remarkably good.

Substituting Equations (5.7) and (5.71) into Equation (5.65), the thermal conductivity of vermiculite can be calculated for different densities at different elevated temperatures. Figure 5.38 compares the calculation results with the results reported in the aforementioned literature. The agreement is considered very good.

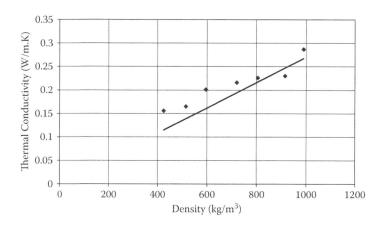

Figure 5.37 Comparison between literature data and calculation results for ambient-temperature thermal conductivity of vermiculite at different densities. (Based on Suvorov, S.A., and Skurikhin, V.V. (2002), High-Temperature Heat-Insulating Materials Based on Vermiculite, *Refractories and Industrial Ceramics*, 43(11–12), pp. 383–389, 2002; and Suvorov, S.A., and Skurikhin, V.V., Vermiculite—a Promising Material for High-Temperature Heat Insulators, *Refractories and Industrial Ceramics*, 44(3), pp. 186–193, 2003.)

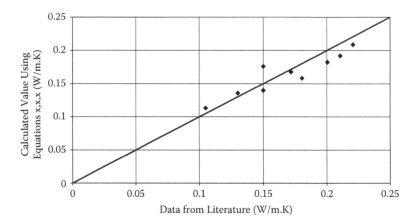

Figure 5.38 Comparison for thermal conductivity of vermiculite between literature data and calculated results using Equations (5.70), (5.71) and (5.65).

Table 5.11 summarises the theoretical thermal property data suggested in this section.

If detailed elevated temperature properties are available from the manufacturer of fire protection material, they should be used. However, the properties in Table 5.11 can be used in performance-based fire engineering calculations if elevated temperature properties are not available.

5.3.3.5 Gypsum

Compared to other fire protection materials, gypsum plaster has attracted more research studies with improved data collation of thermal properties. The data presented next are based on the research by Rahamanian and Wang (2011).

5.3.3.5.1 Density

Gypsum ($CaSO_4.2H_2O$, calcium sulphate dihydrate) has about 3% (by weight) of free water and 21% (by weight) chemically bound water. When exposed to high temperatures, the free water starts evaporation at around 100°C and is lost at around 150°C. The chemically bound water also starts to be lost due to chemical reaction. The chemical reaction happens in two stages:

$$CaSO_4.2H_2O + Q_1 \rightarrow CaSO_4. \tfrac{1}{2}H_2O + \tfrac{3}{2}H_2O$$

$$CaSO_4.\tfrac{1}{2}H_2O + Q_2 \rightarrow CaSO_4 + \tfrac{1}{2}H_2O$$

(5.72)

Table 5.11 Thermal property models for some common generic fire protection
materials

Material	Density ρ, kg/m³	Base value of specific heat, J/kg.K	Thermal conductivity, W/m.K
Rock fibre	155–180	900	$\lambda_{rock\ fibre} = 0.022 + 0.1475\left(\dfrac{T}{1000}\right)^3$
Mineral wool	165	840	$\lambda_{min\,eral\ wool} = 0.03 + 0.2438\left(\dfrac{T}{1000}\right)^3$
Calcium silicate	Various	900	$\lambda^* = \lambda_0^* + CT^3$ $\lambda_0^* = 0.23\dfrac{\rho}{1000}$ $C = 0.08 \times \dfrac{(2540 - \rho)}{2540}$
Vermiculite	Various	900	$\lambda^* = \lambda_0^* + CT^3$ $\lambda_0^* = 0.27\dfrac{\rho}{1000}$ $C = 0.18 \times \dfrac{(1000 - \rho)}{1000}$

The first-stage reaction happens at about the same time when the free
water is driven off, during which 75% of the chemically bound water is
lost. There is some uncertainty concerning the temperature range during
which the second reaction occurs, but it is considered that the second stage
is completed when the temperature is about 220°C. Therefore, the change
in density of gypsum may be obtained according to Figure 5.39.

For simplicity without loss of accuracy, a straight line joining the value at
100°C (100%) to the value at 220°C (76%) may be used.

The densities for British Gypsum-produced Fireline gypsum board and
Wallboard gypsum board are 770 and 623 kg/m³, respectively.

5.3.3.5.2 Specific heat

The base value of specific heat of gypsum is 950 J/kg°C as reported by
Mehaffey et al. (1994). Because of the large amount of water driven out
from gypsum under heating, this effect should be included when calculating the specific heat of gypsum.

As explained, gypsum undergoes two phases in which water is lost: the
free water plus 75% of the chemically bound water during the first phase and
the remaining 25% of the chemically bound water during the second phase
of the chemical reaction. Thus, the specific heat-temperature relationship

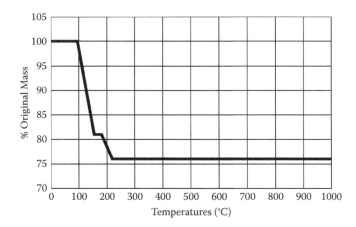

Figure 5.39 Density of gypsum as used in the analysis (percentage of the original density).

should have two peaks above the base value of 950 J/(kg.K). Ang and Wang (2004) conducted a combined heat and mass transfer analysis of gypsum under different heating conditions. They concluded that when calculating these peak values, the energy required to disassociate the water molecules from the gypsum compound as well as the effects of water movement and recondensation in cooler regions should be included.

As a result, the additional average specific heat (J/kg.K) at each dehydration reaction can be expressed by

$$\Delta c = \frac{2.26 \times 10^6}{\Delta T}(e_d f_1 + e_{free})f_2 \qquad (5.73)$$

where

Δc is the average additional specific heat;
e_d is the dehydration water content (percentage by total weight);
e_{free} is the free water content (percentage by total weight);
ΔT is the temperature interval;
f_1, f_2 are correction factors to account for heat of reactions and effects of water movement.

Knowing the heat of reactions, f_1 is easily obtained as 1.28 and 1.42 for the first and second dehydration reactions, respectively. According to Ang and Wang (2004), $f_2 = 1.4$ may be used.

Assuming that the first and second dehydration reactions occur between 95°C and 155°C and 180°C and 220°C, respectively, and assuming a tri-angular distribution of the additional specific heat over the temperature

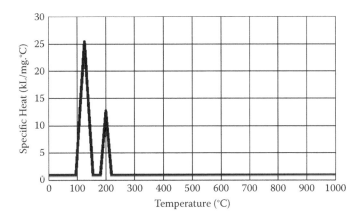

Figure 5.40 Specific heat of gypsum.

intervals, the specific heat-temperature relationship of gypsum may be obtained. Figure 5.40 shows the results.

The peak values at 120°C (first reaction) and 200°C (second reaction) are 25.3 kJ/(kg.K) and 12.7 kJ/(kg.K), respectively.

5.3.3.5.3 Thermal conductivity

Thomas (2002) collated thermal conductivity-temperature relationships for gypsum from different sources, and the results are shown in Figure 5.41.

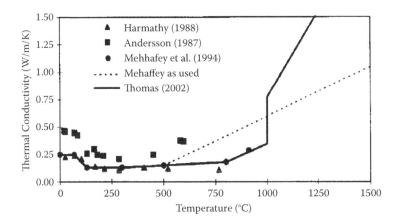

Figure 5.41 Thermal conductivity of gypsum collated by Thomas (2002). (Adapted from Thomas, G., Thermal Properties of Gypsum Plasterboard at High Temperatures, *Fire and Materials*, 26, pp. 37–45, 2002. With permission from John Wiley and Sons.)

The different lines in the figure were used by the researchers, but there was no theoretical basis. The results of the study by Rahamanian and Wang (2011) suggest that it is possible to use the theory of thermal conductivity in porous materials, as given in Section 5.3.3 of this chapter, to explain the variation in thermal conductivity of gypsum. In addition, due to the large amount of water driven out of gypsum, the effect of replacing water by air should be included.

The trends shown in Figure 5.41 may be used to make the following assumptions:

1. The thermal thermal conductivity is constant, up to 95°C before water evaporation, being equal to that at ambient temperature reported by the manufacturer.
2. The thermal conductivity decreases linearly with increasing temperature until 220°C due to evaporation of water. The thermal conductivity of dried gypsum plaster at 220°C may be measured experimentally using the guarded hot plate method (ASTM 1997).
3. The thermal conductivity increases as a function of temperature due to radiant heat transfer within the pores of gypsum.

The ambient temperature thermal conductivity values of the British Gypsum-produced Fireline gypsum board (similar to type X gypsum) and Wallboard gypsum board are 0.24 and 0.19 W/(m.K), respectively. At 220°C, the thermal conductivity values of these two types of dried gypsum plaster have been measured to be 0.12 W/(m.K) and 0.1 W/(m.K), respectively.

Equation (5.60) can be used to obtain the thermal conductivity-temperature relationship of gypsum at high temperatures. To enable this calculation to be made, it is necessary to know the pore size. Microscopic photographs of gypsum have been obtained. Figure 5.42 shows a typical photograph. Figure 5.43 provides the void size distribution in that sample.

Expectedly, the pores in gypsum are not of a uniform size, but there is a range of void sizes with different frequencies. The effective pore size used in Equation (5.60) is the size of uniform pores in a hypothetic gypsum board, which would result in the same thermal conductivity as that of real gypsum with various pores. The effective pore size has been found to be about 0.12 mm.

Figure 5.44 shows the final gypsum thermal conductivity-temperature relationships.

5.3.3.6 Intumescent coating

So far, the thermal properties of fire protection materials have been assumed to be temperature dependent only. For the fire protection materials that have been covered to this point in this chapter, this assumption

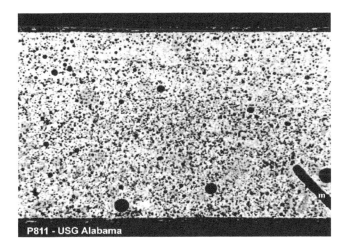

Figure 5.42 Low-magnification photograph of a sample gypsum board. (Courtesy of British Gypsum.)

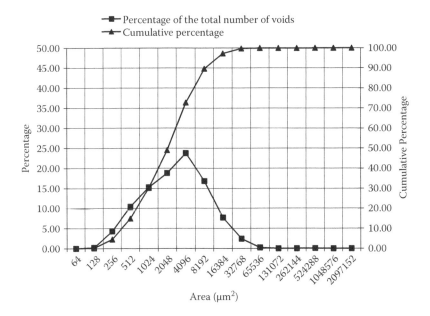

Figure 5.43 Void size distribution of a sample gypsum board.

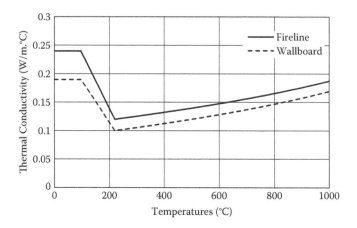

Figure 5.44 Effective thermal conductivity of gypsum.

is acceptable because the materials are inorganic and largely non-reactive under exposure to high temperatures. Intumescent coatings are different. They are organic and highly reactive under fire exposure. It is the strong chemical reactions that result in intumescent coatings to expand many-fold to provide the necessary fire protection. Therefore, it is important that the effects of these chemical reactions are properly considered.

Because the amount of intumescent coatings used is very small relative to the size of the protected structure, heat stored within the coatings amounts to a very small portion of the heat being transferred from fire to the protected structure. Therefore, heat capacitance (density multiplied by specific heat) plays a minor role. For this reason, a nominal density of 1300 kg/m³ and nominal specific heat value of 1000 J/(kg.K) may be used. The focus of this section is thermal conductivity of intumescent coating.

Because intumescent coating changes thickness during heating, one may define two thermal conductivities for intumescent coating: effective thermal conductivity and apparent thermal conductivity. The effective thermal conductivity is the heat conducted per "original" thickness of the intumescent coating per unit degree temperature. The apparent thermal conductivity is the heat conducted per "expanded" thickness of the intumescent coating per unit degree temperature.

In the current methodology of assessing fire protection performance of intumescent coating (EN 13381-8 2010; CEN 2010), the standard fire exposure condition is assumed, and the effective thermal conductivity is quantified. The effective thermal conductivity refers to the original thickness of intumescent coatings. The following equation is used to calculate the thermal conductivity of intumescent coatings. It is merely the inverse of the equation used for calculating the protected steel temperature [see Equation (4.4.5)].

$$\lambda_{p,t} = d_p \times \frac{V}{A_p} \times c_a \times \rho_a \times \frac{1}{\left(T_{f,t} - T_{s,t}\right)\Delta t} \times \Delta T_{s,t} \tag{5.74}$$

Note the second term on the right-hand side of Equation (4.45) in Section 4.5.2 is not included because it makes very small difference when used with intumescent coatings.

This method of quantification is the same as for non-reactive fire protection materials. For non-reactive fire protection materials, because their thermal properties are mainly temperature dependent only, the thermal properties obtained in this way can be used in performance-based fire engineering of structures when different fire exposure conditions are considered. However, for intumescent coatings, the thermal properties obtained under the standard fire exposure condition will most likely not be applicable to other fire conditions owing to the fact that the chemical reactions in intumescent coatings under heating, hence the expansion and thermal properties, will not only be temperature dependent but also be strongly affected by the rate of temperature change.

5.3.3.6.1 Need for a better model

To illustrate this point, Yuan (2009) carried out standard and parametric fire tests. Figure 5.45 shows the furnace temperature-time relationships. Two universal column steel sections, $254 \times 254 \times 132$ UC and $203 \times 203 \times 52$ UC, were used in the parametric fire tests, and section $254 \times 254 \times 89$ UC

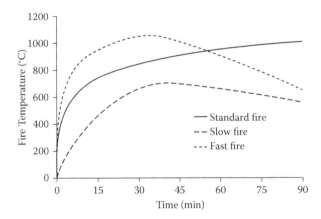

Figure 5.45 Different fire exposure curves.

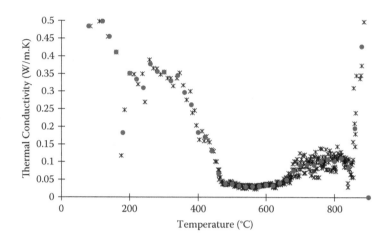

Figure 5.46 Thermal conductivity of intumescent coating extracted from a standard fire test.

was used in the standard fire test. The same intumescent coatings were used, and the coating thicknesses in the three tests were similar, about 0.7 mm.

From the recorded steel temperatures under the standard fire test, the effective intumescent coating thermal conductivity-temperature relationship was obtained using Equation (5.74). The results are shown in Figure 5.46.

Using the effective thermal conductivity of Figure 5.46, steel temperatures were calculated using Equation (5.74) for the steel sections under the parametric fire tests. The results are shown in Figures 5.47 and 5.48 for the two different parametric fire curves.

Clearly, Figures 5.47 and 5.48 show gross inaccuracy in the calculated steel temperature results compared to the experimental results for the two parametric fire conditions. This can be explained by the different recorded ratios of maximum expansion (the maximum expansion thickness divided by the initial thickness) of the intumescent coatings, which are given in Table 5.12. The same intumescent coatings in both the Slow and Fast parametric fire tests expanded much more than in the Standard fire test.

5.3.3.6.2 Modelling the expansion process

To predict the temperatures of intumescent coating protected steel under different fire conditions accurately, the expansion process should be considered. The expanding thickness should be explicitly included, and the apparent thermal conductivity (referred to the expanded thickness) should be used. Research is starting to be undertaken to develop a method for characterising the fire protection performance of intumescent coatings

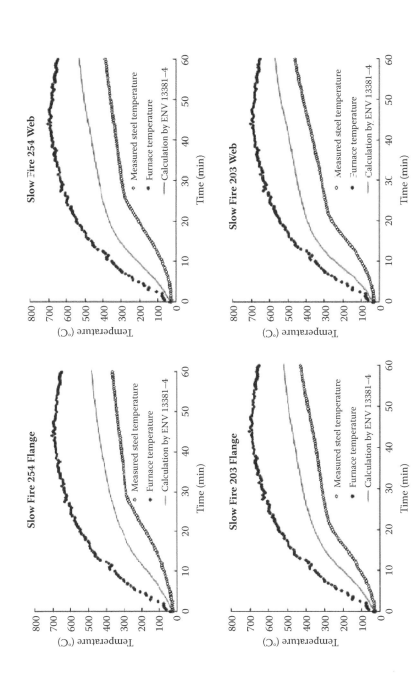

Figure 5.47 Comparison of steel temperatures for the slow parametric fire.

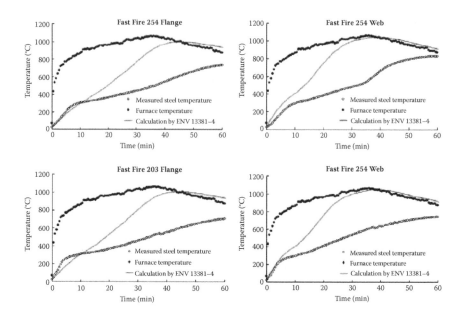

Figure 5.48 Comparison of steel temperatures for the fast fire.

under different fire conditions through understanding the global chemical reactions and the expansion process. As the research is still at an early stage, this section only briefly explains the principal factors that should be included.

Consider a layer of intumescent coating of unit area with thickness x. As it expands under heating, its thickness is $\int_{t_1}^{t_2} \dot{x}dt$, where $\dot{x}$ is the rate of expansion, and t_1 and t_2 are the lower bound and upper bound, respectively, of time during which the expansion of intumescent coatings occurs. Assuming there is a lower-bound temperature (T_1) from which intumescent coatings start to expand and an upper-bound temperature (T_2) at which intumescent coatings expansion stops, then the lower and upper limits of time can be determined by the lower- and upper-bound temperatures.

Table 5.12 Example maximum ratios of expansion of intumescent coatings

	Standard fire	Slow fire	Fast fire
Web	29.5	47.3	46.5
Flange	25.9	37.3	37.5

Because intumescent coatings are very thin, the volume (and hence the thickness) is primarily that of the gas within the pores. If the mass of the gas is m_{gas}, then

$$x = \frac{m_{gas}}{\rho_{gas}} \qquad (5.75)$$

where ρ_{gas} is the gas density.

From ideal gas law,

$$\rho_{gas} = \frac{WP}{RT} \qquad (5.76)$$

where W is gas molecular mass, P is pressure, R is the gas constant and T is the absolute temperature.

Intumescent coatings have three generic chemical components: an organic acid source, an organic blowing agent and a charring material. Gas in intumescent coatings is produced when the blowing agent decomposes. However, some of the gas from the decomposed blowing agent escapes from the intumescent coatings, and it is only the portion remaining that expands the intumescent coatings.

Assuming the amount of decomposed blowing agent is m_2, and assuming the amount remaining in the intumescent coatings is βm_2, then the rate of expansion may be expressed as follows:

$$\dot{x} = \frac{dx}{dt} = \frac{d\left(\dfrac{\beta m_2 RT}{WP}\right)}{dt} = \frac{R\beta}{WP} \bullet \left(T\frac{dm_2}{dt} + m_2\frac{dT}{dt}\right) \qquad (5.77)$$

Thus, to obtain the expanded thickness of the intumescent coatings, the following four values should be obtained: the lower-bound temperature T_1, the upper-bound temperature T_2, the proportionality β and the mass (and its rate of depletion) of the blowing agent.

5.4 CONCLUDING REMARKS

This chapter has presented data on material properties for performance-based fire engineering of structures, including both the mechanical and thermal properties for load-bearing materials and thermal properties for common fire protection materials. Comprehensive information is available for load-bearing materials, which is included in the first part of this chapter.

The second part of this chapter presented the temperature-dependent thermal properties for a range of common fire protection materials, including rock fibre, mineral wool, calcium silicate, vermiculite and gypsum. These materials are non-reactive, and their properties are temperature dependent only. Intumescent coatings are a special case because they are reactive. Their thermal properties are not only temperature dependent but also strongly affected by the fire exposure. This chapter has demonstrated that the intumescent coating thermal properties extracted from the standard fire test, as recommended in EN 13381-8 (CEN 2010), are not appropriate for applications under different fire exposure conditions. Research studies are being conducted to develop a method to characterise intumescent coating performance under different fire conditions.

Chapter 6

Element structural fire resistance design[1]

6.1 DESIGN PRINCIPLES

6.1.1 Basis of element design

As with structural design under ambient temperature, structural fire resistance design may be carried out at three different levels:

1. Using design tables: The design is simple, but the design choice will be conservative and limited.
2. Using analytical models according to basic mechanical principles and good engineering practice. These models are suitable for structural elements with clearly defined loading and boundary conditions.
3. Using numerical models: This approach can be complex and time consuming, but it does allow complex structures to be dealt with and can give very detailed and accurate results.

This chapter focus on the element-based analytical approach.

6.1.2 Structural Eurocodes

This chapter adopts the structural Eurocodes as the basis. The structural Eurocodes were developed from the 1990s and were formally issued in 2008 to supersede the relevant national standards. There are based on ultimate limit states. The design principles, with safety level being achieved by partial safety factors for loading and material/modelling, are described in EN 1990: 2002 (Committee of European Normalization [CEN] 1990–2002, *Basis of Structural Design*). The loading on structures is collected in EN 1991. The fire and mechanical loads are specified in EN 1991-1-2: 2002 (CEN 1991–2002, *Actions of Structures Exposed to Fire*). There is one set of standards for each of the main construction materials, in which Part 1-2

---

[1] A list of notations is provided for this chapter at the beginning of the book.

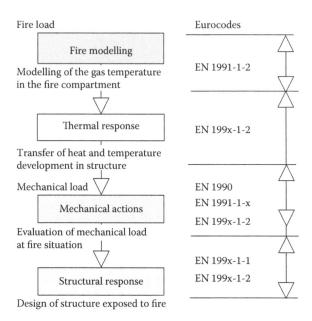

Fire load   Eurocodes

Fire modelling

Modelling of the gas temperature   EN 1991-1-2
in the fire compartment

Thermal response   EN 199x-1-2

Transfer of heat and temperature
development in structure

Mechanical load

Mechanical actions   EN 1990
                                           EN 1991-1-x
                                           EN 199x-1-2

Evaluation of mechanical load
at fire situation

                                           EN 199x-1-1
Structural response   EN 199x-1-2

Design of structure exposed to fire

Figure 6.1 Major steps of fire resistance design and relevant structural Eurocodes.

deals with fire resistance design. Figure 6.1 shows the general arrangement of structural Eurocodes for fire resistance design. The main objective of this chapter is to present the fundamentals of the element-based fire resistance design methods. Therefore, this chapter presents design methods for steel, concrete, composite steel/concrete, timber and masonry structures. General design methods for using other less-common load-bearing constructional materials, such as stainless steel and aluminium, are similar to those for steel structures. Interested readers should refer to the relevant Eurocodes and educational materials (see Lennon et al. 2007, for more detailed information). For steel structures, the reader may consult a number of sources for more worked examples, such as those of Access Steel (http://www.access-steel.com) or DIFISEK⁺ (http://www.difisek.eu) as well as the work of Franssen and Vila Real (2010).

The design value of mechanical material property in the fire situation is described as

$$X_{d,fi} = k_\theta \, X_k / \gamma_{M,fi} \qquad\qquad (6.1)$$

where X_k is the characteristic value of a strength or deformation property for normal temperature design; k_θ is the reduction factor dependent on the material temperature θ; $\gamma_{M,fi}$ is the partial safety factor of the relevant material mechanical property for the fire limit state.

Detailed data of reduction factors (k_θ) for each construction material can be found in the relevant Eurocode for fire resistance design for the particular material (EN 199x-1-2). Some values are summarised in Chapter 5 of this book. Material mechanical properties at elevated temperatures may be sensitive to heating rate and duration. The reduction factors in Eurocodes are derived from large sets of material tests at heating rates and durations similar to those encountered in fire (e.g. for steel between 2 and 50 K/min; see Kruppa et al. 2005).

Fire resistance design is primarily focussed on the ultimate limit state. A serviceability check (e.g. deformation) is not required unless excessive deformation of the structural member adversely affects functions of other members, such as fire protection or separating members.

6.1.3 Design procedures

Verification of fire resistance may be made in the time domain, in the strength domain or in the temperature domain. In verification by time, the required fire resistance time should be equal to or less than the available fire resistance time of the structure. Verification by strength is similar to the familiar approach of structural design under the cold condition. In the temperature domain, the critical temperature in the structure should be lower than the limiting temperature of the structure.

A member-based fire resistance design approach does not consider effects caused by thermal deformation at elevated temperatures. Therefore, the approaches in this chapter only are applicable where these effects may be recognized a priori to be either negligible or favourable; are accounted for by conservatively chosen support and boundary conditions; or are implicitly considered by conservatively specified fire safety requirements.

6.1.4 Thermal loading

When carrying out structural fire resistance design, the first step is to select a fire scenario (thermal loading). Thermal loading may be based on Standard Fire temperature-time curves, other simplified fire models for localised and compartment fire, or output from fire test or numerical simulation models. More detailed information may be obtained from Chapter 3 of this book.

6.1.5 Mechanical loading

In the Eurocode system, fire is considered an accidental loading condition. Therefore, for obtaining the relevant effects of actions $E_{fi,d,t}$ during fire exposure, the mechanical actions should be combined in accordance with EN 1990: 2002 (*Basis of Structural Design*, CEN 1990–2002) for accidental design situations. In Eurocode terminology, the representative value for variable action Q_1 may use the quasi-permanent value $\psi_{2,1} Q_1$ or the frequent value $\psi_{1,1} Q_1$ depending on the nature of the action, where $\psi_{2,1}$ and $\psi_{1,1}$ are the so-called combination factors.

When performing element-based structural fire resistance design, indirect fire actions are not considered, and the effects of actions are the same as at ambient temperature (corresponding to time $t = 0$). These effects of actions $E_{fi,d}$ are applied as constant values throughout the design fire exposure.

As a simplification, the effects of actions may be deduced from those determined in normal temperature design using the following expression:

$$E_{fi,d,t} = E_{fi,d} = \eta_{fi} E_d \qquad (6.2)$$

where E_d is the design value of the relevant effects of actions from the fundamental load combination; $E_{fi,d}$ is the corresponding (required) design value for the fire situation; and η_{fi} is a reduction factor, expressing the ratio of the applied load in fire to the applied load under normal temperature.

For example, if the load combination for fire design is $G_k + \psi_{fi} Q_{k,1}$ and the load combination for normal temperature design is $\gamma_G G_k + \gamma_{Q,1} Q_{k,1}$, then the reduction factor η_{fi} can be taken as

$$\eta_{fi} = \frac{G_k + \psi_{fi} Q_{k,1}}{\gamma_G G_k + \gamma_{Q,1} Q_{k,1}} \qquad (6.3)$$

where $Q_{k,1}$ is the principal variable action; G_k is the permanent action; γ_G is the partial factor for permanent actions for normal temperature design; $\gamma_{Q,1}$ is the partial factor for the leading variable action for normal temperature design; and ψ_{fi} is the combination factor for either quasi-permanent value ($\psi_{1,1}$) or frequent value ($\psi_{2,1}$) as explained in the first paragraph of this section.

As a further simplification for evaluation of structural fire resistance without calculation, a value of $\eta_{fi} = 0.70$ may be conservatively used for concrete structures. For steel, steel/concrete composite, aluminium, and masonry structures, $\eta_{fi} = 0.65$ may be used. For timber structures, $\eta_{fi} = 0.60$. However, in all cases for areas susceptible to accumulation of goods, including access areas, it is recommended to use a value of $\eta_{fi} = 0.70$.

After introducing details of different design calculation methods for different construction materials, this chapter presents a number of design calculation examples.

6.2 CONCRETE STRUCTURES

6.2.1 Basis of design

Due to concrete structural element size, low thermal conductivity of concrete and transport of water during heating, temperature distributions in concrete elements are complex and non-uniform. Numerical heat transfer analysis is often necessary to obtain temperature distributions for fire resistance design. For structural elements exposed to the nominal Standard Fire temperature-time curve, Annex A to EN 1992-1-2: 2005 (CEN 1992–2005b) provides some useful tables of temperature distributions for durations from 30 min to 240 min with moisture content up to 1.6%.

In addition to load bearing, a concrete structural element may have to fulfil the fire resistance requirements for integrity and insulation. For example, the concrete structure may serve as a fire compartment wall or floor.

6.2.2 Simplified methods

Two simplified methods may be used: the 500°C isotherm method and the zone method.

6.2.2.1 500°C isotherm method

The 500°C isotherm method (see Figure 6.2) is based on the hypothesis that concrete at a temperature of more than 500°C is neglected in the calculation of load-bearing capacity, while concrete at a temperature below 500°C is assumed to retain its full strength and stiffness. This method is applicable to reinforced and prestressed concrete cross sections with respect to axial

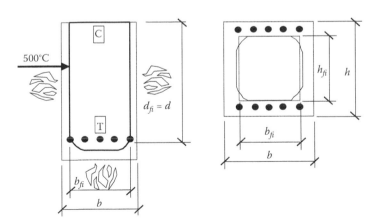

Figure 6.2 The 500°C isotherm method: reduced cross section of reinforced concrete member exposed to fire on three or four sides.

load, bending moment and their combinations. The heating condition may be according to either the nominal Standard Fire temperature-time curve or the parametric fire curve in EN 1991-1-2 (see Chapter 3 of this book for more details) with opening factor $O \geq 0.14$ m$^{1/2}$ and fire load density 200–800 MJ/m^2 floor area. The reinforcement should be taken into account, whether it is inside (where temperature does not exceed 500°C) or outside the effective cross section with reduced values of mechanical properties. The resistance of the cross section is calculated by the normal temperature design method in EN 1992-1-1 (*Design of Concrete Structures*).

The detailed design calculation procedure may be divided into the following steps:

a. Obtaining the 500°C isotherm, according to nominal Standard or parametric fire curve;
b. Finding the new effective (reduced) thickness b_{fi} and height d_{fi} within which the concrete temperature does not exceed 500°C. The isotherm may be simplified to give a rectangular cross section;
c. Finding out temperatures at the centre of gravity of each reinforcing bar (including those outside the effective cross section);
d. Calculating the reduced strength of reinforcement bars;
e. Calculating the resistance of the reduced cross section by the normal temperature design procedure (e.g. for bending resistance, see Figure 6.3);
f. Comparing the effect of design load in fire to the evaluated concrete member resistance from step e.

6.2.2.2 Zone method

In the zone method, a layer (thickness a_z measured from the fire-exposed surface) is reduced from the fire-damaged cross section (see Figure 6.4).

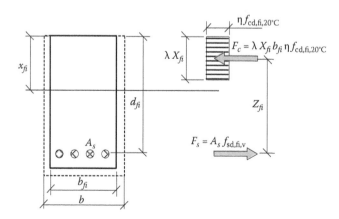

Figure 6.3 Evaluation of bending resistance of reduced cross section.

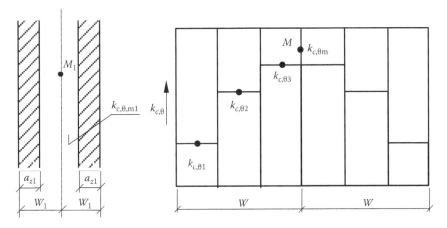

Figure 6.4 Zone method for wall exposed to fire from both sides.

The remaining cross section is assumed to be cold. The zone method is more complex than the 500°C isotherm method but may produce more accurate results.

The zone method is applicable to walls, slabs, beams and columns exposed to nominal Standard Fire temperature-time curve only, but the concrete temperature may be obtained using any means. To obtain the layer thickness a_z, the original cross section is divided into several zones of equal thickness/width that are assumed to be at the same mean temperature with the same corresponding mean compressive strength and modulus of elasticity (e.g. see Figure 6.4).

The design procedure is as follows:

a. The original cross section is divided into several $(n, n \geq 3)$ parallel zones of the same thickness, (e.g. for a rectangular wall, see Figure 6.4);
b. Determine the average temperature θ_i at the centre of each zone;
c. Obtain the reduction factors of concrete for compressive strength $k_{c,\theta i}$ for each zone;
d. The mean value of reduction factor $k_{c,\theta m}$ for the cross section is approximately

$$k_{c,m} = \frac{(1-0,2n)}{n} \sum_{i=1}^{n} k_{c,\theta,i} \qquad (6.4)$$

where n is the number of parallel zones of width w, and m is the zone number;

e. The thickness a_z of the damaged layer, which will be neglected in calculating concrete member capacity, is calculated using the following equations:

- for slabs beams or elements loaded by shear:

$$a_z = w\left[1 - \left(\frac{k_{c,m}}{k_{c,\theta,M}}\right)\right] \tag{6.5}$$

- for columns and walls where a second-order effect should be taken into consideration:

$$a_z = w\left[1 - \left(\frac{k_{c,m}}{k_{c,\theta,M}}\right)^{1,3}\right] \tag{6.6}$$

f. For the reduced cross section, the material properties are based on those at point M.

To reduce the calculation effort for $k_{c,m}$ and a_z, a number of graphs have been prepared in Annex B.2 of EN 1992-1-2: 2005 (CEN 1992–2005b) for slabs, beams and columns of siliceous aggregate concrete exposed to the nominal Standard Fire temperature-time curve.

6.2.3 Spalling

Explosive spalling is unlikely to occur when the moisture content of concrete is less than 3% by weight. For beams, slabs and tensile members, if the moisture content of the concrete is more than 3% by weight, the influence of explosive spalling on load-bearing capacity may be assessed by assuming local loss of cover to one reinforcing bar or bundle of bars in the cross section and then checking the reduced load-bearing capacity of the cross section. Where the axis distance to the reinforcement is 70 mm or more and tests have not been carried out to show that falling off does not occur, a surface reinforcement should be provided. The surface reinforcement mesh should have spacing not greater than 100 mm and a diameter not less than 4 mm.

For high-strength concrete grades C55/67 to C80/95, the same rules may be applied, provided that the maximum content of silica fume is less than 6% by weight of cement. For high-strength concrete grades above C80/95 but not exceeding C90/105, spalling can occur in any situation for concrete exposed directly to fire and at least one of the following methods should be provided:

- a reinforcement mesh with a nominal cover of 15 mm;
- demonstration by testing that no spalling of concrete occurs under fire exposure;
- use of protective layers;

together with inclusion in the concrete mix more than 2 kg/m^3 of monofilament propylene fibres.

6.3 STEEL STRUCTURES

6.3.1 Basis of design

Steel conducts heat very fast, and steel structural cross sections are usually thin. Therefore, it is acceptable to assume uniform temperature distribution in steel cross sections. Simplified methods of calculating steel cross-section temperature are given in Chapter 4 of this book. Steel mechanical properties at elevated temperatures are given in Chapter 5 of this book.

6.3.2 Simple analytical models

The simple models for element fire resistance design are based on calculation methods for normal temperature design taking into consideration reductions in steel mechanical properties.

6.3.2.1 Cross-section classification

At elevated temperatures, the steel modulus of elasticity E and yield strength f_y are reduced. This will affect steel section classification. The relevant value for checking section classification in EN 1993-1-2: 2005 is

$$\sqrt{\frac{E_\theta}{f_{y,\theta}}} = \sqrt{\frac{k_{E,\theta}\,E}{k_{y,\theta}\,f_y}} = \sqrt{\frac{k_{E,\theta}}{k_{y,\theta}}}\sqrt{\frac{E}{f_y}} \cong 0.85\sqrt{\frac{E}{f_y}} \tag{6.7}$$

6.3.2.2 Members in tension

The design resistance $N_{fi,t,\mathrm{Rd}}$ at time t of a tension member with temperature θ may be determined from

$$N_{fi,\theta,Rd} = k_{y,\theta}\,N_{Rd}(\gamma_{M,0}/\gamma_{M,fi}) \tag{6.8}$$

where $k_{y,\theta}$ is the reduction factor for the yield strength of steel at temperature θ_a, reached at time t, and N_{Rd} is the design resistance of the cross-section $N_{pl,Rd}$ for normal temperature design according to EN 1993-1-1: 2005.

6.3.2.3 Members in buckling

When a steel member is exposed to fire, its stiffness is reduced. This may be used to reduce the column effective length. For columns in braced frames, in which each storey comprises a separate fire compartment with sufficient fire resistance, in an intermediate storey the buckling length l_{fi} of a continuous column may be taken as $l_{fi} = 0.5\ L$. In the top storey, the buckling length may be taken as $l_{fi} = 0.7\ L$, where L is the system length in the relevant storey (see Figure 6.5).

The design buckling resistance $N_{b,fi,t,Rd}$ at time t of a compression member with Class 1, Class 2 or Class 3 cross section and uniform temperature θ_a can be determined from

$$N_{b,fi,t,Rd} = \chi_{fi}\, A\, k_{y,\theta} f_y / \gamma_{M,fi} \tag{6.9}$$

where χ_{fi} is the reduction factor for flexural buckling in the fire design situation; $k_{y,\theta}$ is the reduction factor for yield strength of steel at temperature θ_a. The value of χ_{fi} should be taken as the lesser of the values of $\chi_{y,fi}$ and $\chi_{z,fi}$ determined according to

$$\chi_{fi} = \frac{1}{\varphi_\theta + \sqrt{\varphi_\theta^2 - \overline{\lambda}_\theta^2}} \tag{6.10}$$

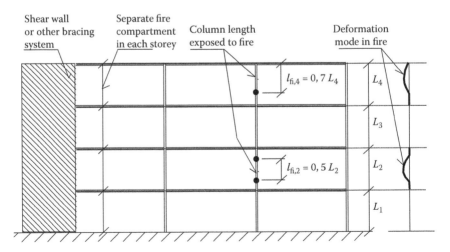

Figure 6.5 Buckling lengths l_{fi} of columns in braced frames in fire.

with $\varphi_\theta = \frac{1}{2}\left(1 + \alpha \overline{\lambda}_\theta + \overline{\lambda}_\theta^2\right)$ and for all steel grades, $\alpha = 0.65\sqrt{235/f_y}$, where f_y is the yield strength of steel at ambient temperature. The relative slenderness $\overline{\lambda}_\theta$ at temperature θ_a is calculated by

$$\overline{\lambda}_\theta = \overline{\lambda}\sqrt{k_{y,\theta} / k_{E,\theta}} \tag{6.11}$$

where $k_{E,\theta}$ is the reduction factor for modulus of elasticity of steel at temperature θ_a.

6.3.2.4 Members in bending

The design moment resistance $M_{fi,\theta,Rd}$ of Class 1, Class 2 and Class 3 cross sections with a uniform temperature θ_a may be determined from

$$M_{fi,\theta,Rd} = k_y\, M_{Rd}\, \frac{\gamma_{M,0}}{\gamma_{M,fi}} \tag{6.12}$$

where M_{Rd} is, for Class 1 and Class 2 cross sections, the plastic moment resistance of the gross cross section $M_{pl,Rd}$ for normal temperature design; for Class 3 cross sections, it is the elastic moment resistance of the gross cross section $M_{el,Rd}$ for normal temperature design, according to EN 1993-1-1: 2005. Here, $k_{y,\theta}$ is the reduction factor for yield strength of steel at temperature θ_a. In case of high shear, the reduced moment resistance allowing for the effects of shear according to EN 1993-1-1: 2005 may be followed. For a member with non-uniform temperature distribution, its bending resistance may be determined from

$$M_{fi,t,Rd} = M_{fi,\theta,Rd}/(\kappa_1 \kappa_2) \tag{6.13}$$

where $M_{fi,\theta,Rd}$ is the design moment resistance of the cross section for a uniform temperature θ_a, which is equal to the maximum temperature θ_a in a critical element of the cross section (usually flange) away from the support. The adaptation factor κ_1 accounts for non-uniform temperature across the cross section. For an unprotected beam exposed to fire on three sides with a composite or concrete slab on side four, $\kappa_1 = 0.70$. For a protected beam exposed to fire on three sides with a composite or concrete slab on side four, $\kappa_1 = 0.85$. The adaptation factor κ_2 is derived for non-uniform temperature distribution along the beam, which may be taken as $\kappa_2 = 0.85$ at the supports of a statically indeterminate beam. In all other cases, $\kappa_2 = 1.0$.

More accurately, the temperature distribution along the beam height may be taken into consideration in plastic design calculations for Class 1 and Class 2 cross sections by dividing the cross section into many parts of approximately equal temperatures and then summing the contributions of each part to the bending moment resistance of the cross section, using

$$M_{fi,t,Rd} = \sum_{i=1}^{n} A_i \, z_i \, k_{y,\theta,i} \, f_{y,i} / \gamma_{M,fi} \qquad (6.14)$$

where z_i is the distance from the plastic neutral axis to the centroid of the elemental area A_i; $f_{y,i}$ is the nominal yield strength f_y for the elemental area A_i.

6.3.2.5 Lateral torsional buckling

The design lateral torsional buckling resistance moment $M_{b,fi,t,Rd}$ at time t of a laterally unrestrained member with a Class 1 or Class 2 cross section may be determined according to

$$M_{b,fi,t,Rd} = \chi_{LT,fi} \, W_{pl,y} k_{y,\theta,com} f_y / \gamma_{M,fi} \qquad (6.15)$$

where $\chi_{LT,fi}$ is the reduction factor for lateral-torsional buckling in the fire design situation; $W_{pl,y}$ is the plastic section modulus about the major axis; and $k_{y,\theta,com}$ is the reduction factor for the yield strength of steel at the maximum temperature in the compression flange $\theta_{a,com}$ reached at time t. The relative slenderness for lateral torsional buckling may be modified from the relative slenderness at ambient temperature using the following equation:

$$\overline{\lambda}_{LT,\theta,com} = \overline{\lambda}_{LT} \sqrt{k_{y,\theta,com} / k_{E,\theta,com}} \qquad (6.16)$$

where $k_{E,\theta,com}$ is the reduction factor for modulus of elasticity at the maximum steel temperature in the compression flange $\theta_{a,com}$ reached at time t. Equation (6.16) is used to calculate the strength reduction factor $\chi_{LT,fi}$ by substituting $\overline{\lambda}_{\theta}$ by $\overline{\lambda}_{LT,\theta,com}$.

6.3.2.6 Class 4 cross sections

Class 4 cross sections are those where local buckling has to be considered. Class 4 cross sections may be designed in the same way as under normal temperature. However, as a gross simplification, the critical temperature of a Class 4 cross section may be conservatively taken as $\theta_{a,cr} = 350°C$.

6.3.2.7 Connection design

At present, information on connection design in code of practice is limited. For example, EN 1993-1-2: 2005 gives two approaches for the design of steel connections. In the first approach, fire protection is applied to the member and its connections; in the second, more detailed approach, it is possible to apply the component approach in EN 1993-1-8: 2005 (*Design of Joints*) together with the mechanical properties of connectors (bolts/welds) at elevated temperatures. Detailed information on connection behaviour and design is provided in Chapter 8 of this book.

6.3.3 Critical temperature

As an alternative to verifying fire resistance in the strength domain, calculations may be performed in the temperature domain. The critical temperature of a structure $\theta_{a,cr}$, which should not be exceeded by the critical temperature rise in the structure, may be calculated using the following equation:

$$\theta_{a,cr} = 39.19 \ \ln\left[\frac{1}{0,9674\mu_0^{3,833}} - 1\right] + 482 \tag{6.17}$$

where μ_0 is the degree of utilisation at time $t = 0$ and must not be taken less than 0.013. For members in bending with Class 1, Class 2 or Class 3 cross sections and for all tension members, the degree of utilization μ_0 at time $t = 0$ may be obtained from

$$\mu_0 = E_{fi,d}/R_{fi,d,0} \tag{6.18}$$

where $E_{fi,d}$ is the design effect of actions for the fire situation; for further information see Section 6.1.5. $R_{fi,d,0}$ is the corresponding load-carrying capacity of the structural member under normal temperature. Conservatively, μ_0 may be obtained from

$$\mu_0 = \eta_{fi}[\gamma_{M,fi}/\gamma_{M0}] \tag{6.19}$$

where η_{fi} is the reduction factor for design load level in the fire situation (see Section 6.1.5), $\gamma_{M,fi}$ is the material partial safety factor for the fire situation, and γ_{M0} is the material partial safety factor for normal temperature design.

6.4 COMPOSITE STEEL AND CONCRETE STRUCTURES

6.4.1 Composite slabs and beams

The fire design of steel and concrete composite structures with partially or fully encased steel elements is covered by EN 1994-1-2: 2005 (CEN 1994–2005b). It is expected that the design of a composite structural element at ambient temperature is according to EN 1994-1-1: 2005 (CEN 1999–2005a). Slabs are expected to be heated from the bottom, columns from all four sides and beams from three sides only if the trapezoidal sheeting of the supported slabs covers at least 85% of the upper flange of the beam.

For composite beams with the concrete section on top of the steel section, either the critical temperature method or the bending moment resistance method may be used.

6.4.1.1 Critical temperature method

The critical temperature method is applicable to symmetric cross sections of a maximum depth $h = 500$ mm and a minimum slab depth h_c not less than 120 mm, used in conjunction with simply supported beams exclusively subject to sagging bending moments. The temperature of the steel cross section is assumed to be uniform. The advantage of this method is that the critical temperature θ_{cr} may be determined from the load level $\eta_{fi,t}$ applied to the composite cross section and from the strength of steel at elevated temperatures f_y without calculating the bending resistance of the cross section. The critical temperature is a function of the load level in the fire design at time t:

$$\eta_{fi,t} = \frac{E_{fi,d,t}}{R_d} = \frac{\eta_{fi} \, E_d}{R_d} = \frac{R_{fi,d,t}}{R_d} \tag{6.20}$$

where $E_{fi,d,t}$ is the design effect of actions in the fire situation, R_d is the design resistance for normal temperature design, and E_d is the design effect of actions for normal temperature design. The ultimate limit state is reached when the design resistance in the fire situation $R_{fi,d,t}$ has decreased to the level of the design effect of actions in the fire situation $E_{fi,d,t}$. The load level may be used to derive an expression to obtain the required level of steel yield strength $f_{a\max,\theta cr}$ at elevated temperatures as follows:

$$\eta_{fi,t} = \frac{R_{fi,d,t}}{R_d} \cong \frac{f_{a\max,\theta cr}}{\dfrac{f_{ay,20°C}}{\gamma_M}} \cong \frac{f_{a\max,\theta cr}}{f_{ay,20°C}} \frac{1}{0,9} \tag{6.21}$$

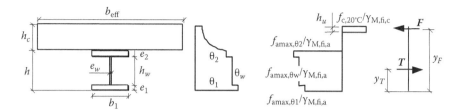

Figure 6.6 Calculation of sagging moment resistance.

The steel yield strength value can be used to obtain the critical temperature of the composite beam.

6.4.1.2 Bending moment resistance model

When using the plastic analysis, the composite cross section may be divided into a number of areas, each with the same temperature. An example is shown in Figure 6.6.

If the plastic neutral axis of the composite cross section is in the concrete slab, the resultant tension force is from the steel cross section, and it may be calculated using the following equation:

$$T = [f_{amax,\theta1}(b_1\ e_1) + f_{amax,\theta w}(h_w\ e_w) + f_{amax,\theta2}(b_2\ e_2)] / \gamma_{M,fi,a} \tag{6.22}$$

The thickness of the compressive zone in the concrete slab h_u is determined from equilibrium $T = F$, giving

$$h_u = T / (b_{eff}\ f_{c,20°C}\ / \gamma_{m,fi,c})$$

The sagging moment resistance is

$$M_{fi,Rd^+} = T\ (y_F - y_T) \tag{6.23}$$

This calculation model may be used for composite slabs with profiled steel sheet if h_u is replaced by h_{eff}.

6.4.2 Composite columns

Composite column design may be carried out using the same method as for steel columns, provided the cross-sectional resistance and the flexural stiffness are calculated using the properties of the composite cross section. Furthermore, the strength reduction factor is calculated using column buckling curve c instead of Equation (6.10). This simple calculation model should only be used for columns in braced frames.

6.5 TIMBER STRUCTURES

6.5.1 Basis of design

Fire resistance of timber structural elements of sufficiently large cross section can be calculated based on predictable charring rate. The boundary between the charred timber and carbonized timber is assumed at 300°C.

6.5.2 Charring depth

The charring depth is the distance between the outer surface of the original member and the position of the char line and may be calculated from the time of fire exposure and the relevant charring rate. Under the Standard Fire exposure, an approximate linear relationship between the fire duration and the charring depth has been found to exist from numerous fire tests on wood and wood-based materials. Therefore, the charring depth is approximately

$$d_{char,0} = \beta_0\, t \tag{6.24}$$

where β_0 is the one-dimensional design charring rate under the Standard Fire exposure (see Table 6.1), and t is the time of fire exposure in minutes.

Charring should be considered for all surfaces of the timber element directly exposed to fire. For surfaces protected by fire-protective claddings, other protection materials or other structural members, charring is assumed to be delayed by a time duration t_{ch}. For fire-protective claddings consisting of one or several layers of wood-based panels or wood panelling,

Table 6.1 Typical one-dimensional design β_0 and notional β_n charring rate according to EN 1995-1-2: 2005

Material		β_0 mm/min	β_n mm/min
Softwood and beech	Glued laminated timber with a characteristic density of ≥ 290 kg/m³	0.65	0.70
	Solid timber with a characteristic density of ≥ 290 kg/m³	0.65	0.80
Hardwood	Solid or glued laminated hardwood with a characteristic density of 290 kg/m³	0.65	0.70
	Solid or glued laminated hardwood with a characteristic density of ≥ 450 kg/m³	0.50	0.55

the time to the start of charring t_{ch} of the protected timber member may be taken as

$$t_{ch} = \frac{h_p}{\beta_0}$$
(6.25)

where h_p is the total thickness of the protective panel.

6.5.3 Simple analytical models

Two methods may be used to evaluate the required fire resistance of timber structural members: the reduced cross-section method and the reduced properties method.

6.5.3.1 Reduced cross-section method

In addition to the char depth, it is assumed that a layer of timber inside the char line of thickness $k_0 d_0$ has zero strength and stiffness, while the strength and stiffness properties of the remaining cross section are unchanged (Figure 6.7). Therefore, the effective cross section is obtained by reducing the initial cross section by the following effective charring depth:

$$d_{ef} = d_{char,n} + k_0 d_0$$
(6.26)

with d_0 = 7 mm, $d_{char,n}$ is determined according to Equation (6.24). For unprotected surfaces and if the time of Standard Fire exposure $t < 20$ min, then $k_0 = t/20$. For longer fire exposure durations, $k_0 = 1.0$.

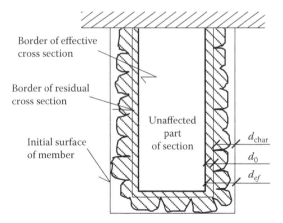

Figure 6.7 Definition of residual cross section and effective cross section.

Due to evaporation of water inside the effective cross section, the design strength and stiffness properties of the effective cross section should be calculated with a modification factor $k_{mod,fi} = 1.0$.

6.5.3.2 Reduced properties method

As an alternative to the reduced section method, the reduced properties method may be applied to rectangular cross sections of softwood exposed to fire on three or four sides and round cross sections exposed to fire along their whole perimeter. The residual cross section should be determined according to the charring depth given in Equation (6.26). A modification factor $k_{mod,fi}$ is used to replace the modification factor for normal temperature design k_{mod} given in EN 1995-1-1: 2005. The design strength and modulus of elasticity at elevated temperature may be evaluated from

$$f_{fi,d} = k_{mod,fi} k_{fi} \frac{f_k}{\gamma_{M,fi}} \tag{6.27a}$$

$$E_{fi,d} = k_{mod,fi} k_{fi} \frac{E_{k,05}}{\gamma_{M,fi}} \tag{6.27b}$$

$$E_{fi,d} = k_{mod,fi} \frac{E_{mean}}{\gamma_{M,fi}} \tag{6.27c}$$

where $k_{mod,fi}$ is the modification factor for fire, which takes into account the influence of temperature and humidity; $\gamma_{M,fi}$ is the partial factor for timber in fire; f_k is the characteristic strength at ambient temperature; $E_{k,05}$ is the characteristic modulus of elasticity at ambient temperature for 5% fractile; and E_{mean} is the average modulus of elasticity at ambient temperature. Here, k_{fi} is the coefficient for transferring the model from characteristic value to average value. For solid timbers, $k_{fi} = 1.25$. For glued-laminated timber and wood-based panels, $k_{fi} = 1.15$. For connections with fasteners in shear with side members of steel and for connections with axially loaded fasteners of wood and wood-based panels, $k_{fi} = 1.05$. The modification factor for fire $k_{mod,fi}$ may be determined in the following way:

- For bending strength:

$$k_{mod,fi} = 1.0 - \frac{1}{200} \frac{p}{A_r} \tag{6.28a}$$

- For compression strength:

$$k_{\mathrm{mod},fi} = 1.0 - \frac{1}{125}\frac{p}{A_r} \tag{6.28b}$$

- For tension strength and modulus of elasticity:

$$k_{\mathrm{mod},fi} = 1.0 - \frac{1}{330}\frac{p}{A_r} \tag{6.28c}$$

where p is the perimeter of the fire-exposed residual cross section, and A_r is the area of the residual cross section.

6.5.4 Connections

The fire resistance of timber structures is influenced by the burn-up of its surface. EN 1995-1-2: 2005 (CEN 1995–2005b) gives rules for fire design of connections made with nails, bolts, dowels, screws, split-ring connectors, shear-plate connectors and toothed-plate connectors for resistance up to 60 min. It also gives the best engineering practice related to the structural detailing of walls and floors. For connections with side members of wood, for connections with external steel plates, and for axially loaded screws, the simplified rules and the reduced load method may be used.

6.5.4.1 Simplified rules

The fire resistance of unprotected wood-to-wood connections, where spacing, edge and end distances and side member dimensions comply with the minimum requirements given in Chapter 8 of EN 1995-1-1: 2005, is 15 min. For connections with dowels, nails or screws with non-projecting heads, greater fire resistance periods, but not exceeding 30 min, may be achieved by increasing the timber dimensions by a_{fi} for the thickness of the side members, the width of the side members, and the end and edge distances to the fasteners (see Figure 6.8). The added thickness/width may be calculated from

$$a_{fi} = \beta_n\, k_{flux}\left(t_{req} - t_{d,fi}\right) \tag{6.29}$$

where β_n is the notional charring rate; k_{flux} (= 1.5) is a coefficient taking into account the increased heat flux through the fastener; t_{req} is the required Standard Fire resistance period; and $t_{d,fi}$ is the fire resistance period of the unprotected connection, $t_{d,fi}$ = 15 min.

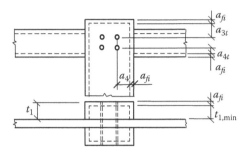

Figure 6.8 Added thickness/distance for fire resistance higher than 15 min.

When the connection is protected by the addition of wood panelling, wood-based panels or gypsum plasterboard, the fire resistance is calculated based on the fire resistance of the unprotected connection after failure of the additional protection. The fixings of the additional protection should prevent its premature failure. The additional protection provided by wood-based panels or gypsum plasterboard should remain in place until charring of the member starts, $t = t_{ch}$.

For joints with internal steel plates of a thickness equal or greater than 2 mm and which do not project beyond the timber surface, the width b_{st} of the steel plates should observe the required conditions.

6.5.4.2 Reduced load method

For Standard Fire exposure, the fire resistance of the unprotected or protected connection in shear should be calculated from its resistance at ambient temperature multiplied by a reduction factor k in accordance with the required fire resistance time.

6.6 MASONRY STRUCTURES

6.6.1 Basis of design

The material properties of masonry vary significantly according to local conditions. The strength and deformation properties of masonry at elevated temperatures may be obtained from stress-strain relationships obtained by tests for a project or from a database. For fire design, a distinction is made between non-load-bearing and load-bearing walls and between separating and non-separating walls. The fire behaviour of a masonry wall depends on the masonry unit material, the clay (calcium silicate, autoclaved aerated concrete or dense/lightweight aggregate concrete, and manufactured stone); the type of unit (solid or hollow, including types of hole, percentage of formed voids, shell and web thicknesses); the type of mortar (general-purpose, thin-layer or lightweight mortar); the relationship of the design load to the design resistance of the wall; the slenderness of the wall; the eccentricity of loading; the density of units; the type of wall construction; and the type and nature of any applied surface finishes.

The fire resistance of masonry walls may be assessed by calculation, taking into account the relevant failure mode in fire exposure, the temperature-dependent material properties, the slenderness ratio and the effects of thermal expansions and deformations by the simplified method of calculation for walls. This is presented next.

6.6.1.1 Isotherm method

In the simplified calculation method, the load-bearing capacity is determined based on the residual cross section of the masonry structure. The method is similar to the 500°C isotherm method for concrete structures. However, for masonry design, two isotherm lines should be considered. At temperatures below 100°C, the masonry mechanical properties are assumed to be unaffected. Above a certain temperature, which is different for different types of masonry, the masonry is assumed to have no strength or stiffness. To use the simplified method for masonry walls and columns under the Standard Fire exposure, the second isotherm temperature is

- 600°C clay units (Group 1S and Group 1 according to EN 1996-1-1: 2005 [CEN 1996–2005a], unit strength f_b from 10 to 40 N/mm², gross density from 1000 to 2000 kg/m³ with general-purpose mortar);
- 500°C for calcium silicate units (Group 1S and Group 1, unit strength f_b from 10 to 40 N/mm², gross density from 1500 to 2000 kg/m³ with thin-layer mortar);
- 500°C for dense aggregate concrete (Group 1, unit strength f_b from 10 to 40 N/mm², gross density from 1500 to 2000 kg/m³ with general-purpose mortar);
- 500°C for lightweight aggregate concrete (Group 1S and Group 1, unit strength f_b from 4 to 8 N/mm², and gross density from 600 for pumice to 1000 kg/m³ with lightweight mortar);
- 400°C for autoclaved aerated concrete (Group 1, unit strength f_b from 2 to 6 N/mm², gross density from 400 to 700 kg/m³ with thin layer of general-purpose mortar).

The design value of the vertical resistance of the wall or column is given by

$$N_{Rd,fi,\theta i} = \Phi \left(f_{d\theta 1} A_{\theta 1} + f_{d\theta 2} A_{\theta 2} \right) \tag{6.30}$$

where Φ is a capacity reduction factor in the middle of the wall obtained from cl. 6.1.2.2 of EN 1996-1-1: 2005 (CEN 1996–2005a), also taking into account the eccentricity $e_{\Delta\theta}$ due to the variation of temperature across masonry; $f_{d\theta 1}$ is the design compressive strength of masonry at temperature θ_1 (the temperature up to which the cold strength of masonry may be used, $= 100°C$); $A_{\theta 1}$ is an area of masonry up to θ_1; $f_{d\theta 2} = c\, f_{d\theta 1}$ is the design strength of masonry in compression between $\theta_1°C$ and $\theta_2°C$; θ_2 is the temperature above which the material has no residual strength; c is a constant obtained from stress-strain tests at elevated temperature; and $A_{\theta 2}$ is the area of masonry between θ_1 and θ_2. The temperature distribution across a masonry section and the temperature at which the masonry becomes ineffective should be obtained from the results of tests or from a database of test results as a function of the time of fire exposure. In the absence of test results, the graphs in Annex C.3 of EN 1996-1-2 (CEN 1996–2005b), an example of which is repeated as Figure 6.9, may be used.

Due to the high-temperature gradient in masonry structures, it is important to take into account the eccentricity created by the temperature gradient $e_{\Delta\theta}$. For use in this simplified calculation method, the eccentricity may be obtained from test results or from the following equation:

$$e_{\Delta\theta} = \frac{1}{8} h_{eff}^2 \frac{\alpha_t (\theta_2 - 20)}{t_{Fr}} \le h_{eff}/20 \qquad (6.31)$$

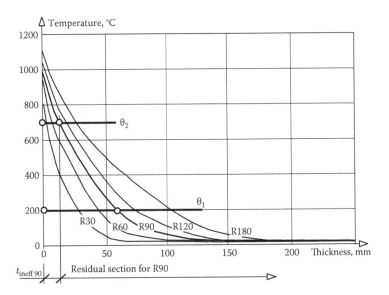

Figure 6.9 The temperature above which masonry is structurally ineffective (θ_2) for autoclaved aerated concrete masonry, gross density 500 kg/m³ ($t_{ineff90}$ is the thickness of the wall that has become ineffective at 90 min), example from EN 1996-1-2: 2005b Figure C.3(f).

where h_{ef} is the effective height of the wall; α_t is the coefficient of thermal expansion of masonry according to 3.7.4 of EN 1996-1-1: 2005; 20°C is the ambient temperature assumed on the cold side; and t_{Fr} is the thickness of the cross section whose temperature does not exceed θ_2.

6.7 ALUMINIUM STRUCTURES

6.7.1 Basis of design

As at normal temperature design, the fire design of aluminium structures is based on the engineering knowledge coming from structural steel design; see EN 1999-1-1: 2007 (CEN 1999–2007a) for more information. Aluminium constructions have low fire resistance because of the low melting point of the alloy, 590°C to 650°C. However, aluminium is reflective and has a low surface emissivity ε_m, which is 0.3 for a clean surface without coating; this compares to a value of 0.4 for stainless steel and 0.7 for carbon steel.

6.7.2 Simple design models

The same principles and relationships as for steel structures are used to calculate the heat transfer into an unprotected and protected aluminium frame (see EN 1999-1-1: 2007; CEN 1999–2007b).

Elements loaded in tension, compression or bending do not have to be checked if the temperature does not exceed 170°C.

The reductions in the modulus of elasticity and strength (defined as 0.2% proof stress) of aluminium alloys are much higher than in steel alloys. For section classification, since the reduction in the strength of aluminium is faster than the reduction in the modulus of elasticity, the same section classification at ambient temperature may be used. Figure 6.10 presents the ratio of the reduction factor for the modulus of elasticity to the reduction factor for the strength of aluminium at elevated temperatures for a variety of aluminium alloys.

Using the properties in Figure 6.10, the column buckling as well as beam lateral-torsional buckling design resistance at elevated temperatures $(N_{b,fi,t,Rd}$ and $M_{b,fi,t,Rd})$ may be evaluated based on the reduced 0.2% proof stress and the same relative slenderness as at ambient temperature.

6.8 FIRE RESISTANCE DESIGN WORKED EXAMPLES

6.8.1 Concrete beam by 500°C isotherm method

Figure 6.11 shows a continuous beam, of cross section 500 × 800 mm. It supports one-way spanning reinforced slabs 250 mm thick spanning 6 m. The concrete grade is C25/30, and the reinforcement grade is B500B

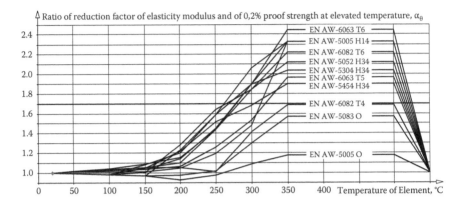

Figure 6.10 Ratio of reduction factor of elasticity modulus to reduction factor of 0.2% proof stress at elevated temperature a_q for aluminium alloys (from Table 1a in EN 1999-1-2:2007).

according to EN 10080. The reinforcement is 6 Ø R25 in span and 6 Ø R32 at supports. The cover to the longitudinal reinforcement is 45 mm. The required fire resistance is R240.

6.8.1.1 Normal temperature design: ultimate limit state

The characteristic values of the actions are

$$g_k = 70 \text{ kN/m, } q_{k1} = q_{k2} = q_{k3} = 70 \text{ kN/m}$$

The partial safety factors for load combinations at the ultimate limit state are summarised in Table 6.2.

Figure 6.12 shows the resulting bending moment envelopes from the load combinations in Table 6.2.

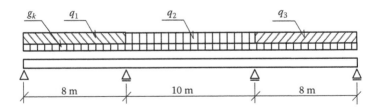

Figure 6.11 Actions on continuous concrete beam.

Table 6.2 Partial safety factors for
load combinations at
ultimate limit state

Load combination	g	q_1	q_2	q_3
1	1.35	1.5	0	0
2	1.35	1.5	1.5	0
3	1.35	1.5	1.5	1.5
4	1.35	0	1.5	1.5
5	1.35	0	0	1.5

6.8.1.2 Design for fire situation

6.8.1.2.1 Mechanical actions and bending moment envelope

The mechanical load at the fire situation is evaluated for partial safety factors $\gamma_{M,fi} = 1.0$, giving the "mechanical action at fire" bending moment envelopes in Figure 6.13. The indirect fire actions due to differential thermal deformations as a result of the temperature gradients across the heated cross section are used to calculate the hogging bending moments (under "influence of temperature" in Figure 6.13). The bending moments are calculated by converting the thermal strain curvatures into negative mechanical strain curvatures at the internal supports. For these calculations, the reduced concrete cross section size is used, and it is assumed that the top of the concrete beam section is at room temperature and the bottom at 500°C for calculating the curvatures. But, the cross section is assumed to be elastic at room temperature for calculating the bending moments. According to the temperature profile in Figure 6.14, the cross-section depth is reduced by 60 mm. Allowing for 15% redistribution of the total bending moments [see cl. 5.5.(4) of EN 1992-1-1 for more detailed information; CEN 1992–2005a], the final design bending moment envelopes are obtained.

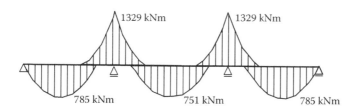

Figure 6.12 Bending moment envelopes (values in kNm).

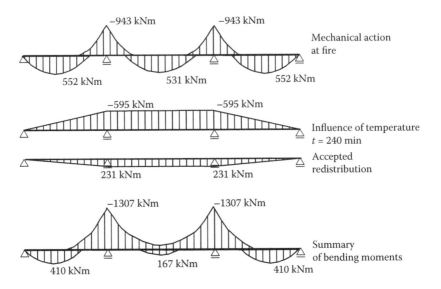

Figure 6.13 Bending moment envelopes for fire resistance design (values in kNm).

6.8.1.3 Verification of fire resistance

6.8.1.3.1 Effective cross section

The position of the 500°C isotherm is $a_{500} = 60$ mm, taken from graph Figure A.10b in EN 1992-1-2: 2005 (CEN 1992–2005b), reproduced in Figure 6.14.

The reinforcing bars in the corners are located outside the 500°C isotherm and have a higher temperature compared to the inside bars. For simplification, they may conservatively be excluded from further calculation see (Figure 6.15).

The four reinforcement bars inside the 500°C contour are at temperature $\theta = 500°C$ (see Figure 6.15). The reduction factor for the reinforcement bar characteristic strength is $k_{s,\theta} = 0.78$.

The cross-section bending resistance is verified as at normal temperature design using the following characteristic strength of reinforcement and concrete:

$$f_{yd,fi} = \frac{k_\theta \, f_{yk}}{\gamma_{M,fi}} = \frac{0.78 \times 500}{1} = 390 \text{ N/mm}^2$$

$$f_{cd,fi} = \frac{k_\varpi \, f_{ck}}{\gamma_{M,fi}} = \frac{1 \cdot 25}{1.0} = 25 \text{ N/mm}^2$$

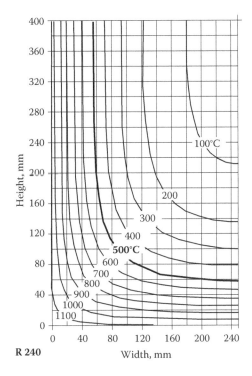

R 240

Figure 6.14 Lower quadrant of temperature profile for a beam $h \times b = 800 \times 500$ mm for fire resistance R 240.

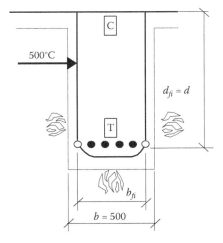

Figure 6.15 Effective cross section according to isotherm 500°C.

The effective depth of the cross section is $d_{fi} = d = 800 - (45 + 0.5 \times 25) = 742.5$ mm. The reinforcement area = 1964 mm². The depth of concrete in compression is

$$x = \frac{A_s \times f_{syd,fi,v}}{b_{fi} \times \lambda \times \eta \times f_{ed,fi,20°C}} = \frac{1964 \times 10^{-6} \times 390}{0.8 \times 0.5 \times 25} = 0.077 \text{ m}$$

Note the top concrete is in compression, so the concrete width is not reduced. The factors λ (= 0.8) and η (= 1) are taken at room temperature design according to EN 1992-1-1: 2005 (CEN 1992–2005a).

$$M_{Rd,fi} = A_s \times f_{syd,fi,v} z_{fi} = A_s \times f_{syd,fi,v} \times (d_{fi} - 0.5 \times \lambda \times x_{fi})$$

$$= 1964 \times 10^{-3} \times 390 \times (0.7425 - 0.5 \times 0.8 \times 0.077) = 545 \text{ kNm}$$

$$M_{fi,Rd} = 545 \text{ kNm} > M_{d,fi} = 410 \text{ kNm}$$

The cross section in the span is satisfactory.

6.8.1.3.2 Cross section at supports

The 500°C isotherm position is the same as shown in Figure 6.14. The design cross section is shown in Figure 6.16.

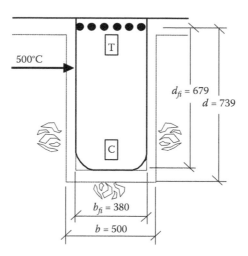

Figure 6.16 Isotherm 500°C for section at support.

The isotherm 500°C, $a_{500} = 60$ mm, reduces the concrete cross section width to $b_{fi} = 380$ mm (see Figure 6.16). The effective height of the reduced cross section is

$$d_{fi} = 800 - 45 - 32/2 - 60 = 679 \text{ mm}$$

The top reinforcement is at the position where there is no reduction in its characteristic strength, giving

$$f_{yd,fi} = \frac{k_\theta \, f_{yk}}{\gamma_{M,fi}} = \frac{1 \times 500}{1} = 500 \text{ N/mm}^2$$

and

$$f_{cd,fi} = \frac{k_\varpi \, f_{ck}}{\gamma_{M,fi}} = \frac{1 \times 25}{1.0} = 25 \text{ N/mm}^2$$

Reinforcement area = 4826 mm²

Concrete depth in compression: $x = \dfrac{4826 \times 10^{-6} \times 500}{0.8 \times 0.38 \times 25} = 0.318 \text{ m}$

$$M_{fi,Rd} = 4826 \times 10^{-3} \times (0.679 - 0.5 \times 0.8 \times 0.318) = 1331 \text{ kNm}$$

$$M_{fi,Rd} = 1\,331 \text{ kNm} > M_{d,fi} = 1\,307 \text{ kNm}$$

Therefore, the cross section at support is satisfactory.

6.8.2 Concrete wall by zone method

Figure 6.17 shows a wall of thickness 300 mm with normal-weight concrete with siliceous aggregate, reinforced by R14/150 bars with 40-mm cover. The wall is exposed to fire from both sides, and the required fire resistance is 120 min.

The zone method is utilised by dividing the cross section into five zones, $n = 5$. Figure 6.18 shows the temperatures at the middle of each zone θ_i, the average compressive strength reduction factor $k_{ci,\theta i}$ and the section compressive strength reduction factor $k_{c,\theta,M}$ at point M.

The mean strength reduction factor is

$$k_{c,m} = \frac{(1 - 0.2/n)}{n} \sum_{i=1}^{n} k_{c,\theta} = \frac{(1 - 0.2/5)}{5}(0.22 + 0.76 + 0.95 + 1 + 1) = 0.75$$

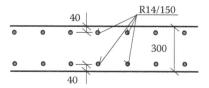

Figure 6.17 Wall cross section.

The thickness of the damaged layer is (see Figure 6.19)

$$a_z = w\left[1 - \left(\frac{k_{c,m}}{k_{c,\theta,M}}\right)^{1,3}\right] = 150\left[1 - \left(\frac{0.75}{1}\right)^{1,3}\right] = 47\,\text{mm}$$

6.8.3 Unprotected steel beam

The beam is uniformly loaded and restrained against lateral torsional buck-ling by the presence of a concrete slab on the top (compression) flange.

The beam, shown in Figure 6.20, is fabricated from a hot-rolled IPE section. The required fire resistance time is 30 min.

Material properties
    Steel grade: S 275
    Yield stress: $f_y = 275$ N/mm²
    Density: $\rho_a = 7850$ kg/m³

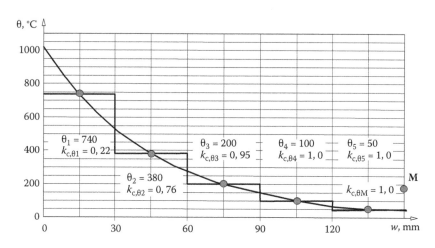

Figure 6.18 Temperature distribution and compressive strength reduction factors for one-half of the symmetrical cross section in Figure 6.17.

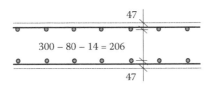

Figure 6.19 Reduced cross section of the wall according to the zone method.

Loads
 Permanent action: $g_k = 4.8$ kN/m
 Variable action: $q_k = 7.8$ kN/m

Mechanical actions for normal temperature design
Design load value:

$$v_d = g_k \; v_G + q_k \; v_Q = 4.8 \times 1.35 + 7.8 \times 1.5 = 18.18 \text{ kN/m}$$

Bending moment and shear force

$$M_{Ed} = \frac{1}{8} v_d \, l^2 = \frac{1}{8} \times 18.18 \times 7.4^2 = 124.4 \text{ kNm}$$

$$V_{Ed} = \frac{1}{2} v_d \, l = \frac{1}{2} \times 18.18 \times 7.4 = 67.3 \text{ kN}$$

6.8.3.1 Normal temperature design

Section IPE 300 is selected according to EN 1993-1-1:2005.
 The cross section dimensions are shown in Figure 6.21. The cross section is Class 1.
 The bending moment resistance is

$$M_{pl,Rd} = \frac{W_{pl,y} \, f_y}{\gamma_{M0}} = \frac{628.4 \times 10^3 \times 275}{1.0} = 172.8 \text{ kNm} > 124.4 \text{ kNm} = M_{Sd}$$

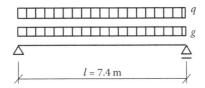

Figure 6.20 Uniformly loaded beam.

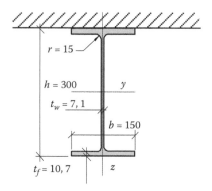

Figure 6.21 Dimensions of cross-section IPE 300.

The shear resistance is

$$V_{pl,Rd} = \frac{A_{V,z}\,f_y}{\sqrt{3}\;\gamma_{M0}} = \frac{2\,568 \times 275}{\sqrt{3} \times 1.0} = 407.7 \text{ kN} > 67.3 \text{ kN} = V_{Sd}$$

6.8.3.2 Design for fire situation

6.8.3.2.1 Mechanical actions for fire design situation

Assuming an office building, the combination factor ψ is $\psi_{1,1} = 0.3$. The reduction factor is:

$$\eta_{fi} = \frac{g_k + \psi_{1,1}\,q_k}{g_k\,\gamma_G + q_k\,\gamma_Q} = \frac{4.8 + 0.3 \times 7.8}{4.8 \times 1.35 + 7.8 \times 1,5} = 0.393$$

6.8.3.2.2 Evaluation of beam temperature exposed to nominal Standard Fire curve

The section factor is calculated for fire exposure from three sides:

$$\left(\frac{A_m}{V}\right)_b = 139 \text{ m}^{-1}$$

Considering the shadow effect (see Figure 6.22), according to EN 1993-1-2:2005, the section factor is

$$\left(\frac{A_m}{V}\right)_{sh} = 0.9 \cdot \left(\frac{A_m}{V}\right)_b = 0.9 \times 139 = 125 \text{ m}^{-1}$$

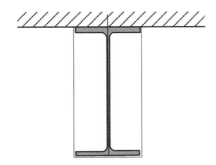

Figure 6.22 Surface of the member exposed to fire and its box value according to EN 1993-1-2:2005.

For an unprotected beam exposed to fire on three sides with a concrete slab on side four, the adaptation factor for non-uniform temperature across the cross section is

$$\kappa_1 = 0.7$$

The degree of utilization at time $t = 0$ may be obtained from

$$\mu_0 = \eta_{fi}\ \kappa_1\ \kappa_2 = 0.393 \times 0.7 \times 1.0 = 0.275$$

The critical temperature is

$$\theta_{a,cr} = 39.19 \ln\left(\frac{1}{0.9674\ \mu_0^{3.833}} - 1\right) + 482$$

$$= 39.19 \cdot \ln\left(\frac{1}{0.9674 \times 0.275^{3.833}} - 1\right) + 482 = 677°C$$

Figure 6.23 shows the temperature increase in the unprotected steel section for different section factors. According to this figure, the fire resistance of the beam is 17 min. This fire resistance is lower than 30 min. Therefore, the cross section is not satisfactory.

6.8.4 Fire-protected steel column

Figure 6.24 shows a steel column continuous over two storeys. The column dimensions (HEB 180) are shown in Figure 6.25. Fire protection is by sprayed vermiculite cement. The required period of fire resistance is 90 min.

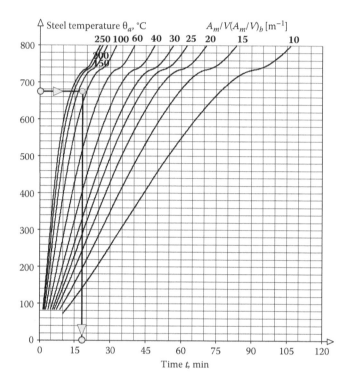

Figure 6.23 Graphic evaluation of fire resistance of the beam.

Material properties
   Steel grade: S 355; yield stress: $f_y = 355$ N/mm²; density: $\rho_a = 7850$ kg/m³

Loads
   Reaction at each floor level due to permanent actions: $R_{G,k} = 185$ kN
   Reaction at each floor level due to variable actions: $R_{Q,k} = 175$ kN

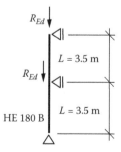

Figure 6.24 Column dimensions and loading and support conditions.

Partial safety factors
$$\gamma_G = 1.35; \gamma_Q = 1.50; \gamma_{M1} = 1.00$$

Data for fire calculation
Material properties of fire protection: sprayed vermiculite cement
Thickness d_p = 20 mm, density ρ_p = 550 kgm^{-3}; specific heat c_p =
1100 Jkg^{-1}K^{-1}; thermal conductivity λ_p = 0.12 Wm^{-1}K^{-1}.

Mechanical actions at normal temperature
The design value of load in the lower part of the column is

$$N_{Ed} = 2\,R_{Ed} = 2\left(R_{G,k}\,\gamma_G + R_{Q,k}\,\gamma_Q\right) = 2\times(185\times1.35 + 175\times1.5) = 1024.5\,\text{kN}$$

6.8.4.1 Normal temperature design

The column cross section is Class 1 (see Figure 6.25).
The buckling length of the column is

$$L_{cr,y} = L_{cr,z} = 3.5\,\text{m}$$

Buckling perpendicular to the z axis is critical.
The elastic critical load for normal design according to EN 1993-1-1:
2005 (CEN 1993–2005b) is

$$N_{cr} = \frac{\pi^2\,E\,I_z}{L_{cr,z}^2} = \frac{\pi^2\times210000\times1363\times10^4}{3500^2} = 2306\,\text{kN}$$

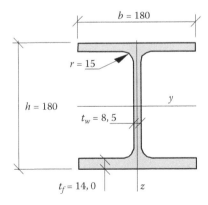

Figure 6.25 Dimensions of column cross section.

The non-dimensional slenderness is

$$\bar{\lambda}_z = \sqrt{\frac{Af_y}{N_{cr}}} = \sqrt{\frac{6530 \times 355}{2306 \times 10^3}} = 1.003$$

Select column buckling curve c for hot-rolled I-sections with h/b ratio < 1.2, giving α = 0.49. Therefore,

$$\Phi = 0.5\,(1 + \alpha\,(\bar{\lambda}_z - 0.2) + \bar{\lambda}_z^2) = 0.5 \times (1 + 0.49 \times (1.003 - 0.2) + 1.003^2) = 1.$$

$$\chi_z = \frac{1}{\Phi + \sqrt{\Phi^2 - \bar{\lambda}_z^2}} = \frac{1}{1.200 + \sqrt{1.200^2 - 1.003^2}} = 0.538$$

The column design resistance at ultimate limit state is

$$N_{b.Rd} = \chi_z \frac{A f_y}{\gamma_{M1}} = 0.538 \times \frac{6525 \times 355}{1.0} = 1246.2 \text{ kN} > 1024.5 \text{ kN} = N_{Ed}$$

The section is satisfactory at normal temperature.

6.8.4.2 Design for fire situation

6.8.4.2.1 Mechanical actions for fire design situation

Assuming an office building, the combination factor ψ is taken as $\psi_{2,1} = 0.3$. The reduction factor for the design load level is

$$\eta_{fi} = \frac{G_k + \psi\,Q_k}{G_k\,\gamma_G + Q_k\,\gamma_Q} = \frac{185 + 0.3 \times 175}{185 \times 1.35 + 175 \times 1.5} = 0.464,$$

giving

$$N_{fi,Ed} = \eta_{fi} \cdot N_{Ed} = 0.464 \times 1024.5 = 475.0 \text{ kN}$$

6.8.4.2.2 Evaluation of fire gas temperature

The Standard Fire temperature-time curve is used of evaluation of the fire gas temperature.

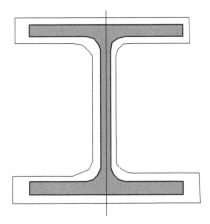

Figure 6.26 Evaluation of section factor A_p/V for fire-protected cross section.

6.8.4.2.3 Evaluation of column temperature

The dotted line in Figure 6.26 indicates the perimeter of the column cross section exposed to fire. The section factor is calculated as follows:

$$\frac{A_p}{V} = \frac{4\,b + 2\,(h - t_w - 4\,r) + 2\,\pi\,r}{A} = \frac{4 \times 180 + 2 \times (180 - 8.5 - 4 \times 15) + 2 \times \pi \times 15}{6\,525}$$

$$= 0.159 \text{ mm}^{-1} = 159 \text{ m}^{-1}$$

Using the simplified temperature calculation of Equation (4.45) in Chapter 4 of this book for protected steel section and temperature interval of 30 s, the column temperature can be calculated. Table 6.3 lists the results from a few steps. Figure 6.27 shows the evolution of the gas and steel temperatures.

From the results shown in Table 6.3 and Figure 6.27, the steel temperature at time $t = 90$ min is $\theta_a = 554°C$.

6.8.4.3 Verification in the strength domain

6.8.4.3.1 Classification of the cross section at elevated temperature

The slenderness of the flange in compression (see Figure 6.28) is

$$\frac{c}{t_f} = \frac{70.75}{14.0} = 5.05$$

Table 6.3 Calculation of steel temperature

Minutes	Seconds	T, min	θ_g °C	c_a J/kg°C	ϕ	$\Delta\theta_{a,v}$ °C	$\theta_{a,v}$ °C
	0	0	20.0	440			20.0
	30	0.5	261.1	440	0.557	0.0	20.0
I	00	1.0	349.2	440	0.557	0.0	20.0
I	30	1.5	404.3	440	0.557	0.0	20.0
2	00	2.0	444.5	440	0.557	0.0	20.0
2	30	2.5	476.2	440	0.557	0.7	20.7
3	00	3.0	502.3	440	0.557	1.4	22.0
88	00	88.0	1002.6	700	0.350	2.1	545.5
88	30	88.5	1003.5	702	0.349	2.1	547.6
89	00	89.0	1004.3	704	0.348	2.1	549.7
89	30	89.5	1005.2	706	0.347	2.1	551.7
90	00	90.0	1006.0	708	0.346	2.1	**553.8**
90	30	90.5	1006.8	710	0.345	2.1	555.8

Note: A selected value is marked by a bold number.

The limit for Class 1 is 9ε. For the fire design situation, ε is taken as 0.85 of that used for normal design. Thus, the limit for S355 is

$$9 \times 0.85 \times 0.814 = 6.22$$

The limit is not exceeded. Therefore, the flange is Class 1.
    The slenderness of the web in compression is

$$\frac{d}{t_w} = \frac{122.0}{8.5} = 14.35$$

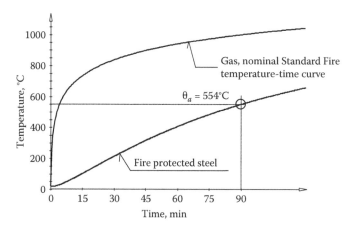

Figure 6.27 Steel and fire gas temperature-time curves.

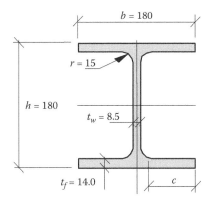

Figure 6.28 Classification of the cross section based on plate slenderness.

The limit for Class 1 is 33ε. For the fire design situation, ε is taken as 0.85 of that used for normal design. Thus, the limit is

$$33 \times 0.85 \times 0.814 = 22.8$$

The limit is not exceeded. Therefore, the web is Class 1.

Therefore, the whole section at elevated temperature is Class 1.

According to Figure 5.14 of this book, the steel reduction factor at temperature $\theta_a = 554°C$ is

$$k_{y,\theta} = 0.613 \text{ and } k_{E,\theta} = 0.444$$

Provided that the column forms part of a braced frame and the fire resistance of the concrete slab separating the floors is not less than the fire resistance of the column, the buckling length is reduced to

$$L_{cr,y,fi} = L_{cr,z,fi} = 0.7 \times L = 0.7 \times 3.5 = 2.45 \text{ m}$$

The modified critical buckling load at normal temperature is then given by

$$N_{cr} = \frac{\pi^2 E I_z}{L_{cr,z}^2} = \frac{\pi^2 \times 210000 \times 1363 \times 10^4}{2450^2} = 4706 \text{ kN}$$

The modified non-dimensional slenderness at normal temperature is given by

$$\bar{\lambda} = \sqrt{\frac{A \times f_y}{N_{cr}}} = \sqrt{\frac{6530 \times 355}{4706.3 \times 10^3}} = 0.702$$

The non-dimensional slenderness at elevated steel temperature θ_a is

$$\bar{\lambda}_\theta = \bar{\lambda}\sqrt{k_{y,\theta}/k_{E,\theta}} = 0.702\sqrt{0.613/0.444} = 0.825$$

The imperfection factor α is equal to

$$\alpha = 0.65\sqrt{\frac{235}{f_y}} = 0.65\times\sqrt{\frac{235}{355}} = 0.53$$

The column strength reduction factor is

$$\phi_{z,\theta} = 0.5\left(1+\alpha\,\bar{\lambda}_{z,\theta}+\bar{\lambda}_{z,\theta}^2\right) = 0,5\times\left(1+0.53\times0.825+0.825^2\right) = 1.058$$

giving

$$\chi_{z,fi} = \frac{1}{\phi_{z,\theta}+\sqrt{\phi_{z,\theta}^2-\bar{\lambda}_{z,\theta}^2}} = \frac{1}{1.058+\sqrt{1.058^2-0.825^2}} = 0,581$$

The column design resistance at elevated $\theta_a = 554\degree C$ is

$$N_{b,fi,\theta,Rd} = \chi_{z,fi}\,A\,k_{y,\theta}\,\frac{f_y}{\gamma_{M,fi}} = 0.581\times6\,525\times0.613\times\frac{355}{1.0} = 825.0\text{ kN}$$

The design effect of actions is

$$N_{fi,Ed} = 475.0\text{ kN and } N_{b,fi,\theta,Rd} \ge N_{fi,Ed}$$

Therefore, the section is satisfactory for the fire design situation.

6.8.5 Composite beam by bending moment resistance model

This example illustrates the fire-resistant design of a simple supported composite beam forming part of the floor structure of a multistorey storage.

The beam (see Figure 6.29) is fabricated from a hot-rolled HEB 180 section of S235 with concrete slab C20/25 of effective width 2 m and is to be designed to achieve R30 fire resistance at nominal Standard Fire temperature-time curve without the use of fire protection. At normal temperature design, the beam is loaded by a design permanent load of 10 kN/m and a design variable load $q_d = 14$ kN/m.

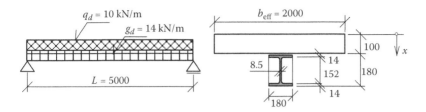

Figure 6.29 Unprotected partially encased composite beam.

6.8.5.1 Normal temperature design

Bending moment

$$M_{Ed} = \frac{1}{8}(q_d + g_d) \times 5^2 = \frac{1}{8}(10+14) \times 5^2 = 75 \text{ kNm}$$

Cross-section resistance:
  Material properties:
    Concrete: $f_{ck} = 20 \text{ N/mm}^2$ and $f_{cd} = 0{,}85 \times 20/1.5 = 11.3 \text{ N/mm}^2$
    Steel: $f_{ay} = 235 \text{ N/mm}^2$ and $f_{yd} = f_y/\gamma_a = 235/1.00 = 235 \text{ N/mm}^2$;
      HEB 180: $A_a = 6530 \text{ mm}^2$
Position of neutral axes:

$$x = \frac{A_a\, f_{yd}}{b_{eff}\, f_{cd}} = \frac{6530 \times 235}{2000 \times 11.3} = 67.9 \text{ mm}$$

Bending moment resistance:

$$M_{pl,Rd} = 6530 \times 235 \times (90 + 100 - 67.9/2) = 239.5 \text{ kNm}$$

The beam is satisfactory:

$$M_{Ed} = 72 \text{ kNm} < M_{pl,Rd} = 239.5 \text{ kNm}$$

6.8.5.2 Design for fire situation

At the fire situation, the bending moment is lower then the normal temperature design situation. Conservatively, a simplified value of $\eta_{fi} = 0.70$ may be used for areas susceptible to the accumulation of goods such as storage, according to load category E given in EN 1991-1-1: 2002 (CEN 1991–2002). The bending moment at the fire situation is

$$M_{Ed,fi} = 0.7 \times 75 = 52.5 \text{ kNm}$$

For the calculation of the beam bending moment resistance, the beam cross section is divided into separate parts.

6.8.5.2.1 Concrete slab

The lower part of the thickness (30 mm) is not taken into account; see Table D.5 of EN 1994-1-2: 2005 for the temperature distribution in a solid slab of 100-mm thickness composed of normal-weight concrete and not insulated. The rest of the slab may be assumed to have unreduced material properties.

6.8.5.2.2 Lower flange

Section factor:

$$\left(\frac{A_m}{V}\right)=\left(\frac{2\cdot(b_f+t_f)-t_w}{b_f\cdot t_f}\right)=\left(\frac{2\times(180+14)-8.5}{180\times14}\right)=0.1509\ \text{mm}^{-1}=150.9\ \text{m}^{-1}$$

The increase in the temperature of the steel section is calculated by a step-by-step procedure (see the worked example in Section 6.8.3). For a nominal Standard Fire temperature-time curve and at 30 min, the steel temperature is $\theta_{a,1}=815.8°C$

The reduction factor for the yield strength of steel (see Figure 5.14) is $k_{y,\theta}=0.10$.
The reduced design strength is $f_{y,\theta,1}=k_{y,\theta}\times f_{y,\theta}=0.10\times235=23.5\ \text{N/mm}^2$.

6.8.5.2.3 Beam web

Section factor:

$$\left(\frac{A_m}{V}\right)=\left(\frac{2\cdot h_w}{h_w\cdot t_w}\right)=\left(\frac{2\times152}{152\times8.5}\right)=0.2353\ \text{mm}^{-1}=235.3\ \text{m}^{-1}$$

The increase of temperature of the steel section is calculated by a step-by-step procedure.

The steel temperature at time 30 min is $\theta_{a,2w}=832.3°C$.
The reduction factor for yield strength is $k_{y,\theta}=0.09$
The reduced design strength is $f_{y,\theta,w}=k_{y,\theta}\times f_{y,\theta}=0.09\times235=21.2\ \text{N/mm}^2$.

6.8.5.2.4 Upper flange

Section factor:

$$\left(\frac{A_m}{V}\right) = \left(\frac{2 \cdot t_f + b_f - t_w}{b_f \cdot t_f}\right) = \left(\frac{2 \times 14 + 180 - 8.5}{180 \times 14}\right) = 0.0792 \text{ mm}^{-1} = 79.2 \text{ m}^{-1}$$

The steel temperature at time 30 min is $\theta_{a,2} = 741.2°C$.
The reduction factor for the yield strength of steel is $k_{y,\theta} = 0.18$.
The reduced design strength is $f_{y,\theta,2} = k_{y,\theta} \times f_{y,\theta} = 0.18 \times 235 = 42.3 \text{ N/mm}^2$.

The final cross section and the the reduced design strengths are shown in Figure 6.30. The partial safety factor for all materials is $\gamma_{M,fi} = 1.0$. The tension force in the steel part is

$$T = (f_{a\max,\theta1} \cdot b_1 \cdot e_1 + f_{a\max,\theta w} \cdot b_w \cdot e_w + f_{a\max,\theta2} \cdot b_2 \cdot e_2)/\gamma_{M,fi,a}$$

$$= (23.5 \times 180 \times 14 + 21 \times 2 \times 152 \times 8.5 + 42.3 \times 180 \times 14)/1.0 = 333.8 \text{ N}$$

The thickness of the compressed part of the slab from $F = T$ is

$$h_u = T/(b_{eff} \cdot f_{c,20°C}/\gamma_{M,fi,c}) = 333.8/(2000 \times 20/1.0) = 8.3 \text{ mm}$$

The bending moment resistance is

$$M_{fi,Rd^+} = T \ (y_F - y_T) = 333.8 \times [(180 + 100 - 8.3) - 180/2] = 60.7 \text{ kNm}$$

$$\geq M_{Ed,fi} = 52.5 \text{ kNm}$$

Therefore, the composite beam is satisfactory for R30.

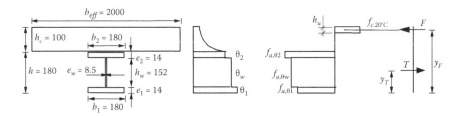

Figure 6.30 Distribution of temperatures and stresses in the composite section.

6.8.6 Solid timber beam

The design beam is made of soft solid timber of cross section 180×220 mm. The required fire resistance is 60 min. The beam span is 5.0 m. The total mechanical load for the ultimate limit state design at normal temperature is $g_d + q_d = 6.5$ kN/m. The ratio of the characteristic variable load to the characteristic permanent load is $Q_{k,1}/G_k = 1.0$.

The bending moment at normal temperature is

$$M_{Ed} = \frac{(g_d + q_d)\, \ell^2}{8} = \frac{6.5 \times 5^2}{8} = 20.31\ \text{kNm}$$

The ratio of the variable load to the permanent load is

$$\xi = Q_{k,1}/G_k = 1.0$$

The reduction factor for fire design is

$$\eta_{fi} = (\gamma_{GA} + \psi_{1,1}\, \xi)/(\gamma_G + \gamma_{Q,1}\, \xi) = (1.0 + 0.2 \times 1.0)/(1.35 + 1.5 \times 1.0) = 0.42$$

The design bending moment at fire situation is

$$M_{fi,d} = \eta_{fi}\, M_d = 0.42 \times 20.31 = 8.53\,\text{kNm}$$

6.8.6.1 Using the reduced properties method

The coefficient for the 20% fractile of strength is $k_{fi} = 1.25$ (from Table 2.1 of EN 1995-1-2: 2005 [CEN 1995–2005b]).
The partial safety factor for the material at fire situation $\gamma_{m,fi} = 1.0$.
The design notional charring rate under the Standard Fire exposure for solid timber $\beta_n = 0.8$ mm/min (see Table 3.1 of EN 1995-1-2: 2005 [CEN 1995–2005b]).
The charring depth is

$$d_{char} = \beta_n\, t = 0.8 \times 60 - 48\,\text{mm}$$

The residual width is

$$b_r = b - 2\, d_{char} = 180 - 2 \times 48 = 84\,\text{mm}$$

The residual height is

$$h_r = h - d_{char} = 220 - 48 = 172\,\text{mm}$$

The elastic modulus for the residual cross section exposed to fire by three sides is

$$W_r = \frac{b_r \times h_r^2}{6} = \frac{84 \times 172^2}{6} = 414 \cdot 10^3 \text{ mm}^3$$

The area of the residual cross section is

$$A_r = b_r \, h_r = 0.084 \times 0.172 = 1.4 \times 10^{-2} \text{ m}^2$$

The perimeter of the fire-exposed residual cross section is

$$p = b_r + 2 \, h_r = 0.084 + 2 \times 0.172 = 42.8 \cdot 10^{-2} \text{ m}^2$$

The modification factor for fire for the reduced properties method is

$$k_{\text{mod},fi} = 1.0 - \frac{1}{200} \frac{p}{A_r} = 1,0 - \frac{1}{200} \frac{42.8 \times 10^{-2}}{1.4 \times 10^{-2}} = 0.85$$

The design bending strength is

$$f_{m,fi,d} = k_{\text{mod},fi} \, k_{fi} \frac{f_{m,k}}{\gamma_{M,fi}} = 0.85 \times 1.25 \frac{22}{1.0} = 23.4 \text{ N/mm}^2$$

The design bending normal stress is

$$\sigma_{m,fi,d} = \frac{M_{fi,d}}{W_r} = \frac{8.53 \times 10^6}{414 \times 10^3} = 20.6 \text{ N/mm}^2$$

Evaluation by stresses

$$\sigma_{m,fi,d} = 20.6 \text{ N/mm}^2 \leq k_{\text{crit}} \, f_{m,fi,d} = 1.0 \times 23.4 = 23.4 \text{ N/mm}^2$$

Therefore, the beam is satisfactory for a fire resistance rating of 60 min. Note that the utilisation factor is $\sigma_{m,fi,d} / f_{m,fi,d} = 20.6/23.4 = 0.88$.

6.8.6.2 Using the reduced cross-section method

The beam was further checked using the reduced cross-section method, and the results were more conservative compared to the reduced properties method. The reduced cross-section calculation method results are presented next.

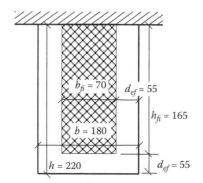

Figure 6.31 Reduced cross section of a beam exposed to fire from three sides.

The modification factor for fire $k_{mod,fi} = 1.0$;
The coefficient for the 20% fractile strength or stiffness property for solid timber $k_{fi} = 1.25$;
The partial safety factor for fire $\gamma_{m,fi} = 1.0$;
The charring rate under Standard Fire exposure for solid timber $\beta_n = 0.8$ mm/min;
The depth of layer with assumed zero strength and stiffness $d_0 = 7$ mm;
$k_0 = 1.0$ (the surface of the beam is not protected and $t \geq 20$ min).

The effective charring depth is

$$d_{ef} = \beta_n\, t + k_0\, d_0 = 0.8 \times 60 + 1.0 \times 7 = 55\, mm$$

The reduced width (see Figure 6.31) is

$$b_{fi} = b - 2\, d_{ef} = 180 - 2 \times 55 = 70\, mm$$

The reduced height is

$$h_{fi} = h - d_{ef} = 220 - 55 = 165\, mm$$

The modulus of elasticity for the beam with the reduced cross section exposed is

$$W_{fi} = \frac{b_{fi} \cdot h_{fi}^2}{6} = \frac{70 \times 165^2}{6} = 318 \times 10^3\, mm^3$$

The design strength in fire in bending is

$$f_{m,fi,d} = k_{\mathrm{mod},fi}\, k_{fi}\, \frac{f_{m,k}}{\gamma_{M,fi}} = 1.0 \times 1.25\, \frac{22}{1.0} = 27.5\,\mathrm{N/mm^2}$$

The design bending normal stress is

$$\sigma_{m,fi,d} = \frac{M_{fi,d}}{W_{fi}} = \frac{8.53 \times 10^6}{318 \times 10^3} = 26.8/\mathrm{mm^2} < f_{m,fi,d} = 27.5\ \mathrm{N/mm^2}$$

Therefore, the beam is satisfactory for 60 min of fire resistance.

Note that the utilisation factor is $\sigma_{m,fi,d}/f_{m,fi,d} = 26.8/27.5 = 0.97$. However, $0.97 > 0.88$, so the reduced cross-section method is more conservative than the reduced properties method.

6.8.7 Fire-protected timber column

This worked example demonstrates the fire design of a timber column in a multistorey building using both the reduced properties method and the reduced cross-section method.

The column is made of soft solid timber of cross section 100/100 mm and protected by Oriented Standard Board (OSB) plates of density 550 kg/m³. The column height is 3.0 m. The mechanical load is $N_{Ed} = 40.0$ kN. The required fire resistance rating is 30 min.

Mechanical action
Assume a ratio of variable load to permanent load of 2, that is,

$$Q_{k,1}/G_k = 2.0$$

The reduction factor for fire design is

$$\eta_{fi} = (\gamma_{GA} + \psi_{1,1}\, \xi)/(\gamma_G + \gamma_{Q,1}\, \xi) = (1.0 + 0.5 \times 1.0)/(1.35 + 1.5 \times 1.0) = 0.46$$

The design force for the fire situation is

$$N_{fi,d} = \eta_{fi}\, N_{Ed} = 0.46 \times 40.0 = 18.4\,\mathrm{kN}$$

Time of delay for charring due to OSB plate protection
For OSB protective panel of thickness $h_p = 20$ mm, the design notional charring rate for density 450 kg/m³ in one-dimensional charring under the Standard Fire exposure $\beta_{0,450,20} = 0.9$ mm/min (from Table 3.1 in EN 1995-1-2: 2005 [CEN 1995–2005b]).

The coefficient for charring rate for $\rho_k = 550$ kg/m³:

$$k_p = \sqrt{\frac{450}{\rho_k}} = \sqrt{\frac{450}{550}} = 0.9$$

The coefficient for panel thickness, which is $h_p = 20$ mm, is:

$$k_h = \sqrt{\frac{20}{h_p}} = \sqrt{\frac{20}{20}} = 1.0 \leq 1.0$$

The charring rate is

$$\beta_{0,\rho,t} = \beta_{0,450,20}\ k_\rho\ k_h = 0.9 \times 0.9 \times 1.0 = 0.81 \text{ mm/min}$$

The time to the start of charring of the protected member is

$$t_{ch} = \frac{h_p}{\beta_{0,450,20}} = \frac{20}{0.81} = 24.5 \text{ min}$$

The failure time of the protection is

$$t_f = t_{ch}$$

6.8.7.1 Using the reduced cross-section method

The modification factor for fire is $k_{mod,fi} = 1.0$.
The coefficient for timber fire design is $k_{fi} = 1.25$.
The partial safety factor for timber at fire situation is $\gamma_{m,fi} = 1.0$.
The design notional charring rate under the Standard Fire exposure for
solid timber $\beta_n = 0.8$ mm/min.
The depth of layer with assumed zero strength and stiffness is $d_0 = 7$ mm.
The charring time is

$$t_{fi,\,req} - t_{ch} = 30 - 24.5 = 5.5 \text{min}$$

Coefficient k_0:

$$k_0 = \frac{t_{fi,req} - t_f}{20} = \frac{30 - 24.5}{20} = 0.3$$

The effective charring depth is (see Figure 6.32)

$$d_{ef} = 2\beta_n\,t + k_0\,d_0 = 2 \times 0.8 \times 60 + 0.3 \times 7 = 11 \text{ mm}$$

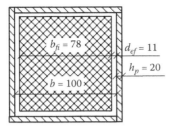

$b_{fi} = 78$

$d_{ef} = 11$

$h_p = 20$

$b = 100$

Figure 6.32 Reduced cross section of a column exposed to fire from all sides.

The effective width is

$$b_{fi} = b - 2\, d_{ef} = 100 - 2 \times 11 = 78\,\text{mm}$$

The buckling length of a continuous column in a non-sway frame is $\ell_{efi} = 0.5\, L$.
The slenderness ratio is

$$\lambda = \frac{\ell_{ef}}{i} = \frac{3\,000}{0.289 \times 78} = 133$$

The critical bending stress is

$$\sigma_{c,crit} = \pi^2\, \frac{E_{0,05}}{\lambda^2} = 3.14^2\, \frac{6\,700}{133^2} = 3.7\,\text{N/mm}^2$$

The relative slenderness ratio is

$$\lambda_{rel} = \sqrt{\frac{f_{c,0,k}}{\sigma_{c,crit}}} = \sqrt{\frac{20}{3.7}} = 2.3$$

The instability factor is

$$k = 0.5\left[1 + \beta_c\, (\lambda_{rel} - 0.5) + \lambda_{rel}^2\right] = 0.5\,[1 + 0.2\,(2.3 - 0.3) + 2.3^2] = 3.3$$

$$k_c = \frac{1}{k + \sqrt{k^2 - \lambda_{rel}^2}} = \frac{1}{3.3 + \sqrt{3.3^2 - 2.3^2}} = 0.17$$

The design compressive strength is

$$f_{c,0,fi,d} = k_{mod}\, k_{fi}\, \frac{f_{c,0,k}}{\gamma_{M,fi}} = 1.0 \times 1.25 \times \frac{20}{1.0} = 25\,\text{N/mm}^2$$

The reduced area is

$$A_{fi} = b_{fi}^2 = 78^2 = 6 \times 10^3 \, \text{mm}^2$$

The buckling resistance is

$$N_{c,Rd,fi} = k_c \cdot A_{fi} \cdot f_{c,0,d,fi} = 0.17 \times 6 \times 10^3 \times 25 = 25.5 \times 10^3 \, \text{kN}$$

To check buckling resistance,

$$N_{fi,d} = 18.4 \times 10^3 \, \text{kN} < N_{c,Rd,fi} = 25.5 \times 10^3 \, \text{kN}$$

The utilisation factor $N_{fi,d}/N_{c,Rd,fi} = 18.4/25.5 = 0.72$.
Therefore, the column is satisfactory for 30 min of fire resistance.

6.8.7.2 Using the reduced properties method

The coefficient for the 20% fractile strength $k_{fi} = 1.25$.
The partial safety factor for fire is $\gamma_{m,fi} = 1.0$.
The design notional charring rate under the Standard Fire exposure for
   solid timber is $\beta_n = 0.8$ mm/min.
Charring depth is

$$d_{char} = 2 \, \beta_n \, t = d_{char} = \beta_n \, t = \ 2 \times 0.8 \times 5.5 = 9 \, \text{mm}$$

Effective width is

$$b_{fi} = b - 2 \, d_{ef} = 180 - 2 \cdot 9 = 82 \, \text{mm}$$

Slenderness ratio is

$$\lambda = \frac{\ell_{ef}}{i} = \frac{3\,000}{0.289 \times 82} = 127$$

Critical bending stress is

$$\sigma_{c,crit} = \pi^2 \, \frac{E_{0,05}}{\lambda^2} = 3.14^2 \, \frac{6\,700}{127^2} = 4.1 \, \text{N/mm}^2$$

The relative slenderness ratio is

$$\lambda_{rel} = \sqrt{\frac{f_{c,0,k}}{\sigma_{c,crit}}} = \sqrt{\frac{20}{4.1}} = 2.2$$

The instability factor is

$$k = 0.5\left[1 + \beta_c \, (\lambda_{rel} - 0.5) + \lambda_{rel}^2\right] = 0.5 \, [1 + 0.2 \, (2.2 - 0.3) + 2.2^2] = 3.1$$

$$k_c = \frac{1}{k + \sqrt{k^2 - \lambda_{rel}^2}} = \frac{1}{3.1 + \sqrt{3.1^2 - 2.2^2}} = 0.19$$

The area of the residual cross section is

$$A_r = b_{fi}^2 = 0.082^2 = 67 \times 10^{-2} \text{ m}$$

The perimeter of the fire-exposed residual cross section is

$$p = 4 \, b_{fi} = 4 \cdot 0.082 = 32.8 \times 10^{-2} \text{ m}$$

The modification factor for exposure of the fire-protected cross section is $30 - 24.5 = 5.5$ min:

$$k_{\mathrm{mod},fi,20} = 1.0 - \frac{1}{125} \frac{p}{A_r} = 1.0 - \frac{1}{125} \frac{32.8 \times 10^{-2}}{67 \times 10^{-4}} = 0.6; k_{\mathrm{mod},fi,,0} = 1.0$$

$$k_{\mathrm{mod},fi,5,5} = 1.0 - \frac{(1.0 - 0.6) \times 5.5}{20} = 0.9$$

The design compressive strength is

$$f_{c,0,d,\,fi} = k_{\mathrm{mod},fi} \, k_{fi} \, \frac{f_{c,0,k}}{\gamma_{M,fi}} = 0.9 \times 1.25 \, \frac{20}{1.0} = 19 \, \mathrm{N/mm^2}$$

The buckling resistance is

$$N_{c,Rd,fi} = k_c \cdot A_r \cdot f_{c,0,d,fi} = 0.19 \times 6.7 \times 10^3 \times 22.5 = 19.1 \times 10^3 \text{ kN}$$

To check buckling resistance,

$$N_{fi,d} \, 18.4 \times 10^3 \text{ kN} < 19.1 \times 10^3 \text{ kN}$$

The utilisation factor is $N_{fi,d}/N_{c,Rd,fi} = 18.4/19.1 = 0.96$.
Therefore, the column is satisfactory for 30 min of fire resistance.
   Note, however, that the utilisation factor 0.96 is greater than 0.72 from using the reduced cross-section method.

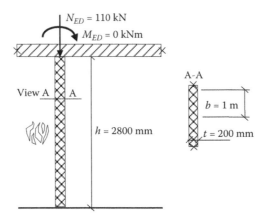

Figure 6.33 Geometry and loading condition of a load-bearing and separating wall.

6.8.8 Masonry wall loaded by fire on one side

A load-bearing and separating wall (see Figure 6.33) is designed from the autoclave aerated concrete masonry solid units P2, category I, 500 kg m^{-3} gross density, of size 200 × 599 × 249 mm with mortar for thin joints M2.5 in the third class of execution control. It is loaded by centric force, $F_{Ed} = G_{Ed} + Q_{Ed} = 110$ kN. At normal temperature, the load combination is calculated according to Equation (6.10b) in European standard EN 1990–2002. The required fire resistance for fire exposure from one side is R90.

6.8.8.1 Normal temperature design

6.8.8.1.1 Calculation of masonry strength

Solid units and mortar:

$$f_b = \delta \cdot \eta \cdot f_u = 1.25 \times 1.0 \times 2.0 = 2.5 \text{ MPa and } f_m = 2.5 \text{ MPa}$$

The masonry strength in compression, for masonry with thin joints, execution control class 3, is

$$f_d = \frac{f_k}{\gamma_M} = \frac{K \cdot f_b^{0.85}}{\gamma_M} = \frac{0.8 \times 2.5^{0.85}}{2.2} = \frac{1.74}{2.2} = 0.79 \text{ MPa}$$

Resistance at normal temperature
Cross-section area:

$$A = 0.2 \times 1.0 = 0.2 \text{ m}^2$$

Weight of 1 m² of wall:

$$G_{0d} = 1.35 \times 0.2 \times 1.0 \times 5.0 = 1.35 \text{ kNm}^{-2}$$

At the top and bottom, the eccentricity is $e_i = 0.05 \cdot t$. The reduction factor at the top and bottom of the wall is $\Phi_i = 0.9$, giving the wall resistance of

$$N_{Rd,i} = 0.9 \times 0.2 \times 0.79 \times 10^3 = 142.2 \text{ kN} > N_{Ed,i} = 110 + 1.35 \times 2.8 = 113.8 \text{ kN}$$

At the middle, the wall is influenced by its slenderness. For $\frac{h_{eff}}{t} = \frac{0.75 \times 2.8}{0.2} = 10.5$, the reduction factor within the middle height of the wall is $\Phi_m = 0.775$, and the wall resistance is

$$N_{Rd,m} = 0.775 \times 0.2 \times 0.79 \times 10^3 = 122.45 \text{ kN} > N_{Ed,m} = 110 + 1.35 \times 0.5 \times 2.8$$

$$= 111.9 \text{ kN}$$

The wall is satisfactory at normal temperature.

6.8.8.2 Design at fire situation

The illustrative design at fire situation is prepared using the simple model based on Figure C.3(g) in European standard EN 1993-1-2 (see Figure 6.9). For more accurate design, the results of further tests are required. At the fire limit state, the design value of the vertical load applied to a wall should be less than or equal to the design value of the vertical resistance of the wall such that

$$N_{Ed,fi} \leq N_{Rd,fi,\theta i}$$

The load reduction factor for fire design for the acting design force in the wall is $\eta_{fi} = 0{,}65$. Therefore,

$$N_{Ed,fi,i} = \eta_{fi} \cdot N_{Ed,i} = 0.65 \times 113.8 = 74.0 \text{ kN}$$

$$N_{Ed,fi,m} = \eta_{fi} \cdot N_{Ed,m} = 0.65 \times 111.9 = 72.7 \text{ kN}$$

The design value of the vertical resistance of the wall is

$$N_{Rd,fi,\theta i} = \Phi \cdot \left(f_{d,\theta 1} \cdot A_{\theta 1} + f_{d,\theta 2} \cdot A_{\theta 2} \right)$$

For a wall made of autoclave aerated concrete masonry solid units, the temperatures are $\theta_1 = 200$ °C, $\theta_2 = 700$ °C, and according to Figure 6.9, the thicknesses are $t_{700\ °C,90} \approx 20$ mm and $t_{200\ °C,90} \approx 60$ mm. The design compressive strength of the masonry at less than or equal to θ_1

$$f_{d\theta 1} = \frac{1.74}{1.0} = 1.74 \text{ MPa}$$

The design compressive strength of the masonry between θ_1 and θ_2 may be evaluated as $f_{d\theta 2} = c \cdot f_{d\theta 1}$. Conservatively, $c = 0$. Therefore, the design resistance is

$$N_{Rd,fi,\theta i} = \Phi \cdot f_{d,\theta 1} \cdot A_{\theta 1}.$$

The area of the masonry up to temperature θ_1 in the case of fire exposure from one side only is

$$A_{\theta 1} = (0.20 - 0.06) \cdot 1.0 = 0.14 \text{ m}^2$$

For the evaluation of factor Φ, the wall thickness may be assumed to be

$$t = 0.20 - 0.04 = 0.16 \text{ m}$$

For the cross section at the bottom,

$$\frac{e_{mk}}{t} = \frac{0.005}{0.16} = 0.031 < 0.05 \text{ and } \frac{h_{eff}}{t} = \frac{0.75 \times 2.8}{0.16} = 13.125, \text{ so that}$$

$$\Phi_i = 0.9 \text{ and } \Phi_m = 0.605$$

For fire design, the eccentricity due to the fire load should be added:

$$e_{\Delta\theta} = \frac{1}{8}(0.75 \times 2.8)^2 \times \frac{9 \times 10^{-6}(700 - 20)}{0.2 - 0.02} = 0.0187 \text{ m} \le \frac{h_{eff}}{20} = \frac{0.75 \times 2.8}{20} = 0.105 \text{ m}$$

$$e_{mk} = \frac{h_{eff}}{450} + e_{\Delta\theta} = 0.005 + 0.0187 = 0.0237 \text{ m}$$

$$\frac{e_{mk}}{t} = \frac{0.0237}{0.16} = 0.148 \quad \text{and} \quad \frac{h_{eff}}{t} = 13.125 \text{ so that}$$

$$\Phi_i = 0.9 \text{ and } \Phi_m = 0.605$$

The fire resistance of the wall at the bottom is

$$N_{Rd,fi,\theta i,i} = 0.9 \times 0.14 \times 1.74 \times 10^3 = 219.2 \text{ kN} > N_{Ed,fi,i} = 74.0 \text{ kN}$$

The resistance of the wall in the middle is

$$N_{Rd,fi,\theta i,m} = 0.523 \times 0.14 \times 1.74 \times 10^3 = 127.4 \text{ kN} > N_{Ed,fi,m} = 72.7 \text{ kN}$$

Using the approximate assumptions, the wall is satisfactory for fire exposure from one side for 90 min.

Chapter 7

Global modelling of structures in fire

7.1 INTRODUCTION

As a result of the introduction of performance-based approaches to design, it is now possible for designers to treat fire loading in the same manner as any other form of load. However, for this to happen it must be possible for designers to predict with confidence how a structure will respond to fire. Considerable research effort has been dedicated in recent years to providing the knowledge needed for this, and much progress has been made. It turns out that structural behaviour in fire in all but the simplest cases is much more complex than analyses based solely on loss of material strength due to heating, such as those based on the Standard Fire Test, can predict (e.g. Usmani et al. 2001). A finding is that analyzing structural elements, such as beams and columns, in isolation and with idealized supports (as is common at ambient temperature) in a fire analysis is insufficient if an understanding of the fire resistance of entire structures is desired. For accurate results to be produced, either the behaviour of whole structures or the behaviour of large parts of structures with appropriate boundary conditions must be considered. As a result, in all but the most straightforward cases, numerical analyses are required to predict accurately the strength and behaviour of structures in fire.

This chapter aims to explain the need for, and the requirements of, global modelling of heated structures. It has three parts. Firstly, it explains why analyzing and designing structures for fire loading are particularly challenging tasks in structural engineering. It considers in detail a simple example that highlights the importance in any thermostructural analysis of various phenomena not normally considered at ambient temperature. This section also notes possibly unconservative aspects of using the Standard Fire Test that are often ignored. Secondly, it discusses modelling techniques that are appropriate to structural fire problems. The advantages and drawbacks of various numerical schemes are considered, and practical methods of ensuring that numerical models are both efficient and accurate are identified.

Finally, consideration is given to realistic examples and how output from numerical models of heated structures should be handled.

7.2 NATURE OF GLOBAL STRUCTURAL BEHAVIOUR IN FIRE

7.2.1 Requirements

At ambient temperature, the "actions" on a structure typically result from statistically extreme combinations of wind and gravity loading. Such actions are forces and are (or can reasonably be assumed to be) invariant with time. As a result, the stresses caused in a structure by each load case can be regarded as constant, and it is straightforward to design for sufficient strength. Simplifications in structural analyses may often be made, mainly because most commonly used structural materials are very stiff, which means deflections under normal loads remain small. In most structures, small deflections are also ensured by serviceability requirements since the users of buildings are unwilling to accept perceptible movements. This means that geometric non-linearity can often be neglected in analyses. Moreover, it is usually possible to assume either linear-elastic or rigid-plastic material behaviour, further simplifying the analytical process by removing the difficulties of handling material non-linearity in calculations. This simplification is even possible for concrete, which has non-linear stress-strain behaviour, by the use of equivalent stress blocks.

The situation at elevated temperatures is very different for several reasons. The actions on a heated structure are primarily temperatures, or more fundamentally the heat fluxes that result from exposure of the structure to hot gases and radiation. These produce heating as a fire grows and cooling when fuel is being expended or as a result of firefighting. Since not all parts of a structure heat at the same rate, and because structural elements expand when heated, interference stresses are produced. Whereas the design stresses in a structure at ambient temperature may be considered as invariant, this is not the case in a heated structure because thermal equilibrium will not occur during a typical fire. The interplay between thermal expansion, restrained by this expansion and the large deflections commonly present in fire conditions, also result in variation of the stresses within structural members during a heating-cooling cycle. There is no reason why the greatest stresses should occur simultaneously with the peak of either the applied heat flux or the structural temperatures. A further complication is that neither heating nor cooling will occur simultaneously in all parts of a structure. This implies that stresses and material strengths may be increasing in some areas but decreasing in others at any point during a fire event.

High temperatures also affect structural materials' mechanical properties, with key factors being loss of linearity, strength, modulus and a clear yield point, as discussed in Chapter 5. These changes mean that not only the stresses within a heated structure change with time but also the structure's strength, and this must be considered during analyses. A second consequence of heating is thermal expansion. If this is restrained in any way, even by equilibrating the different free expansions due to a temperature distribution in a single cross section, large stresses will result. Thermal expansion also frequently causes large deflections to be present in heated structures. As these deflections are caused by the changing free lengths of heated members, it is not necessarily the case, as at ambient temperatures, that they indicate impending failure. Indeed, it may be the case that large deflections allow thermally induced stresses to be relieved. However, large deflections do mean that it is necessary to account for the effects of geometric non-linearity in structural analysis if accurate results are to be produced.

This discussion shows that to obtain an accurate prediction of structural behaviour at a high temperature, it is necessary to consider in analyses the following factors that may typically be excluded or disregarded under ambient conditions: material non-linearity, geometric non-linearity, and time- and temperature-varying strength. If a structure is to be designed to resist fire using a performance-based approach, it is necessary to ensure that it has sufficient strength and fulfils other design requirements during the entire period it is exposed to temperatures above ambient. The complex and time-varying nature of both stresses and strengths in heated structures mean that it is not possible to identify the most critical set of applied temperatures in the same way as the most critical load cases can often be identified at ambient temperature. Fire loading is a rare example in structural engineering in which all these phenomena need to be considered simultaneously to predict behaviour accurately.

The complexity of the behaviour of heated structures has not traditionally been recognised in fire safety design calculations because assessing fire resistance has almost always been done with reference to the Standard Fire Test, which assumes that individual elements of a structure may be considered in isolation from each other. In other words, structural fire design has tended to assume static determinacy member by member. In these conditions, high-temperature strength calculations only need to account for loss of material strength to obtain a reasonably accurate critical temperature. However, almost all real structures contain a degree of redundancy, and simplistic calculations will not provide the accurate estimates of strength that are needed for performance-based design.

7.2.2 An illustrative example

To highlight the importance of the various phenomena considered in the receding section to even very simple structures in fire, Gillie (2009)

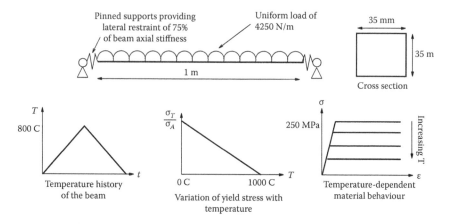

Figure 7.1 The benchmark problem provided by Gillie (2009) for numerical modelling of structures in fire. Note that the Young's modulus of the material is taken to be constant with temperature. (From Gillie, M., Analysis of Heated Structures: Nature and Modelling Benchmarks, *Fire Safety Journal*, 44(5), pp. 673–680, 2009. With permission from Elsevier.)

discussed the benchmark problem shown in Figure 7.1. It consists of a beam cross section, 1 m long and 35 mm square, that is subject to uniform heating and then cooling, as well as a constant distributed gravity load. The material behaviour is elastoplastic with a temperature-dependent yield stress, as shown in the figure. The coefficient of thermal expansion is taken to be $1.2 \times 10^{-5}/°C$ and is independent of temperature. The beam is fixed vertically, free to rotate at its ends and subject to a lateral restraint stiffness equal to 75% of its axial stiffness. The benchmark is not intended to represent a structure that might exist but rather to provide a tightly defined problem that allows a comparison of analyses that include or exclude various phenomena to be undertaken.

Figure 7.2 shows the predicted deflections of this structure when it is analysed using full material and geometric non-linearity and when it is analysed with various other assumptions. It is clear that the omission of any of the phenomena highlighted in the previous section results in different predictions. The most significant difference occurs when the effects of geometric non-linearity are omitted from the analysis, so that the $P - \delta$ effect of the axial restraint force is not included. In contrast to beams at ambient temperature, when geometric non-linearity is routinely ignored in analysis because its effects are generally minimal, this example shows that it is vital that it is included in fire analyses if accurate predictions are to be made. Predicted deflections for a geometrically linear analysis are twice as great as for the non-linear case because the stabilising effects of catenary tension are not accounted for. It is also notable that in the geometrically linear case deflections continue to

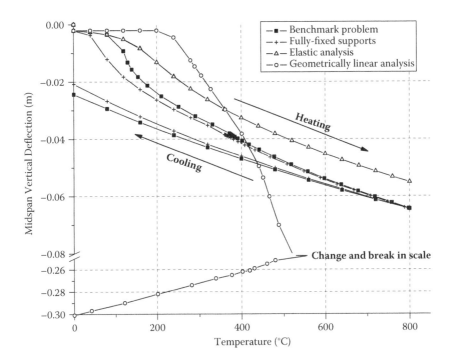

Figure 7.2 Predicted deflections for a simple heated structure with differing phenomena modelled. (After Gillie, M., Analysis of Heated Structures: Nature and Modelling Benchmarks, *Fire Safety Journal*, 44(5), pp. 673–680, 2009. With permission from Elsevier.)

increase on cooling, a counterintuitive finding that results from the beam yielding in tension and losing bending strength more rapidly through this mechanism than its increase due to cooling. Figure 7.2 also shows the effects of omitting the effects of plasticity and of varying the support conditions, both of which are shown to have a significant effect. The effect of the support conditions on the structural response is significant because of the assumption of simple support that is made in Standard Fire tests.

7.3 ANALYTICAL METHODS FOR STRUCTURES IN FIRE

7.3.1 Introduction

The types of analysis needed to determine structural behaviour in fire will depend on the problem being solved, the level of accuracy required and the available input data. Decisions must be made about the physical phenomena that need to be represented and about the numerical schemes that

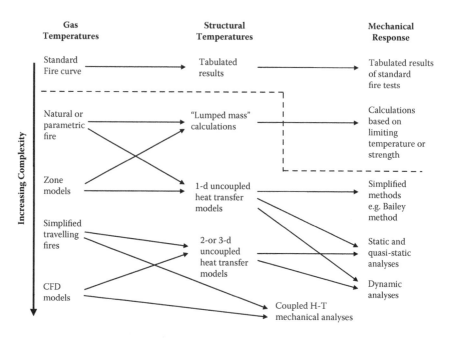

Figure 7.3 Relationship between analysis methods for design of heated structures. Methods above the dashed line are prescriptive in nature; those below suitable for performance-based design. Arrows indicate appropriate combinations of calculation methods.

are to be used to represent them. It will always be necessary to calculate the stresses and other mechanical quantities that develop in a structure. In addition, depending on the information available and design approach being taken, it may be necessary to calculate the gas temperatures and then the resulting structural temperatures that will occur. The relationships between the various kinds of analysis that may be required are shown in Figure 7.3. The following discussion focuses on methods of calculating mechanical response (means of estimating gas temperatures are discussed in Chapter 3 and calculating structural temperatures in Chapter 4). It is also assumed that computer-based analyses are being used because simpler methods of analysis are fully discussed in other publications (e.g. Bailey 2002b). The structural analysis may be either decoupled from or coupled with heat transfer analysis.

7.3.1.1 Uncoupled heat transfer analyses

The temperatures within the structure are calculated for the duration of the fire (and for as long afterwards as necessary). Afterwards, the mechanical analysis is performed in a completely uncoupled fashion. An uncoupled

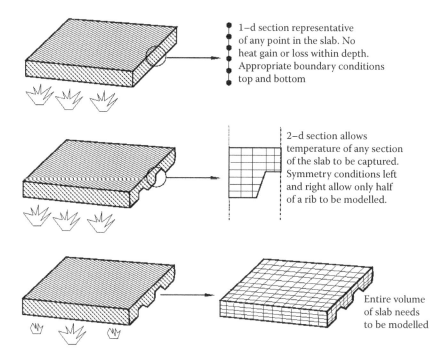

1–d section representative of any point in the slab. No heat gain or loss within depth. Appropriate boundary conditions top and bottom

2–d section allows temperature of any section of the slab to be captured. Symmetry conditions left and right allow only half of a rib to be modelled.

Entire volume of slab needs to be modelled

Figure 7.4 Suitable discretization schemes for heat-transfer analyses in three scenarios. Top, a slab of uniform thickness heated by a uniform fire. Middle, a ribbed slab heated by a uniform fire. Bottom, a ribbed slab heated by a non-uniform fire.

approach has the advantage of allowing different models and software to be used for heat transfer and structural analyses. Since many methods of modelling compartment fires assume a uniform gas temperature within the compartment, it is often possible to use a one- or two-dimensional heat transfer analysis to determine structure temperatures, even in large structural members such as slabs (Figure 7.4). Even if the same (undeformed) geometry is used for thermal and mechanical analyses, the optimum mesh densities for the two analyses are usually different in the two cases, so adopting a sequential approach can improve calculation efficiency and accuracy.

7.3.1.2 Fully coupled analyses

A fully coupled thermomechanical analysis is rarely required for building fire problems because there is normally only a weak two-way coupling between heating and mechanical response. However, there are circumstances when full coupling will be needed, for example, if spalling of concrete or other damage that changes the exposure of the structure is being modelled. It may also be the case that local fires that do not flash over cause roof structures to fail locally, which may alter the fire ventilation conditions and therefore change the nature of the fire. In some cases, the geometry of the structure may alter significantly as a result of heating, and this may in turn alter the compartment's thermal boundary conditions, resulting in two-way coupling. The input data required for the heat transfer aspects of using a fully coupled approach is the same as for uncoupled modelling, except for the choice of finite elements, which need to have both mechanical and thermal degrees of freedom, and their meshing, which has to ensure accuracy across both regimes.

One potential advantage of a coupled approach, even when it is not strictly necessary, is that only one analytical process is required, so that no data transfer from a heat transfer analysis to a structural analysis need be undertaken.

7.3.2 Mechanical analyses

Various kinds of mechanical analysis available for modelling heated structures are shown in Figure 7.5, together with the phenomena that they are able to represent. The most significant difference between analysis types is the degree to which time has physical meaning. Time is effectively ignored in many non-computer-based types of analysis. As a result, such analyses are only able to calculate stresses for one set of loads and temperatures (typically peak structural temperatures are of most interest). Numerical models allow time to be represented in various ways, as discussed in the following sections.

7.3.2.1 Static analyses

As for ambient temperature finite element analyses (FEAs) of structures, static analyses are the most common method for analysing heated structures. "Time" in static analyses is a non-physical solution parameter; it allows a sequence of loads to be applied to a numerical model and the model's equilibrium state to be calculated throughout this sequence. This means that, although the order in which actions (mechanical or thermal) are applied affects the analytical predictions, the (non-physical) time over which they are applied is arbitrary. Increasing the time period of an analysis will not affect the predictions, provided that the relative times at which loads are applied remain the same. This means that time-dependent phenomena, such as inertial forces, cannot be represented in such analyses. Since inertial effects are not modelled, static analyses are only appropriate for predicting structural behaviour where structural movement is slow; they are not suitable for collapse modelling in which the stability of equilibrium is lost, and large inertial forces can be developed.

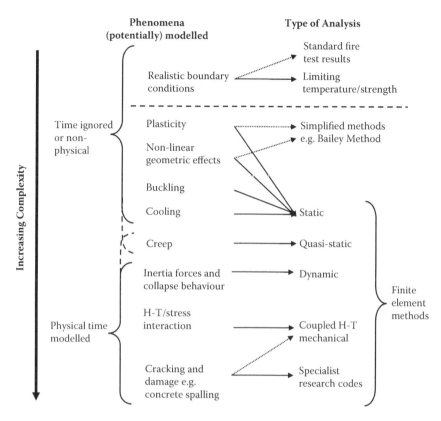

Figure 7.5 Phenomena modelled by various structural analysis methods for heated structures. Solid arrows indicate the simplest analysis method by which a phenomenon may be represented. Generally, more complex analysis methods will include all phenomena represented by simpler methods. Dotted arrows indicate that a phenomenon may be represented in a limited manner. Methods above the dashed line are prescriptive in nature; those below are suitable for performance-based-design.

Typically, a static analysis of a heated structure assumes that mechanical loading is constant, while thermal loading varies. This represents most fire scenarios reasonably accurately because the self-weight and live loads on a structure will generally be present prior to a fire and remain roughly constant during heating. Normally, such an analysis is broken into two parts: during the first part, mechanical loads are applied, and during the second part, these are maintained while thermal loads are introduced. Thermal loads can vary over the non-physical time used in the analysis and may include cooling.

Static analyses of non-linear problems typically use a solution method based on the well-known Newton-Raphson method for solving non-linear problems. However, this method is not good at obtaining solutions when the stability of equilibrium is lost and structural softening or snap-through buckling (in reality a dynamic event) occurs. Softening and buckling behaviour at ambient temperature are sometimes handled by using an "arc-length" algorithm (Zienkiewicz and Taylor 2000) to solve for load and displacement simultaneously. For analyses with varying temperatures, the use of arc-length methods is generally not possible. This is because the applied loads on a structure remain constant during thermal loading, which is predefined. Thus, it is not possible for an arc-length search to be undertaken in either load-deflection or load-temperature space. This can cause difficulties for analysing heated structures and often either results in assumptions being made about material behaviour that may not be entirely justified to allow a Newton-Raphson-based solution to be used (e.g. it is often assumed that concrete behaviour in tension is ductile) or forces the use of dynamic analysis to obtain solutions. Another possible solution is by introducing a small amount of pseudodamping to the structure to enable the numerical simulation to go through a stage of temporary loss of static equilibrium (Dai et al. 2010a; Elsawaf et al. 2011). Despite the shortcomings, static analyses are widely used to develop understanding of structural behaviour at elevated temperatures and, if used with care, will produce useful results rapidly.

7.3.2.2 Quasi-static

It is possible to introduce some time-dependent effects, such as creep, into quasi-static analyses for which inertial effects are still not included. This may be desirable in the analysis of heated structures because creep effects are accelerated at high temperatures and may need to be modelled. Huang et al. (2006) recommend modelling creep when considering the behaviour of heated steel frames. However, many authors have been able to replicate well the behaviour of experimentally heated structures using numerical models without modelling creep numerically. This may, in part, be due to creep effects being included in some form in the material input data used in the analysis. For instance Eurocode stress-strain data for steel includes an allowance for high-temperature creep by means of its use of transient test results.

7.3.2.3 Dynamic analyses

In dynamic analyses, time has a proper physical meaning, and inertial forces can be captured. There are broadly two types of numerical scheme used for dynamic FEAs: "explicit" time integration and "implicit" time integration. Both solve the dynamic equilibrium equations incrementally through time, and both have been used successfully for modelling structures in fire.

Explicit dynamic analyses use the known (explicit) state of a numerical model at the end of one incremental time step to calculate its state at the next time step. These numerical schemes are mathematically stable, provided that the time steps used are less than a certain size, which is related to the smallest element size in the model and to the material density (in reality, to the highest relevant natural frequency). Consequently, in detailed modelling they can require a large number of very small time steps and be computationally expensive. However, by judiciously artificially increasing the density of the material being modelled ("mass scaling"), and thereby increasing the stable time increment, structural fire-engineering problems in which inertial forces are small can still be solved using an explicit algorithm in an reasonable (real-world) time. This is particularly useful if the real requirement is for results that represent equilibrium states of the structure, and it is not important to represent accurately transient states if static equilibrium is lost and the structure moves freely. Collapse behaviour can often be modelled with this kind of analysis, although once inertia forces become predominant, details of the velocities and accelerations of structural components will not be accurate if mass scaling has been used. A detailed discussion of the use of explicit numerical schemes to model collapse behaviour of multistorey structures in fire with the ABAQUS (ABAQUS, 2007) finite element package was undertaken by Flint (2005), who also presented a parametric study that considered the effects on modelling accuracy of mass scaling and many other parameters. Since dynamic explicit numerical models will generally reach a solution of some kind, even if it is not a physically meaningful one, without the convergence problems associated with static analyses, it is recommended that very careful benchmarking is undertaken of such models. This process may include running a secondary static analysis until convergence problems are encountered and comparing the results with those predicted by the explicit dynamic analysis up to the same point.

Implicit dynamic analyses solve the dynamic equilibrium equations by direct integration in an iterative manner to estimate the solution at the next time step. This process requires large matrices to be inverted at every time step, which is computationally intensive; however, there is no mathematical limit on the size of time step that may be used in such an analysis. This form of numerical scheme has been less frequently used for structural fire-engineering applications, but Varma et al. (2008) adopted it successfully when analysing a multistorey steel frame.

7.4 PRACTICAL MODELLING TECHNIQUES

7.4.1 Software

The complexity of even the simplest structural fire problems means that a numerical analysis will be needed if realistic predictions of deflections and stresses are to be produced. To date, the finite element method has been used almost exclusively to achieve this, and it seems likely that FEA will remain the only realistic choice for the foreseeable future, although simplified methods are available for checking strength and fire resistance for design purposes. However, there are choices to be made over the nature of the code to be used. Finite element codes suitable for structural fire analyses can be broadly divided into two categories: commercial general-purpose codes such as ABAQUS (ABAQUS 2007), Ansys (2007) and others; and research-based codes such as Vulcan (Vulcan, 2008), ADAPTIC (ADAPTIC 2008) and Safir (Franssen 2005). Commercial codes have the advantages of being able to handle larger problems than research codes, access to a large range of element types, a range of solution algorithms, and a wider range of capabilities outside fire engineering. This means that if a structure needs to be analysed for several loading conditions, perhaps fire and seismic loading, only one model should be needed. Commercial programs are, however, costly and normally restrict the ability of the analyst to extend or alter the code. This "black box" aspect can be frustrating if numerical convergence is not achieved, but the reasons for this non-convergence cannot be fully investigated. By contrast, research-based codes tend to be much cheaper and more focused in terms of their suitability for typical structural fire situations; the analyst may have access to the source code and thus be able to adapt it according to need.

7.4.2 Representation of concrete material behaviour

Chapter 5 presents a detailed discussion of the existing engineering models of material properties, for all levels of modelling, of structures in fire conditions. This section is not intended to restate this information but to discuss the numerical implementation of aspects of concrete material characteristics that may affect practical thermostructural modelling.

Concrete is a much more complex and variable material than steel, and even at ambient temperatures numerical modelling of concrete structures is usually less accurate than modelling of steel structures. It is important, therefore, that the limitations of any model used for modelling high-temperature concrete structures are recognised.

7.4.2.1 Plasticity models

Several constitutive models are available for representing the multiaxial plasticity behaviour of concrete; see Chapter 5 for details. These have all been developed for use at ambient temperature and are generally validated against the limited experimental evidence that exists for the multiaxial behaviour of concrete (e.g. Kupfer et al. 1969). Their use at elevated temperatures assumes that the same multiaxial behaviour holds, despite lack of direct experimental evidence. A key aspect of the multiaxial behaviour of concrete is that it is not independent of hydrostatic pressure, as is assumed by the von Mises yield criterion. The most commonly used yield criterion that accounts for this dependency is the Drucker-Prager model, which is available in most analysis software. The Drucker-Prager model and its variants have been successfully used to model the behaviour of heated concrete structures, but in its normal form, it lacks the ability to model some aspects of concrete behaviour that may be important (e.g. cracking and crushing), and the form of the yield surface means it is only possible to obtain an approximate match with experimental data. As with the von Mises yield criterion for steel, for use at high temperature the Drucker-Prager criterion needs to be made temperature dependent, as discussed in Chapter 5.

A number of more sophisticated yield criteria are available that do not have these shortcomings. Of note is the "damaged plasticity" model implemented in the ABAQUS software and used by a number of authors to represent heated concrete (e.g. Law 2010). This yield criterion combines two surfaces that together are better able to represent the experimentally observed behaviour of concrete than the Drucker-Prager model. The effects of cracking are also captured in the model by allowing strain-softening behaviour to be defined as part of the input. While effective at capturing the cracking behaviour of concrete, this has a side effect of introducing mesh sensitivity into an analysis.

7.4.2.2 Other phenomena

Many numerical codes allow for a range of other phenomena that occur in reinforced concrete to be added to the basic constitutive model, such as cracking and tension softening. While representing such effects at high temperatures is desirable, it is often not possible to determine suitable input parameters due to lack of experimental evidence.

Cracking is the most commonly included phenomenon; it is usual to use a "smeared cracking" approach to represent the cracking behaviour of concrete numerically. This represents cracks by smearing the equivalent strain over whole elements and so does not fully represent the localised, discrete nature of cracks that may occur in concrete structures. It is an approach that has been shown to be fairly accurate for modelling the overall behaviour of heated steel-concrete composite structures, but its accuracy when used to represent other forms of construction that use concrete in fire is currently not clear. Recent developments in this area include the so-called extended finite element method (XFEM) (e.g. Sukumar et al. 2000), which allows for finite elements with shape functions that are discontinuous and is therefore able to represent discrete cracks. To date, however, this approach has not been adopted in structural fire engineering.

The interaction between concrete in tension and reinforcement is complex and is normally represented by "tension softening". It is a complex phenomenon that is difficult to capture accurately in numerical models, and attempts to do so can lead to numerical instabilities. How best to model this behaviour is currently an active research area.

7.4.3 Buckling

Buckling in structures at ambient temperature is an instability phenomenon in the sense that the stability of equilibrium is lost, either by a bifurcation into alternative equilibrium paths or by the occurrence of a limit load beyond which the stiffness of the system changes from positive to negative, and a "downhill" equilibrium path is generated. Particularly in the latter case, deflections grow rapidly in an uncontrolled dynamic fashion ("run away"), with the possibility that a new stable equilibrium configuration is encountered before the structure is considered to have collapsed. This runaway process happens particularly in load-controlled conditions, where the external loading and displacement are not linked by any stiffness relationship. Where load is applied using a displacement control, it may be possible to avoid dynamic runaway, but this condition is usually only available in experimental test conditions. However, in fire conditions thermal expansion of members against restraint can generate buckling, which is effectively displacement controlled. A simple example of this is shown schematically in Figure 7.6, which shows a fixed-ended beam subject to uniform heating.

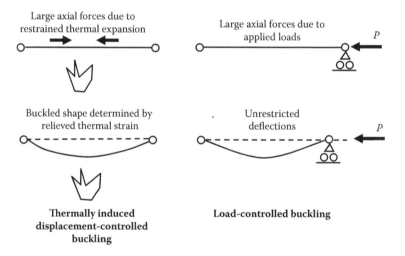

Figure 7.6 Comparison of the effects of displacement- and load-controlled buckling. Thermally induced displacement-controlled buckling does not imply imminent collapse and may in fact serve to reduce mechanical strains in certain structural components by providing a means by which geometric lengthening may occur.

As heating occurs, the axial force in the beam grows until the critical buckling load, given that the material's tangent stiffness has degraded to some extent at the elevated temperature, is reached. Lateral deflection then develops rapidly because it has a quadratic relationship with axial deflection; further thermal expansion simply causes extra lateral deflection, while the axial force gradually reduces. Since the stress-strain curve is curvilinear, so that some permanent compressive strain has occurred, cooling from the buckled state results in reduced deflection but not a return to the original straight configuration. Buckling of this type occurs in many forms of heated structure because of the restraint that is usually provided to thermal expansion. It can either be beneficial (by relieving thermal strains) or deleterious, either because it can indirectly lead to collapse (Röben 2010; Flint 2005) or because of integrity failure by breach of compartmentation.

7.4.4 Orthotropic floor slabs

Models of heated structures frequently require orthotropic floor slabs, of the kind commonly used in steel-concrete composite construction, to be modelled. Such floor slabs are quite complex three-dimensional (3-D) structural elements that usually include ribs running in one direction, formed by profiled steel decking below the concrete; anticracking reinforcing steel mesh within the continuous depth; and complex temperature fields through the concrete thickness. Discretising these elements for numerical analyses in a manner that does not result in an overly large numerical model, while still capturing the required structural phenomena, requires care.

Apart from some early grillage models (Sanad et al. 2000), almost all analysts to date have used shell finite elements to represent such floor slabs. Grillages lack an aspect of slab behaviour that is vital for geometrically non-linear analyses, namely, the very high in-plane shear stiffness, which is inherent in flat plates and which is automatically present in any solid representation of the slab. Shell elements have the important benefit of being able to model bending and membrane behaviour much more efficiently in computational terms than solid elements and, unless great detail is required, appear to be the most logical choice for structural modelling of heated slabs. However, it is not straightforward to model the ribs that run in a single direction under the slabs when using shell elements. This is firstly because of the geometric orthotropy produced by the ribs and secondly because of the complex temperature fields that develop in and between the ribs. Various approaches have been taken in dealing with these complexities. The most robust in terms of representing the geometric and thermal complexities is that developed by Gillie et al. (2001b), who programmed FEAST, a stress-resultant based method of modelling the bending resistance of orthotropic shells. However, this approach represented non-linear elastic (rather than true plastic) material properties, so later workers have selected an approach from several logical alternatives, depending on the purpose of the analysis and the degree of precision required:

- The most accurate approach without using solid "brick" elements uses beam elements of general segmented cross section (often referred to as fibre elements) of T shape to represent the rib portion and part of the continuous upper part of the slab, together with layered flat shell elements to connect these together (Yu et al. 2008b) (see Figure 7.7). If the vertical spacing of beam element segments coincides with that of the shell element layers, then a consistent two-dimensional thermal analysis can be performed. This has been found to work well for most purposes (Lamont et al. 2006). The accuracy of the thermostructural analysis then depends mainly on the coarseness of the mesh and degree of through-thickness resolution available in the finite elements selected.
- An approximation of orthotropic slabs as isotropic flat slabs is often used for design purposes using layered flat shell elements. The decision regarding the slab thickness to be assumed can be based on the degree of conservatism considered necessary as a compensation for the assumption of isotropy. A full-depth flat slab is clearly and demonstratively unconservative, but the conservatism provided by treating the slab as simply the continuous upper part above the ribs, even maintaining its offset from the composite steel beams, may be such that member selection or fire protection is uneconomic.

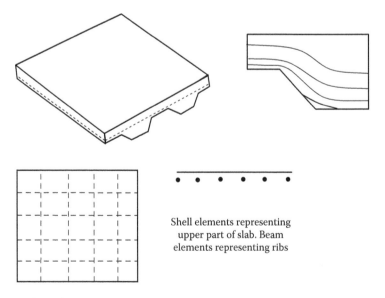

Shell elements representing upper part of slab. Beam elements representing ribs

Figure 7.7 Simplification of a ribbed slab.

- A more reasonable extension of this method is either to use an isotropic slab with its lower face at the mean level of the lower faces of the ribs and troughs. An even more reasonable version is provided by Eurocode 4 Part 1-2, in which the mean level is weighted with respect to the relative mean widths of the ribs and troughs; this gives a better representation particularly for reentrant decking where the width of troughs is usually much less than that of the ribs.
- It is relatively easy to construct a flat slab element in which the orthotropy is represented by factoring the stiffnesses in either direction and at any temperature by their ratios to the full-depth bending stiffness at ambient temperature. This "effective stiffness method" clearly has arbitrary aspects, but evidence from parametric studies (Yu et al. 2008b) suggests that it performs well, at least for slabs with trapezoidal decking.
- Although very expensive in computational terms, the most effective way of ensuring that the analysis is accurate for any combination of heating and slab details is to use solid brick elements to model the whole slab volume. This enables both 3-D thermal and thermostructural analyses using the same mesh of elements. However, the time implied both for preparation and for computation when dense 3-D mesh is used currently dictates that this should generally only be used for models of very limited extent in practical design analysis. More important, however, it can be used for validation and sensitivity assessment of the simpler methods discussed.

Shell elements in most finite element codes are able to handle different material properties through the depth of the shell by treating the shell depth as a series of connected layers. This means it is possible to represent a layer of anticracking rebar mesh within a slab with ease by specifying a layer of steel of equivalent area at the appropriate depth. This is usually done by assigning orthogonal unidirectional properties to two thin adjacent layers, representing the bars in either direction. A single such layer can also be used to represent additional layers of rebar; these are most often used in the ribs when the longer slab spans are used between secondary beams. It is generally assumed that steel decking on the underside of slabs should be ignored in fire analyses because it will rapidly reach almost the same temperature as the gases in the fire compartment and hence have negligible strength. In certain cases, such as heating from local fires, this assumption may not be applicable, in which case representing the decking using a further layer of steel within shell elements may be considered. However, steel decking will so often debond from the concrete below a slab, even under local heating, and cease to act compositely, so its effects should only be included in analyses after considerable thought and rarely in design situations. In any case, it is unlikely that steel decking will act compositely in both directions, given the ease with which it can "unfold" in its weak direction.

7.4.5 Units

Most finite element codes do not require the user to adopt particular units but allow any consistent set to be used. The most commonly adopted internationally is the SI system. If only the "base" units of this system (metres for length, grams for mass, newtons for force, seconds for time) are used, then all input and output will be consistent, although some input and output will have orders of magnitude that are unfamiliar to engineers and inconvenient for specifying structural layouts and details. If multiples of the base units are used without care (e.g. millimetres for length, as is common in structural engineering) in analyses of heated structures, then confusion can result.

In static mechanical analyses, no time-dependent phenomena are modelled, so forces result solely from the mechanical properties of the structure and equilibrium. This fact has useful consequences. If in a static mechanical analysis millimetres are used as the length unit, the force unit should strictly be one thousandth of a newton because the definition of force has a length dimension within it (MLT^{-2}). However, because forces are being calculated from the stiffness of the structure via

$$F = kx = (\text{Force/Length})*\text{Length} \qquad (7.1)$$

in which all length terms are eliminated from the the right-hand side, force will in fact be output in newtons, providing that material moduli are entered in units of newtons and millimetres, as is also common in structural engineering. Hence, it is possible to use newtons and millimetres, an inconsistent set of units, in a static analysis but to avoid inconsistent results because of the simplified methods by which forces are being calculated. This sleight of hand is widely used in structural engineering practice and in ambient-temperature finite element work without being explicitly acknowledged. It can also be used for purely mechanical high-temperature static analyses without problems. However, as a general principle, it is better practice to convert from engineering units to consistent units before entering data and then to reconvert results to engineering units after the FEA is complete.

In thermal analyses, using inconsistent units for convenience is no longer reasonable because time-dependent phenomena are being modelled, and incorrect results can easily be generated. A danger lies in attempting to take an existing mechanical finite element model sized in millimetres that has produced correct results and using it to perform a heat-transfer analysis. This can initially result in gross error, but perhaps more important, even if the requirement for truly consistent units is identified, this may make the use of the model for heat-transfer analysis very complex, as physical quantities will need to be entered in unusual units. In general, it is simplest to use plain SI units throughout, even if this goes against usual modelling practice in structural engineering.

In dynamic mechanical analyses, time derivatives are involved, and forces are calculated from dynamic equilibrium:

$$F = m\ddot{x} + c\dot{x} + kx \tag{7.2}$$

The time derivatives contain both time and length dimensions that are not eliminated as in static analyses, so again it is no longer possible to use millimetres, newtons per square millimetre (N/mm^2; megapascal) and newtons to enter data as the predicted forces would be incorrect. The most practical option is again to use plain SI units.

Example 7.1

To illustrate the previous discussion, suitable approaches to modelling a real structure are presented and discussed. The structure analysed is the Cardington test frame, a composite steel-concrete structure designed and used specifically to test a largely unprotected building of a typical commercial type subjected to fire. The Cardington frame was one of the very few full-scale structures that have ever been tested in fire while heavily instrumented. Consequently, it has been intensively analysed in the past (Bailey et al. 1996; Elghazouli et al. 2000; Gillie et al. 2001a), and the accumulated experimental data allow real and numerically predicted responses to be compared easily. Full data including deflections, strain, rotations and temperatures during a sequence of six fire tests are freely available online (http://www.mace.manchester.ac.uk/project/research/structures/strucfire/, 2011). The behaviour of the Cardington frame during a localised fire affecting one secondary beam, the "restrained beam test", is discussed here.

THE STRUCTURE AND TEST

The Cardington frame was an eight-storey building, constructed in 1994 and tested during 1995–1996, with five 9-m composite secondary beam bays across its frontage, supporting almost identical composite floors based on 305 × 165 × 40 UB primary and 610 × 228 × 101 UB secondary steel downstand beams, arranged as shown in Figure 2.3 in Chapter 2. Concrete composite slabs were cast over the beams on trapezoidal metal decking with the ribs running perpendicular to the 9-m secondary beams. The slab's overall thickness was 130 mm, with a concrete strength of 47 N/mm^2 and a single layer of A142 (142 mm^2 per metre run) anticrack mesh. Total gravity loading on all floors amounted to 5.48 kN/m^2. The tested secondary beam is indicated as number 1 in Figure 2.3. Fire loading during the test was simulated by gas burners along its entire 9-m span, except for 500-mm lengths at each end, which were protected, as were the connections themselves. The floor slab was also heated to 1 m from the steel beam on either side.

NUMERICAL IDEALIZATION

To model the mechanical behaviour of the fire test with a reasonably sized finite element model, a representative area of one floor of the structure was represented numerically, as shown in Figure 7.8. Symmetry boundary conditions were applied along all edges of this area to simulate the surrounding structure and to take advantage of the symmetrical nature of the heated area. Although symmetry conditions on the three unheated sides of the area do not strictly represent the surrounding structure, they are sufficiently distant from the heated area to have only a minimal effect on the predictions. The steel beams were modelled using linear beam finite elements and the concrete slab using linear four-node shell elements with an embedded layer of steel at the appropriate depth to represent the reinforcing mesh. A significant approximation was made by ignoring the ribs on the undersurface of the slab and taking the thickness to be a uniform 70 mm. This is justified by the accuracy of the results, but a more detailed model could have been adopted using the techniques discussed. As is usual when modelling heated composite floor slabs, the metal decking was ignored in the analysis. The column in the centre of the area considered was not modelled explicitly but instead represented by a fully fixed boundary condition.

Temperature-dependent stress-strain behaviour for both steel and concrete in compression was taken from the Eurocodes (Committee of European Normalization [CEN] 1992–2005b, 1993–2005b, 1994–2005b). Key values from the test data that allow the stress-strain curves to be determined from the codes are shown in Table 7.1. The tensile strength

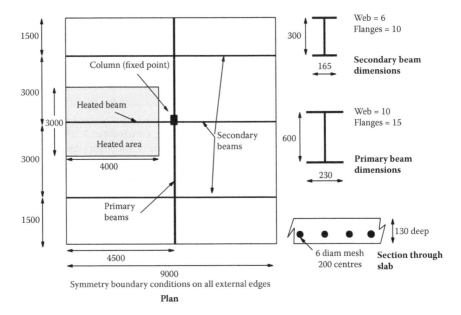

Figure 7.8 Details of the Cardington frame and of the area of the structure modelled numerically.

Table 7.1 Material properties used for modelling the Cardington restrained beam test

Material	E, GPa	Poisson's ratio, (−)	Coefficient of thermal expansion, °C	Yield stress, MPa	Ultimate stress, MPa
Mild steel	210	0.3	1.35×10^{-5}	300	400
Reinforcing steel	210	0.3	1.35×10^{-5}	450	460
Concrete	—	0.25	9×10^{-6}	—	47

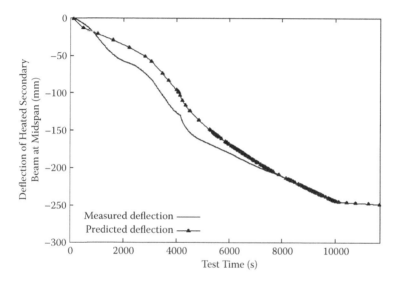

Figure 7.9 Comparison of predicted and measured midspan deflections of the Cardington restrained beam test.

of concrete was taken as 5% (2.25 MPa) of the compressive strength, and ductile behaviour in tension was assumed to aid numerical convergence. This assumption of ductility can be partly justified by the tension-softening behaviour of reinforced concrete, but it remains a simplification of the true behaviour.

A static analysis was chosen so all time-dependent phenomena were ignored. First, the gravity load of 5.48 kN/m² was applied to the slab, then temperature loading was added with the gravity loading held constant.

Figure 7.9 shows a comparison of the predicted and measured deflection of the midspan of the heated beam. A good correlation is apparent, particularly if the measured temperatures were used.

Chapter 8

Steel and composite joints

8.1 INTRODUCTION

The details of joints used in steel-framed and composite construction vary widely and depend on a variety of factors, such as the basic design philosophy and assumptions of the framing system. If a "simple" or "gravity" frame is used, then the assumption is that beams are simply supported, with a separate bracing system or structural core resisting horizontal forces on the frame. In this case, the basic role of beam-end joints is to carry the vertical end reactions of the beams. If the frame is designed to carry horizontal loads without a separate bracing system, then the beam-to-column joints are designed as either rigid or semirigid and must carry combinations of moment and vertical force. In either case, the joints adopted will have both moment and vertical force resistance to some extent, but their design details will be different.

The fire resistance of joints must be at least the same as for the connected members. This means that beam-to-column joints should be able to transmit the internal forces during the whole fire resistance time. When passive fire protection is used on the members, this requirement is generally considered to be fulfilled if the same thickness of fire protection is applied to the joints. Because of this, it is usually said that beam-to-column joints do not present a major problem because, due to the concentration of material, the temperature of the joint tends to be lower than that of the connected members; therefore, their strength increases relative to that of the beams to which they are connected. However, this is only superficially logical. The forces transmitted through a joint change massively during the course of heating by a fire. In fact, in framed structures they will usually change considerably because of the effects of restrained thermal expansion interacting with thermal gradients across members and the temperature-related reductions in strength and stiffness of the steel and concrete.

It is worth at this stage giving some attention to the use of the terms *joint* and *connection*. In most previous work and in research papers, these have been considered almost synonymous, but in the development of

Eurocode 3 EN 1993-1-8 (Committee of European Normalization [CEN] 1993–2005c), they have been separately defined. The term *joint* is used to refer to the *whole region* where the members intersect. It thus includes the plates, bolts and welds, which facilitate the "connection" of the individual elements, and parts of the connected elements in the immediate vicinity, such as the column web and beam ends. Thus, *connection* is used to refer to the details of particular connected zones. In broad terms, this chapter uses this nomenclature, although the distinction is not always clear-cut.

8.2 TYPICAL JOINTS AT AMBIENT TEMPERATURE AND IN FIRE

8.2.1 "Simple" joints

For simple braced frames, the joints are assumed only to transfer beam-end vertical shear forces into the columns. Such joints should possess very low rotational stiffness so that the moments induced in the columns are caused only by the eccentricity of the reaction forces from the column centre lines. Ambient-temperature design procedures for three standard simple joints (Figure 8.1) are given in the Steel Construction Institute (SCI; 1992, 1993) "Green Books".

The flexibility of such joints is purely rotational. In horizontal terms, they may be required to carry a nominal "tying force" to fulfil the robustness requirement of avoiding progressive collapse as a result of a local structural failure. In the fire condition, such joints are required to perform a much more extensive function. Joint rotation is very much higher, initially because the differential expansion of the steel and concrete of the floor system causes thermal bowing towards the fire and eventually because the loss of bending stiffness and strength of the exposed steel beams at very high temperatures causes the floors largely to hang in catenary tension between supports. In terms of horizontal force, the joints are subjected to

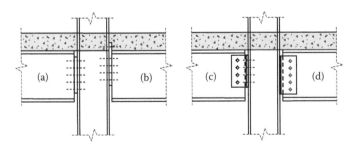

Figure 8.1 Standard simple joint types: (a), (b) partial-depth end-plates; (c) web cleat; (d) fin plate.

high compressive force in the early stages when overall expansion of the beam members overrides the effective shortening due to deflection. As temperatures increase, the development of catenary action in the beams causes the horizontal joint force component to reverse, and a high tensile force develops. If temperatures continue to rise, as in the BS 476 standard fire (British Standards Institution [BSI] 1990a), this tensile force eventually declines gradually as the tensile strength of the steel beam is reduced by increasing temperatures. Finally, in the cooling phase, the joint is subjected to very high tension as members, which have effectively shortened in heating, contract.

The temperature distribution in a joint zone is usually considerably lower than that of the members it connects, especially if some of these members are unprotected, because of the local concentration of material and relative lack of exposed surfaces. As a result of this, it is possible to generate moments, even in joints such as these that are designed as simple hinges for ambient temperature, which can very effectively help to resist the deflection of the connected beams. It has previously been thought that this moment resistance in fire can be utilised in a simple fire-engineering calculation for beams, with a residual resisting moment at the beam ends. However, observations of local buckling due to compression stresses in fire have suggested that more caution needs to be applied until further research has been done.

8.2.1.1 Observed behaviour of "simple" joints in fire tests

The standard joint details shown in Figure 8.1 would not all be expected to behave identically in fire. Partial-depth end-plates of type (a) were used extensively in the Cardington fire tests and showed considerable evidence of local buckling in the beam lower flanges, in combination with shear buckling of the beam webs, as shown in Figure 8.2.

In cooling, a partial failure of these connections was often observed, in which the end-plate material adjacent to the welds on one side of the beam

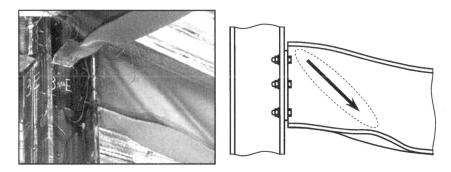

Figure 8.2 Local buckling observed in Cardington fire tests.

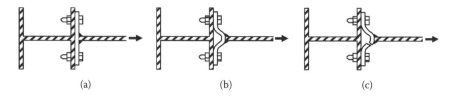

Figure 8.3 Partial fracture of end plates in cooling.

flange fractured, as illustrated in Figure 8.3. In no case did both sides of the end-plate fracture, indicating that the increased flexibility produced by the fracture of one side was enough to allow the remaining connection to perform in a ductile fashion.

This is seen in the aftermath of most accidental fires in which either steel or composite beams in internal regions of a floor have been subject to the restraint to their thermal expansion provided by continuous areas of slab surrounding these beams. As the steel beam temperature increases, this restraint causes high axial compression to grow rapidly in the lower flange and web of the downstand beam, while the expansion is prevented but most of the steel strength remains and deflections are relatively low. The essential limit to this phase comes as the steel strength declines rapidly between 400°C and 700°C, and the axial compression falls. During this phase, plastic buckling occurs in the lower flange and web near the connections, where the axial compressive stresses due to restraint are enhanced by bending compression stresses due to hogging rotation. When the beam strength reduction becomes very high, it loses most of its bending strength so that deflections increase rapidly, the compressive force reduces rapidly, eventually changing to tension as the steel section begins to carry its loads mainly in catenary action.

This behaviour is illustrated in Figure 8.4, which shows the axial force changes in the unprotected steel downstand of a composite beam tested at Cardington.

It is notable that in heating the tensile force at very high temperatures approximates the steel strength degradation curve if there is high axial restraint stiffness at the beam ends; however, it is reduced by greater axial flexibility at the beam ends or by higher beam deflections. On cooling from this state, the beam, which has effectively shortened in restrained heating, contracts against its axial restraint and develops a progressively higher tensile force, which again mimics the strength degradation curve. The force generated at this stage is reduced by any axial flexibility, such as low restraint from adjacent structure, low tying stiffness at the end joints, or the flexibility associated with straightening the deflected shape of the beam.

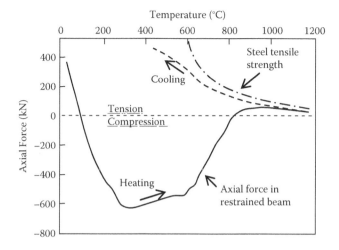

Figure 8.4 Axial force in steel downstand of axially restrained composite beam.

Observations of structural behaviour in natural fires (Wald et al. 2006) and furnace test programmes (Yu et al. 2008a, 2009a, 2009c, 2011) have shown steel joints to fail components such as bolts and end plates because of the high forces induced by the thermal and structural deformations of the connected members.

Cases of partial failure of the type illustrated in Figure 8.3 from different fire tests are pictured in Figures 8.5a and 8.5c. In no case did both sides of the end-plate fracture, indicating that the increased flexibility produced by the fracture of one side was enough to allow the remaining connection to perform in a ductile fashion. The increased flexibility after fracture allowed the tensile force in the cooling beam to be relaxed through deformation of the joint; the remaining connection performed in a ductile fashion and could still transmit the vertical reaction. An alternative behaviour under these conditions is bolt failure, which very often takes the form of thread stripping within the nuts, as shown in Figure 8.5d.

Fin plates were used at Cardington to connect secondary beams to their supporting primary beams. In several cases, it was observed (Figure 8.5b) that the bolts had fractured in shear at the interface between the fin plate and the beam web. This probably occurred as the secondary beam contracted during cooling, but in other cases might happen as it expands during the heating phase. Fin plates rely on steel in direct tension and shear, so they will always behave in a less-ductile fashion than a bending element such as a partial-depth end-plate. However, the rotational stiffness of any end-plate joint is increased considerably when the lower flange of the

Figure 8.5 Cardington and Coimbra joint fractures. (a) and (c) Single-sided fracture of partial-depth end-plate in cooling; (b) shear failure of bolts in fin plate in cooling; (d) thread stripping of nuts.

connected beam makes contact with the face of the column, with a corresponding reduction in rotation capacity.

8.2.1.2 Measured internal forces in joints

In January 2003, seven years after the main series of Cardington fire tests, a seventh full-scale fire test was carried out on the eight-storey steel-framed building with composite floors (Wald et al. 2006). The main purpose of this test was to collect evidence on the behaviour of typical beam-to-column and beam-to-beam connections subjected to a natural fire. The test was carried out in a compartment on the third floor enclosing a floor area 11 by 7 m, as shown in Figure 8.6a.

Heavy fire protection to the columns prevented excessive increase of temperature and allowed measurement of strain gradients, up to 60 min on the third floor and for the whole experiment on the fourth floor, using ambient-temperature strain gauges. The stresses in the columns were used to calculate bending moments. The shear forces in the columns were derived from the bending moments, and finally the horizontal forces transmitted through the beam-to-column joints were calculated. The calculated horizontal forces at various levels are shown during the progress of the fire in Figure 8.6b.

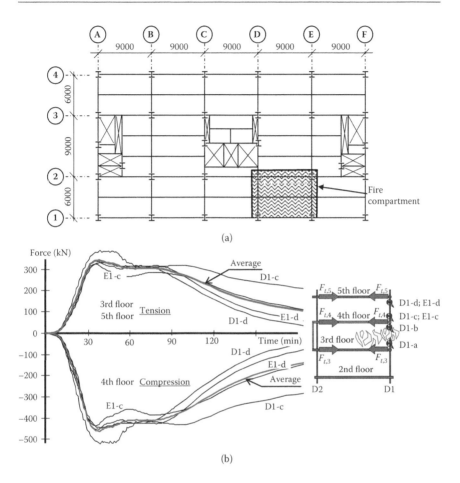

Figure 8.6 (a) Location of Cardington Test 7, on the third floor; (b) horizontal forces calculated in connections at different levels.

8.2.1.3 Variants of simple joints

Many types of simple joint have been used on steel-framed structures, although those discussed have tended to dominate normal fabrication in recent years. For optimal behaviour in fire, a simple joint needs

- High rotation capacity to cope with the large beam deflections experienced in fire;
- Relatively flexible behaviour in horizontal tension, so that catenary tension can be reduced by allowing the effective shortening caused by large deflection;

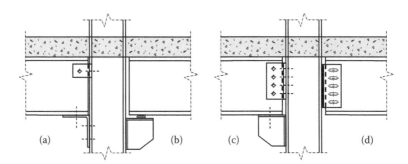

Figure 8.7 Alternative simple beam-column joints. (a) Seating/locating cleats; (b) sliding bearing; (c) seating bracket/web cleat; (d) slotted fin plate.

- Sufficient strength when distorted to resist the catenary tension; and
- Sufficient vertical shear resistance to carry the vertical load component appropriate to the fire limit state at the final steel temperature.

Some suggested details for beam-to-column joints are shown in Figure 8.7. The details shown are variants that may be useful in providing some of the requirements mentioned in particular circumstances. The slotted fin plate shown as type (d) and the seated web cleat (c) can be designed to give the flexibility in rotation and axial movement required to minimise catenary forces, together with the final strength to resist these forces. Type (b) provides high axial movement and rotation capacity but has no tying capacity other than that provided by the slab reinforcement.

8.2.2 Semirigid and rigid joints

Site-welded joints are not favoured in many parts of the world because of the difficulty of achieving the required accuracy and welding quality. Where semirigid or rigid joints are needed, it is usual to use a variant of the end-plate joint, shown in Figures 8.8a and 8.8b. However, flush end-plates may be specified by fabricators for convenience, although they have been designed for ambient-temperature strength as simple shear connections, assuming hinged support. Design that assumes full moment transfer at ambient temperature through rigid joints generally implies either extra shop fabrication and expense in stiffening extended end-plates or site welding. Little research has so far been done on the behaviour of such joints in fire, and it is doubtful whether they can maintain their rigidity in fire when local buckling is likely to occur at the beam end adjacent to the connection.

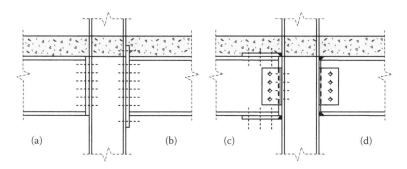

Figure 8.8 Semirigid and rigid joints. (a) Flush end-plate; (b) extended end plate; (c) shop-welded/site-bolted joint; (d) site-bolted/site-welded joint.

8.2.3 Beam-to-beam splice joints

Beams connecting to each other at right angles generally use a subset of the normal range of beam-to-column joint types. However, beams that connect in-line must use quite distinct details; some common types are shown in Figure 8.9.

Little research or testing has been done of such joints, but some logical predictions can be made. Welded moment splices should perform well at high temperatures because the effect of elevated temperature on the weld

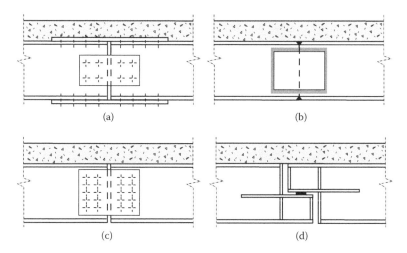

Figure 8.9 Beam-to-beam splice connections: (a) bolted moment splice; (b) welded moment splice; (c) bolted shear splice; (d) sliding support.

and heat-affected zone would not affect these adversely compared with the beam section itself. Bolted moment splices are designed to transmit substantial moments, but it is possible for bolts to lose their preload, because of either thermal expansion or differential softening. Although the connection should carry catenary tension well, it may experience a progressive failure of bolts in transmitting bending moment.

There is now evidence from the Federal Emergency Management Agency (FEMA 2002) that bolted shear splices can be vulnerable to catenary forces in fire if they are not designed for an appropriate tying capacity. Building 5 of the World Trade Center (WTC5) complex in New York, which was subject to a severe fire on 11 September 2001 in the aftermath of the collapse of the Twin Towers, was constructed using prefabricated "column trees" supporting suspended beam spans, as shown in Figure 8.10. The joint shown in Figure 8.11 was covered in fire protection but failed as a result of the effects of fire alone. The upper level of the building was not damaged and remained intact, while several levels below collapsed during the fire. It is probable that the collapse initiated at one level and the impact loads caused the levels below to fail in a progressive manner. The connection utilised a simple shear plate, designed to transfer the vertical shear force along the member. The eccentricity between the bolt rows generates a small moment, but otherwise the bolts and bolt-hole

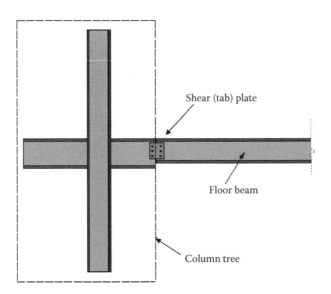

Figure 8.10 Column tree detail from WTC5. (From Federal Emergency Management Agency [FEMA], *World Trade Center Building Performance Study: Data Collection, Preliminary Observations and Recommendations*, FEMA Technical Report 403, Federal Emergency Management Agency, Washington, DC, USA, 2002.)

Figure 8.11 Failed tab plate and column tree edges from WTC5. (From Federal Emergency Management Agency [FEMA], *World Trade Center Building Performance Study: Data Collection, Preliminary Observations and Recommendations*, FEMA Technical Report 403, Federal Emergency Management Agency, Washington, DC, USA, 2002.)

positions are designed for vertical forces only, which explains the "nominal" edge distance between the bolt-holes and the end of the steel plate. Under the catenary forces generated at high deflection, combined with vertical shear, these edge distances are clearly inadequate, and block shear failure ensues.

8.3 CURRENT DESIGN

Joints in steel and composite structures have always had to be designed for the forces that are assumed to be transmitted between the connected members in the load cases considered. In the majority of cases, designers have found it convenient to make the assumption that joints at the ends of beams transmit only the transverse reactions to their supporting members. This has been considered to be a conservative assumption because the supported beams are designed to carry their loads without the benefits of full or partial rotational continuity at their ends, and any rotational stiffness that actually exists is thought to add an implicit extra safety factor that reduces both the sagging bending moments and the beam deflections. Where structural frames are analysed and designed as "continuous" (say, for earthquake resistance), it is generally assumed in analysis that complete rotational continuity exists at all connections, and these are designed to resist the moments transmitted in all the load cases considered; detailing of these connections reflects this requirement. Research conducted over several decades on the real behaviour of typical connection details under rotation has been reflected in the acceptance of "semirigid design", allowing the use of simplified versions of connection rotational stiffness and moment capacity to produce an intermediate solution that can optimise the cost of a structural frame, using much simpler connection details than those for full rigidity, but taking advantage of the control of beam bending moments and deflections that end stiffness gives.

The development of the "component method" as a way of representing the actions of a connection by modelling it as an assembly of uncoupled nonlinear springs has allowed semirigid design to be codified in EN 1993-1-8 (CEN 1993–2005c), covering the ambient-temperature design of steel and composite frames, for the limited range of connection types for which accepted mechanical models of the components exist. At present, this method is only codified to calculate rotational stiffnesses and moment capacities of connections at ambient temperature.

8.3.1 Steel-to-steel connections: Eurocode EN 1993-1-2: 2005 Annex D

EN 1993-1-2 (CEN 1993–2005b) has relatively little to say about joints, in contrast to the highly advanced treatment possible for joints at ambient temperature under its EN 1993-1-8 (CEN 1993–2005c). There is no provision given explicitly for their semirigid behaviour, although the relatively cool temperatures in joint components compared with those in the members they connect make the rotational stiffness of simple joints much more significant in fire than at ambient temperature. Annex D is an "informative" section that deals only with simplified connection temperature calculation and the reduced strength of bolts and welds at elevated temperatures. It does not allow any of the load deflection behaviour to be predicted.

8.3.1.1 Connection temperatures

EN 1993-1-2 provides simplified temperature calculation methods for joint zones. Three degrees of simplification are allowed. The temperature of a connection may be assessed by the incremental linearised methods provided for the temperatures of protected and unprotected members, presented in Chapter 4, using the local section factor (A/V) values of the parts forming the connection.

8.3.1.1.1 Uniform connection temperature

As a simplification of the method, a uniform connection temperature may be calculated using the maximum value of the section factors A/V of the connected steel members.

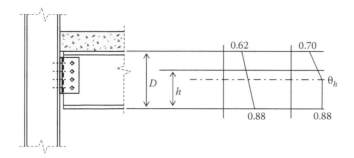

Figure 8.12 Thermal gradient within the depth of a composite connection.

8.3.1.1.2 Non-uniform connection temperature

For beam-column and beam-beam joints in which the beams support any type of concrete floor, the temperature for the connection may be assessed in terms of the temperature of the bottom beam flange at midspan. The beam flange temperature is calculated using the linearised incremental method via a spreadsheet. This is best applied using the section factor for the flange plate itself. When this is determined, the temperature distribution through the connection components is represented as either a linear or a bilinear gradient (Figure 8.12), depending on the depth of the beam:

If the depth of the beam is less than 400 mm

$$\theta_h = 0.88\theta_0[1-0.3(h/D)] \tag{8.1}$$

If the depth of the beam is greater than 400 mm:
When h is less than $D/2$, the temperature

$$\theta_h = 0.88\theta_0$$

When h is greater than $D/2$, the temperature

$$\theta_h = 0.88\theta_0\left[1+0.2(1-2h/D)\right]$$

where θ_h is the temperature at height h of the steel beam, θ_0 is the bottom flange temperature of the steel beam remote from the joint, h is the height of the component being considered above the bottom of the beam (mm), and D is the depth of the beam (mm).

8.3.1.2 Bolted connections

In terms of the member resistance at the connection, there is no requirement to calculate the net section strength in fire because the joint temperature is always lower than that of the member away from the joint. This means that the joint becomes stronger than the member in fire, for any of the normal loading conditions for which it is designed at ambient temperature, although both are weakened.

8.3.1.2.1 Bolts in shear

Bolts in shear may either be of the bearing type, in which the connected parts are assumed to be able to slip over one another without significant friction, or of the "friction grip" type, which use a specified minimum tension in the bolts to generate a frictional resistance.

In the fire case, it is assumed that the heating of friction grip bolts has effectively removed the contact pressure, so that they are assumed to have slipped, and they are treated in the same way as bearing bolts. Bolt strength in general is assumed to degrade with temperature (Figure 8.13) in the fashion given in Table 8.1.

In *single shear* (Figure 8.14), the strength of each bolt is calculated as

$$F_{v,t,Rd} = F_{v,Rd} k_{b,\theta} \frac{\gamma_{M2}}{\gamma_{M,fi}} \tag{8.2}$$

in which

$k_{b,\theta}$ is the strength reduction factor determined for the appropriate bolt temperature from Table 8.1;

γ_{M2} is the partial safety factor at normal temperature; its generic value is 1.25, but it may be changed in National Annexes;

Table 8.1 EC3 strength reduction factors for bolts in shear and tension

Temperature θ_b, °C	Reduction factor $k_{b,\theta}$
20	1.000
100	0.968
150	0.952
200	0.935
300	0.903
400	0.775
500	0.550
600	0.220
700	0.100
800	0.067
900	0.033
1000	0.000

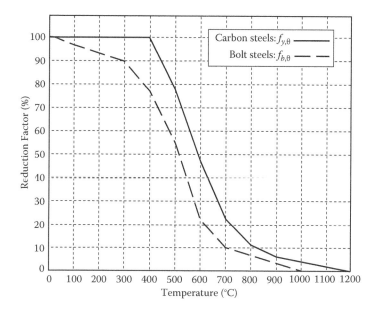

Figure 8.13 EC3 strength reduction for bolt steel, compared with structural carbon steel at high temperatures.

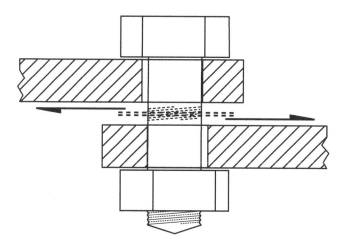

Figure 8.14 Single-shear resistance calculation.

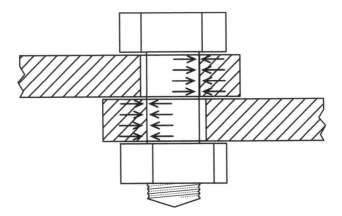

Figure 8.15 Bearing resistance calculation.

$\gamma_{M,fi}$ is the partial safety factor for fire conditions, which has the generic value of 1.0;

$F_{v,Rd}$ is the design shear resistance of the bolt per shear plane, assuming that the shear plane passes through the threads of the bolt, and is given by

$$F_{v,Rd} = \frac{\alpha_v f_{ub} A}{\gamma_{M2}} \qquad (8.3)$$

where $\alpha_v = 0.6$ for bolts of Classes 4.6, 5.6 and 8.8, and A is the gross cross-section area across the threads.

In *bearing* (Figure 8.15), the strength of each bolt is given by

$$F_{b,t,Rd} = F_{b,Rd} k_{b,\theta} \frac{\gamma_{M2}}{\gamma_{M,fi}} \qquad (8.4)$$

in which, apart from the factors defined, $F_{b,Rd}$ is the design bearing resistance of the bolt given by

$$F_{b,Rd} = \frac{k_1 \alpha_b f_u dt}{\gamma_{M2}} \qquad (8.5)$$

where α_b is the lowest of α_d, (f_{ub}/f_u) or 1.0.

Perpendicular to the direction of load transfer:

- For edge bolts, k_1 is the smaller of $(2.8e_2/d_0 - 1.7)$ and 2.5.
- For inner bolts, k_1 is the smaller of $(1.4p_2/d_0 - 1.7)$ and 2.5.

In slotted holes, where the longitudinal axis of the slotted hole is perpendicular to the direction of force transfer, the bearing strength is 0.6 times the bearing resistance of bolts in normal round holes.

8.3.1.2.2 Bolts in tension

The tensile strength of bolts at elevated temperatures is not usually important in calculating the reduced strength of a beam-column or beam-beam joint with respect to the actions for which it is designed at ambient temperature because bolts are rarely used in direct tension in such joints. However, when beams reach high deflections in fire, they lose most of their bending stiffness and strength and hang in catenary between the joints at their ends. At this stage, the tying strength of the joint is probably the key structural property that prevents the floor slabs from collapsing and thus allowing fire to spread vertically to higher storeys.

Both pretensioned and non-pretensioned bolts are treated in the same way. The design strength of a single bolt at elevated temperature is determined from

$$F_{ten,t,Rd} = F_{t,Rd} k_{b,\theta} \frac{\gamma_{M2}}{\gamma_{M,fi}} \tag{8.6}$$

where the ambient-temperature strength $F_{t,Rd}$ is determined from $F_{t,Rd} = \frac{k_2 f_{ub} A_s}{\gamma_{M2}}$, and $k_2 = 0.63$ for countersunk bolts or $k_2 = 0.9$ otherwise. The strength reduction factor $k_{b,\theta}$ for the appropriate bolt temperature is determined from Table 8.1.

8.3.1.3 Welded connections

8.3.1.3.1 Butt welds

The reduced strength of a full-penetration butt weld, for temperatures up to 700°C, is taken as equal to the strength of the weaker part joined, using the appropriate strength reduction factors for structural steel. For temperatures above 700°C, the reduction factors given in Table 8.2 for fillet welds can also be applied to butt welds.

8.3.1.3.2 Fillet welds

The reduced strength per unit length of a fillet weld at elevated temperature is determined from

$$F_{w,t,Rd} = F_{w,Rd} k_{w,\theta} \frac{\gamma_{M2}}{\gamma_{M,fi}} \tag{8.7}$$

Table 8.2 Reduction factors for welds

Temperature θ_b	Reduction factor $k_{w,\theta}$
20	1.000
100	1.000
150	1.000
200	1.000
300	1.000
400	0.876
500	0.627
600	0.378
700	0.130
800	0.074
900	0.018
1000	0.000

in which the strength reduction factor $k_{w,\theta}$ is obtained from Table 8.2 for the appropriate weld temperature. The ambient-temperature strength per unit length $F_{w,Rd}$ is

$$F_{w,Rd} = f_{vw,d}a \tag{8.8}$$

where $f_{vw,d}$ is the design shear strength of the weld, given by

$$f_{vw,d} = \frac{f_u/\sqrt{3}}{\beta_w\gamma_{M2}} \tag{8.9}$$

and f_u is the nominal ultimate tensile strength of the weaker part joined. The correlation factor β_w depends on the weaker steel grade in the parts joined (Table 8.3).

8.3.2 Composite Connections: EN 1994-1-2 cl. 5.4 (CEN 1994–2005b)

8.3.2.1 Connection types covered

Eurocode EN 1994-1-2 cl. 5.4 deals only with shear transfer in connections between composite beams and composite columns and considers only two

Table 8.3 Correlation factors

S235	$\beta_w = 0.8$
S275	$\beta_w = 0.85$
S355	$\beta_w = 0.9$
S420, S460	$\beta_w = 1.0$

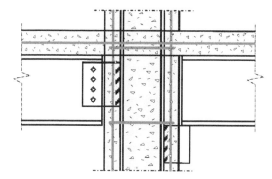

Figure 8.16 Connections to a concrete-encased column.

joint detail types. These are bearing blocks and fin plates (Figure 8.16), both welded to the column face. If bearing blocks are used, they should be detailed to guarantee that the beam cannot slip off these supports during the cooling phase of the fire.

In composite structures, it is very important to guarantee the required level of shear connection between the steel and concrete in the fire situation as well as at ambient temperature. Perhaps the most significant aspect of this part of EN 1994-1-2 is its theme of guaranteeing that the beam-end shear forces can be distributed properly into both the steel and the concrete cross sections of a composite column. This must be done using shear connectors, usually headed studs, or other appropriate detailing, if the connection detail attaches to relatively flexible parts of the steel section of the column, such as flange outstands or the walls of hollow sections. If this is not done, then it is quite easy for these parts to separate from the concrete so that there is no shear transfer between the steel and concrete. Alternatively, the steel and concrete parts must be able to fulfil the fire resistance requirements individually. Shear connectors should not be attached to the directly heated parts of the steel sections. In case of fully or partially encased sections, the concrete must be reinforced (if the concrete encasement has only an insulating function, then nominal steel reinforcement meshes should be sufficient), the concrete cover of the reinforcing bars should be greater than 20 mm and less than 50 mm to prevent spalling of the concrete during the fire.

Under EN 1993-1-2 (CEN 1993–2005b), beam-to-column joints should be designed and constructed to support the applied forces and moments for the same fire resistance time as the member transmitting the actions. For fire-protected members, it is suggested that one way of achieving this is to apply at least the same fire protection as that of the members transmitting the actions. It has been pointed out that this is very superficial and relates badly to the real behaviour of frames in fire.

In the case of a beam that is considered as simply supported for normal temperature design, EN 1994-1-2 suggests that a hogging moment can be developed at the support during the fire because of the beam deflections caused, provided that the concrete slab is reinforced adequately to guarantee its continuity and provided that there can be effective transmission of compression forces generated by restrained expansion through the steel connection. Such hogging moments may always be developed, it is suggested, in fire if the gap between the beam end and the column face is less than 10 mm or, in the case of beam spans over 5 m for fire resistance periods between 30 and 180 min, less than 15 mm.

In the United Kingdom, a fire-engineering calculation method based on the enhancement of the capacity in fire of composite beams whose joints are designed as simply supported at ambient temperature, because of the hogging moments generated at its joints, was published by the Steel Construction Institute (Lawson 1990b). This was widely distributed at the time, but observations of considerable local buckling of the lower flanges of the composite beams next to the connection zones in the Cardington full-scale tests in 1995–1996 cast considerable doubt on the safety of this method. This local buckling derives largely from the compressions caused by restrained thermal expansion. This document is no longer promoted. It is suggested that hogging moments at joints should not be used to enhance the fire resistance capacity of beams unless very specific detailing is used to prevent the possibility of local buckling of the lower beam flange.

8.3.2.2 Connection to concrete-encased column

The practical problem of site connection of steel downstand beams to the concrete-encased column dictates that the only simple connection methods are to use fin plates or bearing blocks that are prewelded to the column face and protrude a sufficient distance from the concrete encasement to support the beam (Figure 8.16).

8.3.2.3 Joints between partially encased beams and columns

The presence of concrete infill between the flanges of I- or H-sections again restricts the connection methods that can be used simply in design, but it is possible to use both types covered in EN 1994-1-2 (Figure 8.17). Additional shear studs should be provided in the vicinity of the connection if unprotected bearing blocks are used because the welds to the column face are exposed to the fire. The shear resistance of these studs should be checked assuming a stud temperature equal to the average temperature of the bearing block. Fin plates can transfer the shear directly into the web of the steel H-section, allowing it to distribute to the concrete, so no such provision is made.

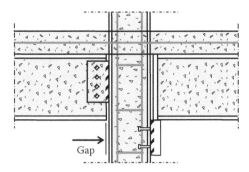

Gap

Figure 8.17 Connections to partially encased H-section column.

For fire resistance classes up to R120, these additional studs are not needed if

- the unprotected bearing block has a minimum thickness of at least 80 mm;
- it is continuously welded on all four sides to the column flange; and
- the upper weld, which is protected against direct radiation, has a thickness of at least 1.5 times the thickness of the surrounding welds and for ambient-temperature design supports at least 40% of the design shear force.

If fin plates are used, the clear gap between beam and column needs no additional protection if it is smaller than 10 mm.

8.3.2.4 Joints between composite beams and composite concrete-filled hollow-section columns

Again, composite beams may be connected to concrete-filled hollow-section columns using either bearing blocks or fin plates (Figure 8.18). The connection details need to be capable of transmitting shear and tension forces by adequate means from the beam to the reinforced concrete core of this composite column type. If bearing blocks are used, the shear load transfer in case of fire should be ensured by additional studs inside the hollow-section column, although these in general need to be inserted from outside the section through drilled holes before welding.

The shear resistance of studs should be checked with a stud temperature equal to the average temperature of the bearing block. It is specified that if fin plates are used, the best way of guaranteeing load transfer to both concrete and steel is for a single plate to penetrate the column and form both connections (Figure 8.19d). This should be welded to both walls of the hollow section.

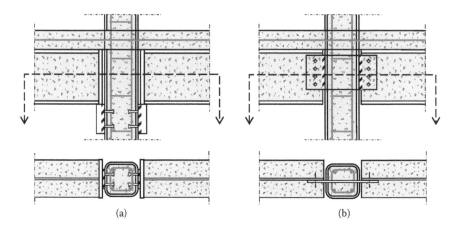

(a)   (b)

Figure 8.18 Connections between composite beams and concrete-filled hollow-section column. (a) Bearing blocks with additional studs; (b) penetrating fin plates.

8.4 DEVELOPMENT OF PERFORMANCE-BASED APPROACHES

As structural fire engineering becomes increasingly based on an understanding of the behaviour of whole structures and their individual parts in fire, the basis for this understanding is a combination of experimental and analytical research. In the case of joints, the motivation for this approach is twofold; on the one hand, designers wish to take advantage of the potential control of beam deflections that can be gained from real joints, and on the other it is necessary to ensure that joints are robust enough in fire to resist progressive structural collapse. At the time of writing, the necessary research into joint behaviour in fire is not sufficiently advanced for detailed design procedures to be developed. Therefore, what follows simply gives a picture of the current situation in research and an indication of the likely directions of this research in future.

8.4.1 Moment-rotation characteristics

Previous data on the real response of joints at elevated temperatures have been gathered from full-scale furnace tests (Lawson 1990a; Leston-Jones 1997; Al-Jabri 1999) on cruciform arrangements and finite element analyses by Liu (1996), which have concentrated exclusively on moment-rotation behaviour in the absence of axial thrusts. From the experimental studies, semiempirical rules were postulated by Al-Jabri et al. (2002), showing the progressive degradation of strength and rotational stiffness, which is illustrated in Figure 8.19.

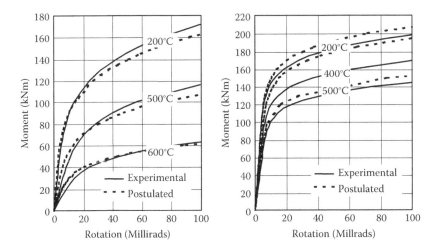

Figure 8.19 Moment-rotation curves for steel-to-steel and composite beam-column joints. (From Al-Jabri, K.S., Burgess, I.W., and Plank, R.J., Prediction of the Degradation of Connection Characteristics at Elevated Temperature, *Proceedings of 3rd European Conference on Steel Structures*, Coimbra, Portugal, pp. 1391–1400, 2002. With permission from Elsevier.)

However, the Cardington full-scale fire tests in particular brought out the fact that, in continuous frames, the behaviour in fire is controlled to a very similar extent by strength degradation at elevated temperatures, by temperature differentials across steel and composite cross sections and by restraint to thermal expansions. In particular, the effect of restraint to expansion means that design procedures based on isolated member behaviour are invalid. For joints, it means that rotational behaviour has to take account of the accompanying axial forces caused by restrained expansion.

8.4.2 Development of component-based methods

As stated, the behaviour of the joints within the overall frame response is greatly affected by the high axial forces created by restraint to the thermal expansion of unprotected beams. The rotational behaviour of a joint is certain to be affected considerably by these axial forces, given the very curvilinear nature of the stress-strain curves for steels, even if local buckling does not take place. Hence, moment-rotation-temperature properties of a joint are not adequate to express the way it will behave as part of a structural frame in fire. If moment-rotation-thrust surfaces were to be generated at different temperatures, this process would require prohibitive numbers of complex and expensive furnace tests for every single joint configuration. The alternative, and more practical, method is to extend the principles of

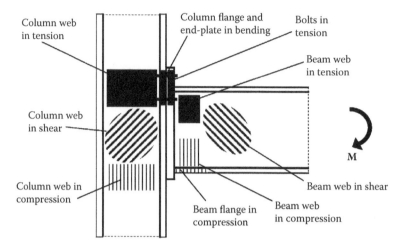

Figure 8.20 The three zones and components in an end-plate joint under moment.

the component method of joint analysis and design to the elevated-temperature situation. The basis of the component method is to consider any joint as an assembly of individual simple zones, each including several components, as shown in Figure 8.20. A steel joint under the action of a member end moment alone can be divided into the three principal zones: the tension, compression and shear zones.

Each of the components is simply a non-linear spring, possessing its own level of strength and stiffness in tension, compression or shear, and each will degrade as its temperature rises. Thus, combinations of moment and thrust are simply different combinations of the horizontal forces in each of these non-linear springs, as shown in Figure 8.21.

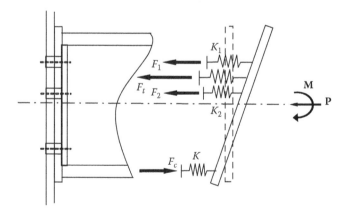

Figure 8.21 Illustration of the action of component springs under moment and thrust.

Research studies by Spyrou et al. (2004a, 2004b) at Sheffield and by Tan et al. (2004) in Singapore began to investigate experimentally and analytically the behaviour of tension and compression zones of end-plate joints at elevated temperatures. Simplified analytical models have been developed for the characteristics of some of the main components of flush and extended end-plates at elevated temperatures, and these have been validated against furnace tests and against detailed finite element simulations. Components that have been evaluated lie within

- the tension zone comprising the end-plate, top bolts and column flange, and
- the compression zone in the column web, both in the absence of high column axial force and when various levels of column axial force are present,

These are sufficient to test the method by using them to regenerate the high-temperature moment-rotation characteristics without axial thrust in the beam or column, which were measured in the earlier furnace tests on cruciform arrangements. The component method has proved very successful in such trials (Figure 8.22), but for practical application, more extensive development is required.

Clearly, each of these component zone studies is to some extent a generic study as well, and it can be anticipated that the models developed will be applicable to other zones with similar characteristics (e.g. the shear panel at the

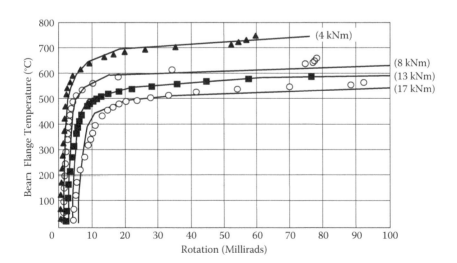

Figure 8.22 Comparison of component-based joint model against moment-rotation characteristics taken from cruciform tests.

beam end should give a model that can be extended to the column web shear panel) and to other joint types that employ some of the same components.

The simplified models have been shown to be very reliable for this common type of joint, although similar methods will need to be developed for other generic joint types. The principles of the component method can be used directly in either simplified or finite element modelling, without attempting to predict the overall joint behaviour in fire, to enable semirigid behaviour to be taken into account in the analytical fire-engineering design of steel-framed and composite buildings.

Significant problems currently need to be solved before component-based connection models can be routinely and reliably used in either simplified analysis or global thermostructural finite element modelling to enable connection response to be taken into account in the analytical fire engineering design of steel-framed and composite buildings. One of the more important aspects is the discontinuous nature of the behaviour of many connection components, which is due partly to elastic reverse straining of components deformed into their inelastic range, partly to the ability of some surfaces to break and regain contact. For example, in Figure 8.23 the top row spring marked 4 represents the stiffness of the column web in compression only; the two rows below this (springs marked 1, 2, 3) represent the stiffnesses of the bolts and end-plate at these levels, which only have an effect in tension.

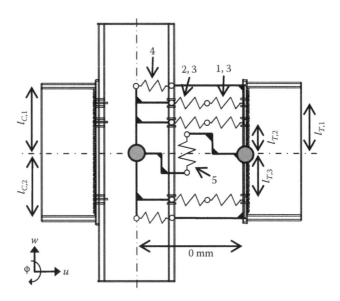

Figure 8.23 Component-based steel joint model including contact-dependent springs. (Courtesy of Block, F.M., Development of a Component-Based Finite Element for Steel Beam-to-Column Connections at Elevated Temperatures, PhD thesis, University of Sheffield, UK, 2006.)

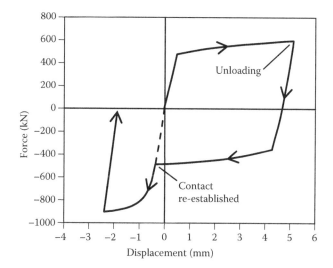

Figure 8.24 Effects of yield and contact on end plate in tying. (Courtesy of Block, F.M., Development of a Component-Based Finite Element for Steel Beam-to-Column Connections at Elevated Temperatures, PhD thesis, University of Sheffield, UK, 2006.).

Figure 8.24 illustrates the complexity of the behaviour that needs to be represented even with very simple connection loading. An end-plate joint is initially loaded to a tying (tension normal to the column face) force of 600 kN, then reverse loaded to a compression of 900 kN, and finally unloaded to zero, showing the rapid change in stiffness when surface contact is remade.

8.4.3 Ultimate survival of joints

It is usually implicitly assumed in fire-engineering design approaches that joints retain their structural integrity, yet evidence from the collapse of the World Trade Center buildings, especially Buildings 5 and 7, and full-scale tests at Cardington indicates that joints may be particularly vulnerable during both heating and cooling. If joint failure occurs, the assumed structural response may not be able to develop fully, and thereby safety levels may be compromised. Furthermore, it is important that the fire should be contained within its compartment of origin; the physical integrity of floors needs to be maintained, even at very high distortions. If joints fail, deformations locally are likely to be increased dramatically. Whilst there may be sufficient redundancy within the structural frame to sustain this, the concrete floor slab has very limited ductility and may not be able to accommodate such deformations without significant cracking, causing loss of compartmentation. Ultimately, joint failure can precipitate local failure of the structural floor system, which may in turn either overload lower floors, causing progressive failure, or may allow the supporting columns to buckle, leading to a much more extensive structural collapse, as appears to have occurred in Building 7 of the World Trade complex in New York. The forces transmitted from one connected member to another across a joint depend on the details of joint behaviour, and without a thorough understanding of this behaviour in fire, it is impossible to predict the whole structural performance accurately. Despite this, little research work has so far been undertaken on failure of steel joints in fire.

The principal structural effects that would normally be considered as "failure" at joints are fracture due to tension and shear and local buckling due to compression and shear. The latter is most likely to occur in parts of the structure that are restrained against thermal expansion. Local buckling of the lower beam flange adjacent to the joint does not in itself constitute a failure in the fire situation but is known to trigger shear buckling in the web as illustrated in Figure 8.2. The diagonal tension field action caused by this shear buckling has the potential to concentrate the shear and tying forces at the top part of the connection, especially when the beam is at high deflection and catenary tension is developing. This could trigger a progressive fracture of the connection from the top downwards, which is a genuine structural integrity failure. Depending on the design details, this could involve failure of the bolts, bolt-holes, welds, beam web or end plate. Even if no fracture occurs during the heating phase, the same progressive fracture can take place during cooling; this was observed after several of the Cardington fire tests. This joint fracture may lead to progressive structural collapse, and its avoidance is defined as "designing the building for robustness"; see Chapter 10 for more detailed discussions.

Having seen the effects of real joint failure, and consequent progressive collapse on a catastrophic scale, in both the Twin Towers and other buildings of the World Trade Center, the ultimate strength of joints in severe fires is clearly a subject that must be high on the research agenda, but at the time of writing, it is not possible to give designers more than the rather sparse guidance provided by the Eurocodes.

8.4.3.1 Structural testing of connections at high temperatures

The only work done so far on the structural testing of connections at high temperatures aspect of joint behaviour in fire has been done recently at the University of Sheffield (Yu et al. 2008a, 2009a–2009d, 2011; Hu et al. 2008, 2009). Steel joints of four different types (flush and partial end-plates, fin plates and web cleats) have been subjected to combinations of high rotation, shear and tying force at high temperatures using the arrangement shown

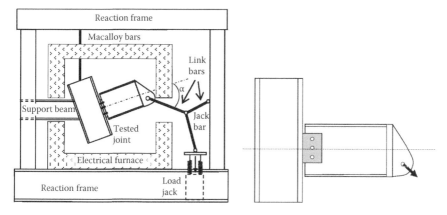

Figure 8.25 Joint robustness tests: (a) test setup; (b) fin-plate specimen.

in Figure 8.25a. The results (Yu et al. 2008a) for fin plates (Figure 8.25b) typically show low ductility, with sudden "brittle" failure (Figure 8.26) by sequential bolt shear at fairly low angles of rotation (Figure 8.27, about 6°), in connections for which ambient-temperature design is based on achieving ductile failure due to bearing.

A large number of high-temperature tests were performed on four different joint types. Fin plates were the least ductile of these; at the other extreme, web-cleat connections were the most ductile, which can be seen in the force-rotation curves shown in Figure 8.28. It can be seen from Figure 8.28 that the ductility of the angle cleats themselves in plastic bending provides a part of this connection type that can be detailed by designers to "tune" the deformation capacity; it can also be seen from the posttest photographs in Figure 8.29 that a number of different failure modes are possible.

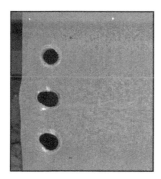

Figure 8.26 A 550°C fin plate test. (a) Deformed beam web holes; (b) sheared bolts.

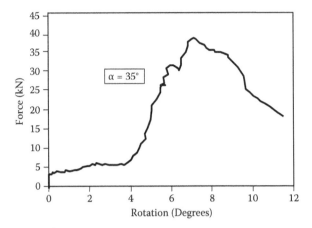

Figure 8.27 Force-rotation curve for fin plate with three Grade 8.8 bolts at 550°C.

8.4.3.2 Component models

To implement component-based connection models of the kind illustrated in Figures 8.20 and 8.22 in global analysis software, the basic requirement is that mathematical models must exist for each of the individual component springs. From the point of view of robustness in fire, the key point is that each component's ultimate strength and deformation capacity should be represented as well as possible. Whilst initial elastic stiffness is considered

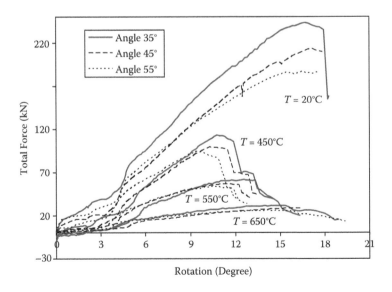

Figure 8.28 Force-rotation curves for web cleats at different temperatures.

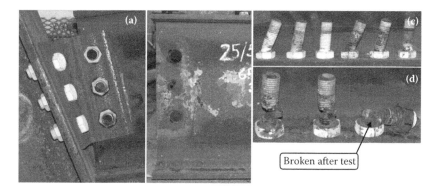

Figure 8.29 A 650°C web-cleat test. (a) Failed connection; (b) beam web deformation; (c) deformed bolts through beam web; (d) deformation of the six bolts to the column flange.

important in ambient-temperature semirigid design of steel-framed structures, in fire it is important that the members should remain connected throughout the design fire scenario. The objective of the series of elevated-temperature tests at Sheffield was to provide experimental data against which component-based connection models could be validated; detailed finite element models were also developed to extend the range of cases beyond the scope of the testing programme.

The characterisation of components as force-deformation-temperature constitutive relationships is still an active research topic. However, because flush end-plate connections have been studied at ambient temperature for several decades, their component modelling is much better established than it is for other details, so the models of various components developed by Yu et al. (2009b) are now summarised. It must be remembered that these components are not intended to form part of a manual design method, but to be assembled into component-based connection elements to be integrated into global numerical modelling.

Figure 8.30 shows a flush end-plate beam-column joint, together with a feasible component representation. The top and bottom springs only act in compression and represent the response of the column web when compressed by the very stiff zone created by the intersection of the beam web and flange with the end-plate. The rows of three springs centred on the bolt groups represent the response of the column web/flange, the bolts and the end plate; the bolt characteristics only come into play in tension. It can be seen that the two closely spaced upper bolt rows have been "grouped" into a single component row to take account of their obvious coupling; this is much more realistic than considering the bolt rows simply to be additive. The central vertical spring can be used to represent shear movement at the connection interface, but in many cases, it should be acceptable to give it an extremely high stiffness.

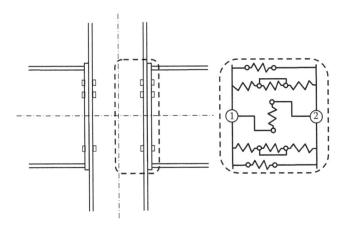

Figure 8.30 Component-based model of an end-plate joint.

The properties of the component springs in the aggregated upper bolt rows and the single lower row are based on a rationalisation of the plate bending behaviour (particularly using yield-line theory) against tee-stub models, as well as a model of bolt behaviour in tension. Both the initial yield and strain-hardening effects are included for both the steel plates and the bolts. Failure and postfailure behaviour of the bolt are defined in simplified terms. Infinite ductility is assumed for steel plate due to a lack of information about plate fracture.

A T-stub is shown in Figure 8.31. When pulled by the tee stalk, two plastic hinges can form on each side of the flange. The axial extension of the bolt is shown as a spring in the half model shown. The plastic work balance principle that the loss of potential of the external force is equal to the plastic work done by the structure gives

$$F\delta = E_{PH1} + E_{PH2} + E_{bolt} \tag{8.10}$$

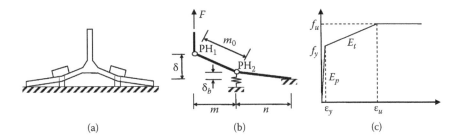

(a)   (b)   (c)

Figure 8.31 Deformation mechanisms possible in a tee-stub model for steel with a simplified trilinear stress-strain curve.

in which E_{PH1} and E_{PH2} represent the internal energies due to the rotation of the plastic hinges PH1 and PH2, respectively, and E_{bolt} is the internal energy generated in the bolt spring.

If the stress-strain curve of steel is simplified to the trilinear general form shown in Figure 8.31c, then the rotational energy generated at either of the hinge positions when the maximum bending stress in the cross section is in one of the three zones of the stress-strain curve is given by the following equations, derived in detail by Yu et al. (2009b). In these equations, ε_m is the maximum strain at the edges of the T-stub plate cross section, which has thickness t, and the length of a plastic hinge is given as $k(t/2)$.

$$E_{PH} = \frac{1}{6k} E\theta^2 \qquad \text{for } \varepsilon_m \leq \varepsilon_y \qquad (8.11)$$

$$E_{PH} = \frac{1}{6k} E_t\theta^2 + (E - E_t)\left[\frac{1}{2}\theta\varepsilon_y + \frac{1}{6}\frac{k^2\varepsilon_y^3}{\theta} - \frac{k}{2}\varepsilon_y^2 \right] \quad \text{for } \varepsilon_m \leq \varepsilon_u \qquad (8.12)$$

$$E_{PH} = \frac{1}{6} E_t k\varepsilon_u^2 + (E - E_t)\frac{1}{2}k\left[\varepsilon_y\varepsilon_u - \varepsilon_y^2 + \frac{1}{3}\frac{\varepsilon_y^3}{\varepsilon_u} \right] + \Omega\left(\frac{k}{\varepsilon_u} - \frac{k^2}{\theta} \right) + \frac{1}{2}f_u(\theta - k\varepsilon_u)$$

$$\text{for } \varepsilon_m > \varepsilon_u \qquad (8.13)$$

where

$$\Omega = (E - E_t)\left(\frac{1}{2}\varepsilon_y\varepsilon_u^2 - \frac{1}{6}\varepsilon_y^3 \right) + \frac{1}{3}E_t\varepsilon_u^3 - \frac{1}{2}f_u\varepsilon_u^2$$

is a parameter depending on the material properties.

If the bolt's force deflection characteristic is also represented by a three-phase linearised elastoplastic material property, with the elastic modulus, tangent modulus, yield strength and ultimate strength denoted by E_b, E_{bT}, f_{by}, and f_{bu}, respectively, then the internal work done by the bolt at any deformation δ_b can then be calculated as

$$E_{bolt} = \frac{1}{2}K_{ba}\delta_b^2 \qquad \text{for } \delta_b \leq d_e \qquad (8.14)$$

$$E_{bolt} = \frac{1}{2}F_{by}K_{ba} + \frac{1}{2}(F_{by} + K_{bT}(\delta_b - d_e)) \cdot (\delta_b - d_e) \quad \text{for } d_e < \delta_b \leq d_y \qquad (8.15)$$

$$E_{bolt} = \frac{1}{2}F_{by}K_{ba} + \frac{1}{2}(F_{by} + F_{bu}) \cdot (d_y - d_e) + F_{bu}(\delta_b - d_y) \quad \text{for } \delta_b > d_y \qquad (8.16)$$

At ambient temperature, bolts fracture abruptly, just beyond their peak resistance. However, at elevated temperatures, tests have shown that bolts

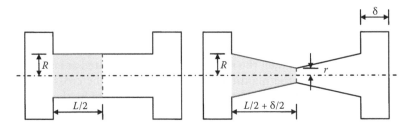

Figure 8.32 Simple model of bolt postpeak necking.

fail much more gradually, with a pronounced necking of the cross section accompanying progressive degradation of their resistance. Because of
these observations, a simple model has been developed to simulate the bolt
behaviour in the postpeak phase.

After a bolt has reached its ultimate strength, its necking is seen to develop
in the free zone between the head and the nut, as shown in Figure 8.32. It
is assumed that the bolt radius changes linearly from its original value R
at the ends of this free zone to a reduced value r at the middle at an axial
extension δ_b. The reduced bolt diameter is determined by assuming that the
volume of the free zone remains unchanged, so that

$$\frac{r}{R} = \frac{1}{2}\sqrt{\frac{12L}{L+\delta} - 3} - \frac{1}{2} \tag{8.17}$$

The reduced bolt resistance, at any extension from the beginning of this
phase, is then

$$F_{bolt} = F_{bu}\left(\frac{r}{R}\right)^2 \tag{8.18}$$

8.4.4 Temperature generation in connections and components

The influence of the accuracy of temperature prediction on the accuracy
of structural modelling is relatively high, as is shown in the comparison of
bolt strength reductions given by temperatures calculated based on different assumptions. Hence it is important that methods of calculating temperatures in joint components, and some degree of certainty about their degree
of conservativeness, should be generated if component-based connection
modelling is to be used in full-structure modelling. Test evidence has been
accumulating in recent years, both from laboratory-based furnace testing and from fire tests in real buildings. However, the only developments
in temperature prediction at this stage have come from a thermal study

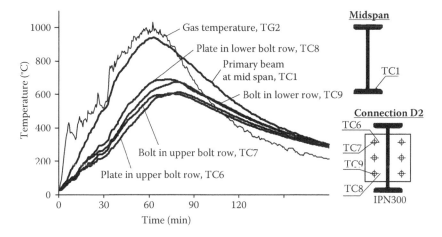

Figure 8.33 Temperatures measured across height of beam-to-column header plate joint in the Ostrava fire test.

carried out (Dai et al. 2007, 2009, 2010b) at the University of Manchester as a part of the joint research project with the University of Sheffield that generated the structural results shown in Section 8.4.3. This included both unprotected and intumescent-protected connections.

8.4.4.1 Experimental evidence

The measured temperatures of an end-plate connection, measured during the Ostrava fire test, performed in 2006 on a building prior to demolition (Wald et al. 2006) are shown in Figure 8.33.

The sensitivity of the prediction may be gauged from the reduction of the bolt tensile resistances based on different thermal assumptions, shown in Figure 8.34.

8.4.4.2 Eurocode-based method for temperatures of unprotected joints

A significant study of temperature development in joints was carried out by Dai et al. (2007, 2009, 2010b) at the University of Manchester. This included thermal testing of four different connection types heated in a gas furnace using the International Organisation for Standardisation (ISO) 834 standard atmosphere curve, rationalised against detailed finite element modelling with ABAQUS. In all cases, a length of beam and a concrete slab carried on the upper flange were included. A simplified method of

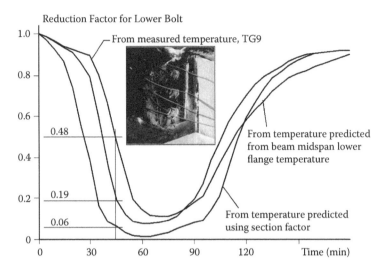

Figure 8.34 Reduction of the bottom-row bolt resistance of the header plate connection using different thermal models, compared to that given by the measured temperature.

predicting temperatures in unprotected connection components was suggested as a result of the study.

In observational terms, the main general conclusions from the testing of all connection types were unsurprising but are useful as a basis for a general view of the thermal conditions at joints:

- Joint component temperatures were all considerably lower than the concurrent beam temperatures remote from the connection zone, and connection temperatures increased with distance away from the connection zone.
- In the vertical sense, temperatures increased with distance away from the concrete slab, with the height of the connection zone being the most important aspect in determining the overall temperature difference. This is due mainly to exchange of radiation between the concrete slab soffit and the steel connection parts.

A major conclusion of the study was that, whether or not the vertical temperature distribution is taken into account, a single section factor is quite adequate to estimate the temperatures of components of a given type and detail of connection. The cross sections used to calculate these section factors are shown for some standard connection types in Figure 8.35.

If the radiation exchange between the slab and a certain location on the connection is considered important, so that the vertical temperature

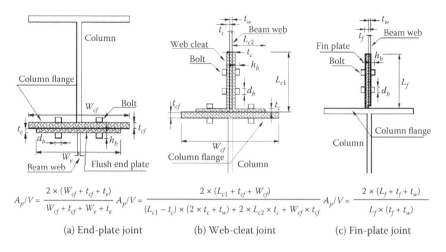

$$A_p/V = \frac{2 \times (W_{cf} + t_{cf} + t_e)}{W_{cf} + t_{cf} + W_e + t_e}$$

(a) End-plate joint

$$A_p/V = \frac{2 \times (L_{c1} + t_{cf} + W_{cf})}{(L_{c1} - t_c) \times (2 \times t_c + t_w) + 2 \times L_{c2} \times t_c + W_{cf} \times t_{cf}}$$

(b) Web-cleat joint

$$A_p/V = \frac{2 \times (L_f + t_f + t_w)}{L_f \times (t_f + t_w)}$$

(c) Fin-plate joint

Figure 8.35 Calculation of section factors for unprotected end-plate, web-cleat and fin-plate joints. (From Dai et al. 2010b.)

distribution is taken into account, the net incident heat flux at this location is given by

$$\dot{h}_{nst,r} = (1 - \Phi)\varepsilon_r\sigma\left[(\theta_r + 273)^4 - (\theta_m + 273)^4\right]$$
$$+ \Phi\varepsilon_r\sigma\left[(\theta_c + 273)^4 - (\theta_m + 273)^4\right] \quad (9.19)$$

in which θ_m is the local steel temperature, θ_r is the effective radiation temperature, and θ_c is the concrete surface temperature; ε_r is the resultant emissivity, and σ is the Stefan-Boltzmann constant; Φ is the configuration factor between the concrete surface and the point of interest in the connection component. Convective heat transfer is a relatively minor effect in the higher-temperature regime, but it is suggested that a convective heat transfer coefficient $\alpha = 10$ W/m²K would be more appropriate than the usual $\alpha = 25$ W/m²K, which is used for midspan zones of the beam, given the slower gas velocities in the connection zone.

The spread of temperatures between the top and bottom of the connections tested was generally fairly low, and a conservative simplified thermal analysis can be done in which the section factors defined in Figure 8.35 are used together with the usual EN 1993-1-2 incremental heating equations for unprotected sections:

$$\Delta\theta_{a,t} = K_{sh}\frac{A_m/V}{c_a\rho_a}\dot{h}_{nst}\Delta t \quad (8.20)$$

This can be used to produce a single temperature for all connection components. The "shadow factor" K_{sh} is set at 1.0 so that the equation becomes

$$\Delta\theta_{a,t} = \frac{A/V}{c_a\rho_a}\dot{h}_{nst}\Delta t \tag{8.21}$$

rather than the EN 1993-1-2 version:

$$\Delta\theta_{a,t} = 0.9\frac{A_{box}/V}{c_a\rho_a}\dot{h}_{nst}\Delta t \tag{8.22}$$

This change makes the method uniform for protected and unprotected members of all types and for joint components. It is necessary in using this version to change the surface emissivity of unprotected steel from 0.7 to 0.5 (which is consistent with prepublication versions of the Eurocode) and to use a fire emissivity of 1.0 so that the resultant emissivity used to calculate radiative flux is 0.5.

8.4.4.3 Extension to joints with inert passive protection

It could be expected, having established from tests and numerical thermal analysis that at least a conservative uniform connection temperature can be predicted using the EN 1993-1-2 incremental method, that the parallel incremental method for passively protected steelwork should be able to be applied using the same section factors. At present, this has not been tested using either profile protection using cementitious sprays or boxed protection using insulating board.

8.4.4.4 Simple analysis of joints with intumescent protection

The joint types that had been tested without protection were also tested by Dai et al. (2010b), using an intumescent coating on the connection and various lengths of the connected members. The popularity of intumescent protection in recent years has largely been based on the fact that reliable processes and coating products have been developed for off-site protection of all the steel sections used. Clearly, this does not include the bolts used for site erection, so a rather unsatisfactory situation exists where bolt heads and nuts have to be painted by hand after erection of the steel frame to obey the principle that all elements of a connection should be protected to the same fire resistance rating. In reality, the film thicknesses delivered by this retrospective process are impossible to guarantee, and it would be a great step forward if bolts could reliably be left unprotected. After testing joints where the exposed parts of bolts were left unprotected, although the connected members were coated with controlled thicknesses of intumescent paint,

Dai et al. (2010b) proposed a simplified design procedure that would allow this to happen in practice. The main general observations were as follows:

- If all the steelwork (excluding the bolts) in the joint assembly were protected, whether the bolts were protected or not had very little effect on temperatures in the protected steelwork, other than the bolts themselves.
- Bolt temperatures were higher when unprotected than when protected, but unprotected bolt temperatures in joints with protection to other steelwork were much lower than bolt temperatures in completely unprotected joints.
- Joint temperatures were unaffected by the protection of an attached beam provided that at least 400 mm of the beam were protected adjacent to the connection.
- If the column alone were protected, only joint components, such as welds, in the immediate vicinity of the column developed noticeable differences in temperature depending on whether the joint assembly was protected or unprotected.

The properties of intumescent coatings vary considerably between different products, depending on chemical composition and their design usage, so the extraction of effective thermal properties for the expanded char layer is always dependent on testing of a particular material on steel sections in standard fire conditions (cf. Section 5.3.3.6). If such testing is done and time-temperature curves are generated for the protected steelwork, then an effective thermal conductivity can be deduced by inverting the EN 1993-1-2 incremental equation for temperature development in profile-protected steelwork:

$$\lambda_{p,t}(t) = \left[d_p \times \frac{V}{A_p} \times c_a \rho_a \times \left(1 + \frac{\Phi}{3} \right) \times \frac{1}{(\theta_t - \theta_{a,t})\Delta t} \right]$$
$$\times \left[\Delta\theta_{a,t} + \left(e^{\Phi/10} - 1 \right) \Delta\theta_t \right] \tag{8.23}$$

The notation here is exactly as defined in Chapter 4, although the thickness d_p is that of the expanded char layer. The term $\phi = \frac{c_p \rho_p}{c_a \rho_a} \times d_p \times \frac{A_p}{V}$, which represents the heat storage in the protective coating itself, is relatively unimportant for intumescent chars.

The process is then as follows:

- Assuming that the temperature of the fire protection material is $\theta_p = (\theta_t + \theta_{a,t})/2$, transform the $\lambda_{p,t}$ (as a function of time) values to λ_p (as a function of temperature) values.

- Calculate a mean λ_{pm} using different λ_p values obtained from various beam or column sections in the same test.
- Use λ_{pm} and the Eurocode Equation (8.24) to calculate temperatures in the protected joint components.

$$\Delta\theta_{a,t} = \left[\frac{\lambda_{p,t} d_p}{c_a \rho_a} \times \frac{A_p}{V} \times \frac{1}{(1+\Phi/3)} \times (\theta_t - \theta_{a,t})\Delta t \right] - \left[\left(e^{\frac{\Phi}{10}} - 1 \right)\Delta\theta_t \right] \qquad (8.24)$$

In calculating component temperatures, the section factors defined in Figure 8.35 for unprotected connections are again used. Effective conductivity values at fire temperatures below about 350°C appear very high and variable, as compared with reasonably uniform values above this temperature. Obviously, the chemical action in expanding the coating takes place during (towards the end of) this initial heating phase, and the final steel temperatures achieved are relatively insensitive to it.

The effectiveness of the proposed method can be gauged from the standard fire test results on four different, completely protected, joint types shown in Figure 8.36.

The predicted temperatures in these graphs assume a protection layer density ρ_p of 1300 kg/m^3 and specific c_p heat of 1000 J/kg K, although it has been pointed out that the parameter ρ_p has very little effect on the temperatures calculated using the EN 1993-1-2 equation. Although the comparisons are based on a relatively small selection of tests and a single intumescent coating product, the correlation seems impressive and without any notable variation. Towards the end of the 60-minute period, the predicted temperatures are in all cases at the high end of the thermocouple readings, which indicates that the method should be conservative if used predictively in design analysis.

If bolts are left unprotected, they can clearly be heated directly by exposure to the fire at their unprotected surfaces. To account for the effect of unprotected bolt surface in an otherwise-protected joint, it is possible to use an "exposure factor" F_e. This exposure factor expresses the unprotected surface area of the bolt heads/nuts as a proportion of the total surface area of the joint assembly. Thus, F_e ranges from 0 for a joint assembly with full protection to 1 for a totally unprotected assembly. It can be assessed in simplified (two-dimensional, 2-D) or more complex (three-dimensional, 3-D) terms. The 2-D, simplified, approach assumes that a bolt head extends across the whole width of the connected plates, so that the exposure factor can be defined simply on the basis of a sectional view through the joint, as shown in Figure 8.37.

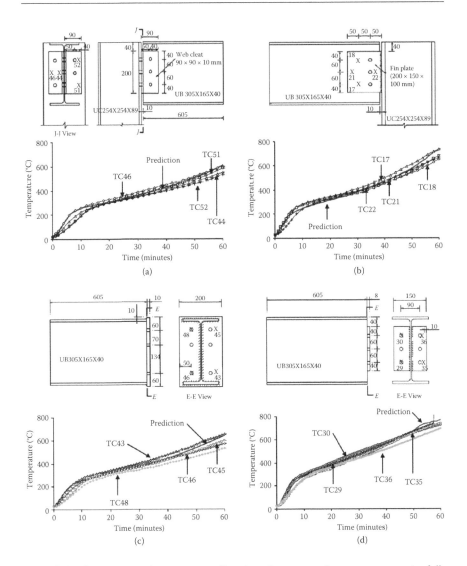

Figure 8.36 Comparison between predicted and measured temperatures in fully protected joint components of different types: (a) web cleat (Test SP1); (b) fin plate (Test SP4); (c) flush end plate (Test SP7); (d) flexible end plate (Test SP9).

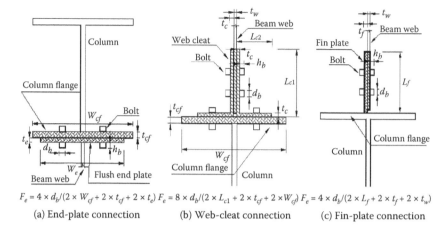

$F_e = 4 \times d_b/(2 \times W_{cf} + 2 \times t_{cf} + 2 \times t_e)$ $F_e = 8 \times d_b/(2 \times L_{c1} + 2 \times t_{cf} + 2 \times W_{cf})$ $F_e = 4 \times d_b/(2 \times L_f + 2 \times t_f + 2 \times t_w)$

(a) End-plate connection   (b) Web-cleat connection   (c) Fin-plate connection

Figure 8.37 Calculation of 2-D (simplified) exposure factors for partially protected end-plate, web-cleat and fin-plate joints where only bolts are unprotected.

The use of the exposure factor is based on consideration of the heat flux per unit area of the unprotected and protected steel surfaces. The aggregate heat flux to the steel section is

$$\dot{h}_{agg} = \dot{h}_{up} F_s + \dot{h}_p (1 - F_s) \tag{8.25}$$

Where $\dot{h}_{up}$ and $\dot{h}_p$ are the heat fluxes to unprotected and protected steel surfaces, respectively. If the concurrent temperatures of the connection completely unprotected and fully protected are θ_{up} and θ_{prot}, respectively, the temperatures of the unprotected bolts may be calculated as

$$\theta_{pp} = \theta_{up} F_s + \theta_p (1 - F_s) \tag{8.26}$$

It must be remembered that, in a protected joint with unprotected bolts, Equation (8.26) only applies to the bolt temperatures. The protected connection component temperature is calculated assuming full protection. When calculating the fully protected temperature θ_p and the completely unprotected temperature θ_{up}, the section factors may be assessed as illustrated in Figure 8.35.

If it is considered necessary to make a more accurate representation of the quantity and dimensions of the bolts and nuts in the joint assembly, it is necessary to use a 3-D representation of the joint instead of the 2-D approximation described. Figure 8.38 shows the dimensions used

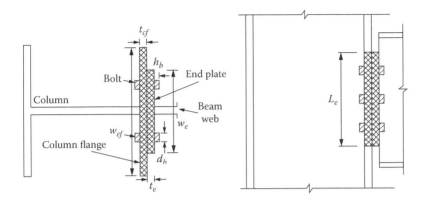

Figure 8.38 Dimensions for calculation of 3-D exposure factors for partially protected end-plate joints in which only the bolts are unprotected.

for calculating the section factor and exposure factor, as an example, for end-plate joints.

The section factor for end-plate joints is then given by

$$\frac{A_p}{V} = \frac{(2W_{cf} + t_{cf} + t_s)L_s + A_{bs}n_b}{W_{cf}t_{cf}L_s + W_s t_s L_s + V_{bo}n_b} \qquad (8.27)$$

The exposure factor that goes together with this section factor is

$$F_s = \frac{A_{bs}n_b}{\left[(2W_{cf} + t_{cf} + t_s)L_s + A_{bs}n_b\right]} \qquad (8.28)$$

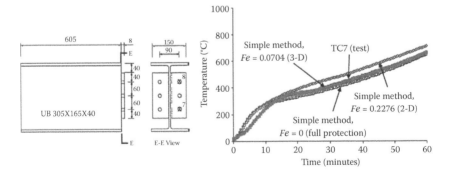

Figure 8.39 Comparison between temperatures of unprotected bolts in an intumescent-protected flexible end-plate connection. Test compared with 2-D and 3-D predictions.

in which n_b is the number of bolts in the joint; A_{bs} is the exposed side surface area of a bolt (approximated as $2\pi d_b h_b$); A_{bs} is the exposed end area of
the two ends of the bolt (approximated as $0.5\pi d_b^2$); and V_{bo} is the volume of
the exposed ends of the bolt and nut (approximated as $A_{bs} h_b$).

The 2-D and 3-D methods of assessing temperatures of exposed bolts in
otherwise totally intumescent-protected connections are compared against
one another and against test results in Figure 8.39 for a flexible (partial-
depth) end-plate connection.

It can be seen from Figure 8.39 that the 3-D method is remarkably accurate compared with test temperatures, while the 2-D method has the advantage of being conservative and therefore secure for design analyses.

Chapter 9

Integrity of compartmentation

9.1 INTRODUCTION

The concept of fire resistance for any structural element that separates one designated fire compartment from another requires that three criteria be satisfied for the required fire resistance period: structural stability (or resistance), insulation and integrity. Integrity in the fire safety context is generally described as the ability of the element to resist the penetration of flames and hot gases so that fire does not spread beyond its compartment of origin. The issue of integrity is generally treated as one of ensuring fire stopping in considering the details of a building and in selecting components and systems that do not provide voids in the separating structure as a fire develops. In testing fire compartmentation systems, the criterion for integrity failure is a very empirical one; a cotton pad held in any position on the unheated side of the component must not ignite during the course of a Standard Fire, continued until its required fire resistance time. Integrity failure can happen because of oversights in the architectural design of a building or very often because of voids specifically created for the passage of building services. In refurbishment operations, it is important to fire-stop gaps created by attaching new cladding to existing structure and where holes are cut in walls and floors for service conduits.

Since the year 2000, different versions of a structural fire-engineering design calculation method for floor slabs in tensile membrane action have emerged; the failure criterion is based on the creation of through-depth cracks in the concrete slab that would allow flames to pass into the storey above. The basic version of this method is therefore presented in this chapter, in the current context, of the integrity of fire compartments.

9.2 ISSUES AFFECTING INTERNAL WALLS

Walls that divide a structurally open-plan floor into fire compartments must be designed to accommodate expected structural movements in a fire without collapse. For walls that conform to the column grid lines, movements of beams, even if unprotected, may be small, and the normal allowance for deflection should be adequate. The prevailing advice is that insulation requirements must be met, and the beams must have protection for 30 or 60 min, with all voids and service penetrations fire-stopped. Beams protected with intumescent coatings require additional insulation because the temperature on the non-fire side is likely to exceed the limits required in BS 476 (British Standards Institute [BSI] 1990a). For walls off grid lines, the deflection of the slabs above these walls may be large, so it is preferable to position fire compartment walls on the grid whenever possible. The deflection allowance may be accommodated by including a sliding, or telescopic, upper portion to the wall. Where potential deflection of the slab could be large, a deformable upper part, using fire-resistant blanket or curtain, may be required, as illustrated schematically in Figure 9.1, or the beam should be protected. It is recommended in SCI (Steel Construction Institute) guidance (Newman et al. 2006) that a deflection allowance of *span*/30 should be provided in walls crossing the middle half of an unprotected beam. For walls crossing the end quarters of the beam, this may be reduced linearly to zero at the beam ends.

A steel beam running above a fire compartment wall is considered part of the wall and is required to have an identical separating function. A steel beam without penetrations thus has integrity, but any service penetrations must be fire-stopped, and any voids above downstand steel sections in composite beams should also be fire-stopped. An unprotected beam in the plane of a compartment wall may not have sufficient insulation and normally requires additional fire protection.

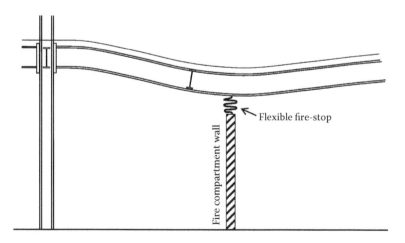

Figure 9.1 Requirement for fire stopping above compartment walls.

It is recommended that compartment walls should be located under beams whenever possible. However, to comply with the insulation criterion for separating elements specified in BS 476, these beams should also normally be protected to provide an additional level of safety in maintaining the integrity of the compartment wall.

9.3 DESIGN OF SLABS FOR INTEGRITY IN FIRE

The remainder of this chapter concerns an important design philosophy and an associated design method that is mainly concerned with establishing the integrity of composite slabs forming the ceilings to fire compartments. It should be understood that it predicts a breach of integrity because its failure criterion is based on a membrane-tension failure of the concrete slab after it has established a yield-line bending failure "mechanism". This tensile failure causes a through-depth crack that is capable of allowing the fire to reignite in the storey above the original fire location. In general, this large crack does not cause even a local structural collapse to happen because the fold lines on the slab do not form a mechanism for large deflection, so additional concrete and reinforcement fractures must happen before a structural resistance failure occurs. To set this method in its proper context, it is really necessary to consider how continuous slab systems, such as composite floors, carry and transmit their loads during a fire scenario in which temperatures may rise at rates and to temperatures that are determined by the conditions of fire load and ventilation. Hence, the temperatures of protected and unprotected steel sections can vary considerably, in both their absolute and relative values, according to the fire conditions. A further result is that the temperature distribution through the concrete slab can vary widely; a short duration but hot fire causes very high bottom-surface temperatures with an extremely cool top surface, whereas a long but cooler fire can produce much more uniform temperatures through the slab.

In real buildings, structural elements form part of a continuous assembly, and building fires may remain localised, with the fire-affected structure receiving significant restraint from cooler areas surrounding it. Even with whole-floor fires, the compatibility conditions between adjacent bays of the structure can significantly restrain the deformations and stresses produced. The real behaviour of these connected structural elements can therefore be very different from that indicated by standard furnace tests. This is not necessarily the case; it is possible to design structural frames that act very much in the fashion of the idealised assumptions normally made in structural design, such as statically determinate support conditions for both beams and one-way-spanning slabs. However, most modern multistorey buildings employ some form of composite construction, at least between the slabs and beams that form their flooring systems. In such systems, the slabs are usually treated in design as spanning one way between adjacent secondary beams but are actually continuous and are made composite with both primary and secondary beams. Practical beam-column and beam-beam joints, which may be treated as perfect hinges in the ambient-temperature design process, usually have a residual stiffness that becomes significant when the connected beams have lost much of their own strength and stiffness at high temperature. Also, a simple steel-work joint at the end of a beam member carrying a composite slab may act together with continuing slab reinforcement at a considerable lever arm above the connection centroid, creating a much stiffer overall joint. Hence, restraint to thermally induced movements may exist in respect of both translations and rotations at the ends of beams and the edges of slabs. There is therefore considerable scope for load sharing and for both advantageous and disadvantageous effects of restraint to rotation and horizontal movement.

9.3.1 Observed behaviour of composite floors: Cardington

The full-scale fire tests on an eight-storey building at Cardington during 1995–1996 (British Steel 1999) made it clear that unprotected steel members forming part of composite floors can have significantly greater fire resistance within real multistorey buildings than when they are tested as isolated members. This appeared to be due to interaction between the heated members within the fire compartment, the concrete floor slabs and the adjacent steel frame structure. The most significant qualitative observation was that in none of the six fire tests conducted was there any indication of runaway failure, which is the eventual outcome of all isolated member tests under International Organisation for Standardisation (ISO) 834 standard fire test conditions. This seemed particularly remarkable since, in some cases, unprotected steel beam temperatures reached over 1100°C, when the steel strength had reduced by well over 95%. Maximum deflections always exceeded *span*/30, and in some cases exceeded the usual test limit of *span*/20. It is worth examining one of the Cardington tests, at least in sufficient detail to reveal some of the structural effects that must be considered in an integrated approach to structural fire engineering. A typical floor plan is shown in Figure 9.2, showing the locations of one of the tests, known as the British Steel Corner-Bay Test. This test was carried out in July 1995 on a compartment 9.98 m wide by 7.57 m deep. The walls of the fire compartment were constructed using lightweight concrete blockwork, the top of which was detached from the slab soffit to allow free deflection of the structure above, with the gap being filled by a flexible insulating blanket. All columns and perimeter beams were wrapped with ceramic

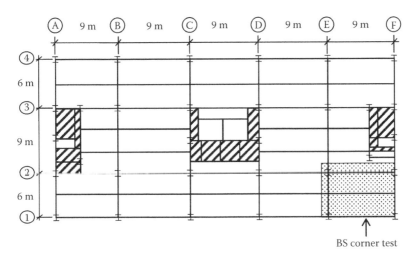

Figure 9.2 Location of the Cardington British Steel corner-bay test and main structural members.

fibre insulation, but all other structural elements were left unprotected. The test was fired using timber cribs, giving an overall fire load of 45 kg/m² to produce a natural fire. During the fire test, the maximum recorded atmosphere temperature in the compartment was 1028°C. The test was numerically modelled by Huang et al. (2002) using both geometrically linear and non-linear slab elements.

The deflection profile of the composite slab at 900°C, as given by the geometrically non-linear slab elements, is shown in Figure 9.3a, including the cracking patterns of the top concrete layer, and the slab membrane tractions (the aggregate principal membrane force vectors) at a steel temperature of 900°C are shown in Figure 9.3b. In this figure, the force vectors are shown at nine points within each slab element, and their magnitudes are proportional to the line lengths. The compressive forces are shown as the dark, thick lines. Tensile forces are shown as grey and are generally of lower magnitude; it should be understood that tension is mainly carried by a small area of anticrack reinforcement. This shows clearly that, in the slab above the fire compartment at this stage, there is effectively a hydrostatic tension field in the central area of the slab, with a ring of compression around the perimeter.

9.3.2 Ways in which steel and composite frames carry loads during a fire

There are clearly only two basic material properties that affect the structural behaviour of steelwork during a building fire. These are the progressive

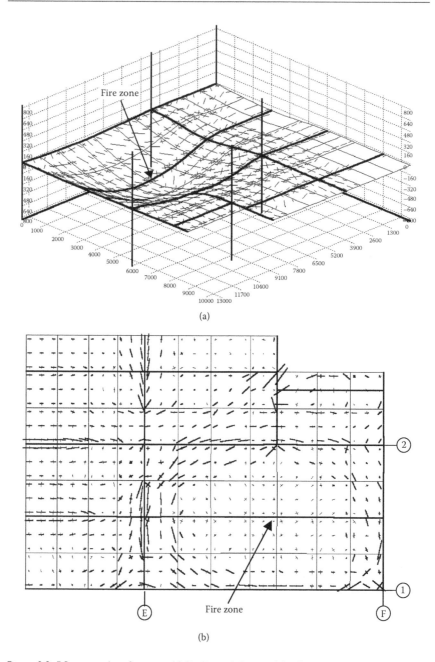

(a)

(b)

Figure 9.3 BS corner-bay fire test: (a) Deflected shape of the fire compartment; (b) principal membrane force vectors. (From Huang, Z., Burgess, I.W., and Plank, R.J., Modelling of Six Full-Scale Fire Tests on a Composite Building, *The Structural Engineer*, 80(19), pp. 30–37, 2002.)

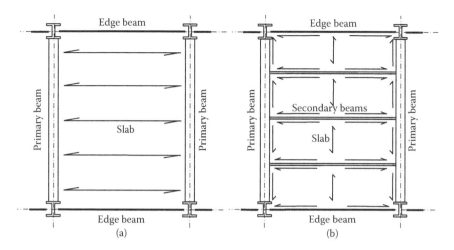

Figure 9.4 Typical load path at ambient temperature: (a) non-composite floor-framing system; (b) composite floor-framing system.

degradation of the stress-strain curves of the steel as its temperature increases and the thermally induced strains that accompany temperature changes. These two basic changes in the steel at any part of the building then interact with the ways in which it is connected to other materials, other parts of the building and its support system to produce structural behaviour that can be very different from that at ambient temperature. The behaviour of multistorey frames in which the floor slabs simply impose load on steel beams (Figure 9.4a) may be quite different from that of composite frames (Figure 9.4b) in which the slabs and steel beams are continuously connected.

This section most directly concerns buildings of composite construction because free-standing slabs need only to be designed for local strength in fire and do not change the structural action of their supporting beams. Both types of framework obviously depend to a large extent on the continued integrity of the joints between members, but the changes in joint forces are much more marked when there is the degree of continuity between heated and cool areas that is afforded by composite flooring systems.

It is clear that a composite floor system such as that shown in Figure 9.4b typically passes through different phases of structural behaviour as temperatures increase in a compartment heated by an intense and long-lived fire.

- Initially, the exposed unprotected steel beams heat rapidly and expand, with little reduction in strength.
- The concrete slab heats more slowly, causing thermal bowing towards the heat source.

- Progressive reduction in steel strength and stiffness then causes very high permanent compressive strains in the steel beams.
- Restraint to thermal expansion, caused by the cool structure surrounding the fire compartment, further increases this compressive straining.

As the steel's strength reduces, so the concrete slab to which it is attached plays an increasingly important role in supporting the floor loads. The characteristics of the slab, together with the way it is supported, will now control the way in which it carries its loads:

- The slab's residual flexural strength may at this stage be great enough for it effectively to carry the load unaided at relatively low deflections, especially if it has been designed with structural reinforcement rather than a nominal anticracking mesh, or if it is thicker than normal (say to control noise propagation in residential buildings).
- If the slab is well supported against vertical deflection along lines that divide it into reasonably square areas, for example by the primary and secondary beams on the column grid lines, then tensile membrane action (Figure 9.5a) can be generated as a load-carrying mechanism. The slab is then forced into double curvature and hangs as a tensile membrane in its middle regions, while a peripheral compressive "ring beam" is generated either around its supported periphery or in its edge beams, as was observed in Figure 9.3b. This forms a self-equilibrating mechanism that supports the slab loading. The particular usefulness of this mechanism is that it requires no horizontal restraint at the edges of each slab panel to maintain the tensile membrane action, provided that vertical support is provided at these edges.
- If temperatures continue to increase, this may eventually end in a real structural collapse if the edge support fails or may simply lead to a tensile slab fracture, either at the edges or within the middle region, that is extensive enough to constitute a compartment integrity failure. The normal failure case for a freely supported individual slab is a single through-depth tensile crack across the central short span, which is the location of the highest tensile membrane forces. Tensile cracking may alternatively occur in continuous slabs along the central parts of their perimeter supporting beams (Figure 9.6a) because the effective lengthening of lines across the slab in its highly deflected central zones, compared with those at or near the support beams, causes a membrane tension in the slab at its edges. This is enhanced by the high hogging curvatures that occur across the supported edges away from the corners. These slab tensile cracking failures do not in themselves lead to collapse but do lead to loss of integrity by allowing paths for the fire to spread vertically.

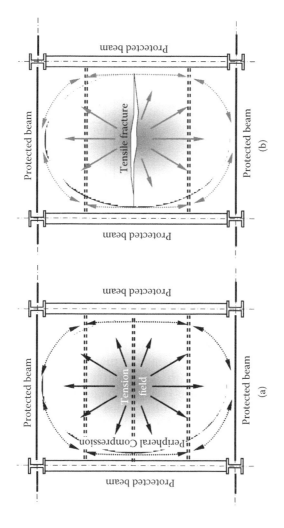

Figure 9.5 Typical composite floor-framing system: (a) tensile membrane action at high deflection; (b) tensile fracture across the shorter span.

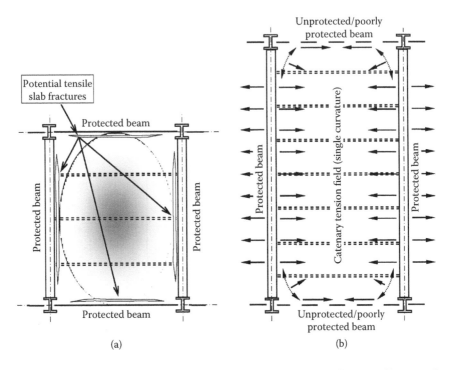

Figure 9.6 (a) Slab tension cracking at support beams at high deflection; (b) uniaxial forces due to catenary action in single curvature.

- If, however, the slab's support is such that it is effectively one-way spanning, which also includes situations where the supported edges form a long, rather than a square, rectangle, then it hangs mainly in single curvature from its longer supported edges. This is *catenary action* (Figure 9.6b), which is distinguished from tensile membrane action by the fact that it is not self-equilibrating, but needs to be anchored in the horizontal sense at the supporting edges of the slab.

In this case, failure may take place due to the tension in the slab or may occur in any of the sequence of structural components that are affected by this tension force. The joints between the composite beams spanning in the direction of the catenary tension and their supporting columns may now fail due to lack of tying capacity at elevated temperature. The failures due to catenary action may therefore be structural and lead to progressive collapse, rather than purely loss of integrity of fire compartments.

9.3.3 Whole-frame fire-engineering design strategies

It has been shown in Section 9.3.2 that composite floor systems may carry loads considerably greater than those at which their individual steel or composite beam members fail by runaway deflection in fire conditions. This may be done by employing tensile membrane action, for which all the edges of slab areas of reasonably square aspect ratio are supported vertically, forcing the deflections of the slab to employ predominantly double curvature so that the flat slab forms a shallow shell with much higher stiffness. For slabs whose support allows near-single-curvature bending to occur, enhancements to strength above that of the flat slab must be due to catenary action, which occurs either when the slab aspect ratio is high or when only some of its edges are supported vertically. In the latter case, horizontal catenary tension forces have to be resisted physically by adjacent structure around the edges of the slab areas. As a self-equilibrating load-bearing mechanism that needs only vertical perimeter support, the use of tensile membrane action to enhance the fire resistance of floors is therefore in general the more attractive prospect for designers.

9.3.4 Background work on slabs at high deflections

During the late 1960s and early 1970s, significant experimental and theoretical research work by Park (1964a, 1964b), Sawczuk and Winnicki (1965), Hayes (1968), Hayes and Taylor (1969) and Brotchie and Holley (1971) was conducted on the behaviour of thin concrete floor slabs when subjected to large vertical displacements. This work showed that concrete slabs at large vertical displacements could support loads considerably greater than those calculated using the well-established yield line approach. The mechanism for supporting the load was shown to be tensile membrane action, which could form within the slab irrespective of whether it was restrained or unrestrained horizontally at its boundaries. For a slab that was unrestrained around its boundaries, compressive in-plane membrane forces were shown to form within the depth of the slab around its perimeter, and these provided the required support to the in-plane tensile membrane forces in the central region of the slab (Figure 9.7). The supporting ring of compression force can only occur if the vertical displacements around the perimeter remain small under increasing load.

Although this early work provided an insight into the behaviour of slabs at large displacements, and some tentative design methods were proposed, no practical use for the research was found, and the work was not developed significantly. However, following the fire tests at Cardington, in which composite floor slabs were used, interest in the behaviour of concrete slabs at large displacements was revived. Numerical studies were done by

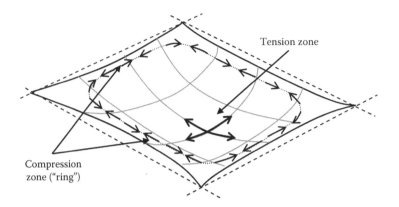

Tension zone

Compression
zone ("ring")

Figure 9.7 Membrane action in a horizontally unrestrained concrete slab.

Huang et al. (2001a, 2001b) on the behaviour, in local and whole-storey fires, of generic frames employing composite floor systems based on a square grid of columns at 9-m spacings with parallel secondary beams 3 m apart. Different patterns of protected beams on the column grid lines were studied, with all other beams unprotected. Placing well-protected beams on the column grid lines is intended to give strong horizontal support to the columns and to provide the requisite vertical edge support to a slab panel within the composite floor so that tensile membrane action can take place. Considerable enhancements to the fire resistance of floor slabs were shown to be possible with sufficient protection on grid line beams, even if other steel beams were unprotected.

9.3.5 Simplified design method based on tensile membrane action

During a fire, large displacements of a structure are acceptable provided that the fire is contained within the compartment of origin so that the risk of fire spread throughout the building is low. A simplified design method was developed initially by Bailey (2000, 2001) at the Building Research Establishment (BRE) in the United Kingdom, based on the simplified models of membrane action of composite floor slabs at large displacements developed in the 1960s, but also subjected to elevated temperatures. By using membrane action, it is possible to design composite floors in which a large proportion of the steel beams within a given floor plate can be left unprotected for a defined fire resistance period. Previously, considering only flexural action, all the supporting steel beams within a given floor plate have generally required some form of passive fire protection to achieve the required fire resistance. The method has been slightly adapted by Clifton et al. (2001) at the Heavy Engineering Research Association (HERA) in New Zealand, but this adaptation is not dealt with here.

The method embodies a basic philosophy that divides the floor plate of a building into square or rectangular slab panels, which are surrounded by protected beams (Figure 9.8). Even in terms of the small-deflection flexural capacity of the slabs, it is easy to see that this arrangement induces the slabs to create yield line failure patterns that lead to the creation of larger-deflection membrane stresses.

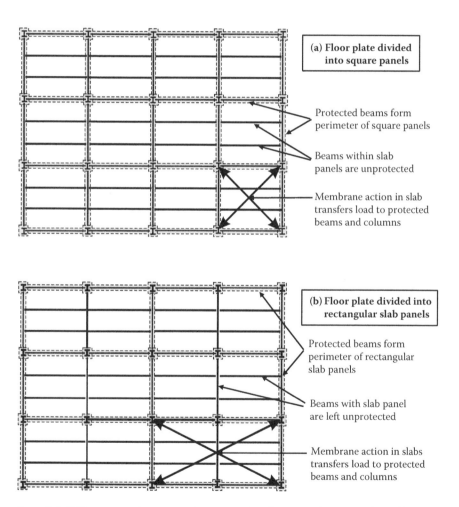

(a) **Floor plate divided into square panels**

Protected beams form perimeter of square panels

Beams within slab panels are unprotected

Membrane action in slab transfers load to protected beams and columns

(b) **Floor plate divided into rectangular slab panels**

Protected beams form perimeter of rectangular slab panels

Beams with slab panel are left unprotected

Membrane action in slabs transfers load to protected beams and columns

Figure 9.8 Different passive protection strategies, leaving a large number of beams unprotected.

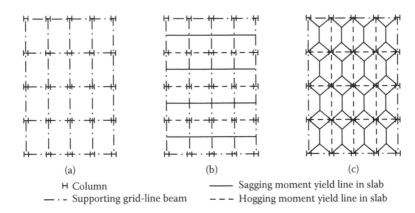

        (a)   (b)   (c)

н Column   ——— Sagging moment yield line in slab
— · — Supporting grid-line beam   – – – Hogging moment yield line in slab

Figure 9.9 Yield line small-deflection hinge patterns that are possible in a composite floor plate.

In a floor plate of the arrangement shown in Figure 9.9a, it is possible for the yield line patterns shown in Figures 9.9b and 9.9c to occur, depending on whether there is adequate support provided by the long-direction grid line beams. In Figure 9.9b, the parallel "concertina" folds do not induce any action other than bending, while in Figure 9.9c the yield line patterns cannot develop without membrane stresses being created.

As seen in the previous section, the method works best when the protected (or otherwise vertically supported) areas are of a reasonably square aspect ratio, so that tensile membrane action, rather than catenary action, is achieved. The secondary beams within each of these panels are left unprotected. During a fire, the unprotected beams lose a significant proportion of their strength and stiffness, which, coupled with thermal effects, results in large vertical displacements of the composite slab. The applied load is then mainly supported by membrane action in the composite slab, effectively transferring the load from the unprotected beams to the protected beams.

The proportion of unprotected beams within a given floor plate will be governed by the size and aspect ratio of the panels and controlled by their membrane capacity. A major assumption that has to be made is that the slab reinforcement over the protected beams is assumed to fracture during the fire due to the combination of high hogging moment and the membrane force that occurs in this area. This is a conservative assumption since if the reinforcement at the edges is assumed to remain intact, the induced tensile membrane forces in the slab, and thus its load-carrying capacity, will be significantly higher. In the design philosophy, the slab panels are assumed to be unrestrained horizontally, so that the peripheral compressive tractions alone provide support to the tension field in the central region.

Although the original version of the design method is based on slabs with isotropic reinforcement, and there is a shortage of published information on slabs with orthotropic reinforcement, the method has since been extended by Bailey (2003) to include the possibility that reinforcement quantities in the orthogonal directions may be unequal. The method considers the behaviour of composite floor slabs and their supporting grillage of steel beams at elevated temperature. Its treatment of the membrane action of the composite slab is based on the observed action of concrete slabs at both ambient and elevated temperatures.

The derivation of the design method to predict the load displacement response of concrete slabs in tensile membrane action has been published for isotropically and orthotropically reinforced slabs by Bailey (2000a, 2003). It is based on the observation from previously published test results (Hayes and Taylor 1969) that a full-depth tension crack eventually forms across the shorter span of the slab. The derivation is not repeated here, but the main equations are reproduced for the isotropically reinforced case; the notation used is mainly defined in Figure 9.10. For ease of presentation, the design method defines the strength displacement relationship in terms of an enhancement factor multiplying the slab's yield line failure load.

This enhancement factor, which increases with the vertical displacement, is given by

$$e = e_1 - \frac{e_1 - e_2}{1 + 2\mu \, a^2} \tag{9.1}$$

in which

$$e_1 = e_{1m} + e_{1b}$$
$$e_2 = e_{2m} + e_{2b} \tag{9.2}$$

where e_{1m} and e_{2m} are enhancement factors on the yield line load due to membrane forces in Elements 1 and 2 (Figure 9.10), respectively. The enhancements e_{1m} and e_{2m} are given by

$$e_{1m} = \frac{M_{1m}}{\mu \, M_0 L} = \frac{4b}{3 + (g_0)_1} \left(\frac{w}{d_1} \right) \left((1 - 2n) + \frac{n\,(3k + 2) - nk^3}{3(1 + k)^2} \right) \tag{9.3}$$

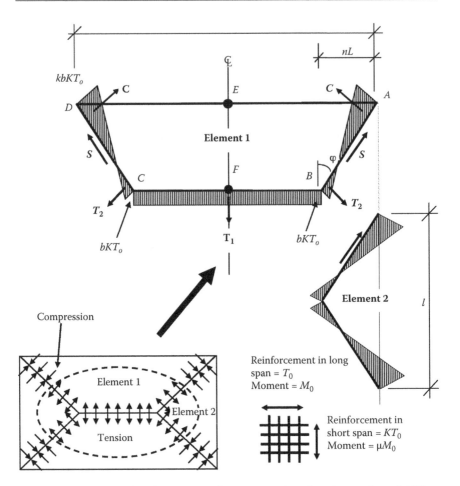

Figure 9.10 Definition of orthotropic reinforcement and in-plane forces along yield lines for a rectangular slab.

and

$$e_{2m} = \frac{M_{2m}}{M_0 l} = \frac{4bK}{3+(g_0)_2}\left(\frac{w}{d_2}\right)\left(\frac{2+3k-k^3}{6(1+k)^2}\right) \tag{9.4}$$

Here, M_{1m} and M_{2m} represent the moments about the support due to membrane forces, which, for a maximum vertical displacement w, are given by

$$M_{1m} = KT_0 L\, b\, w\left((1-2n)+\frac{n\,(3k+2)-n\,k^3}{3\,(1+k)^2}\right) \tag{9.5}$$

and

$$M_{2m} = KT_0 \; lbw \left(\frac{2 + 3k - k^3}{6\,(1+k)^2} \right)$$
(9.6)

where e_{1b} and e_{2b} are enhancement factors on the yield line load due to the increase in bending resistance caused by the membrane forces in Elements 1 and 2 (Figure 9.10), respectively. The enhancement factors e_{1b} and e_{2b} are given by

$$e_{1b} = \frac{M}{\mu \, M_0 L} = 2n \left[1 + \frac{\alpha_1 b}{2} (k-1) - \frac{\beta_1 b^2}{3} (k^2 - k + 1) \right] + (1-2n)(1 - \alpha_1 b - \beta_1 b^2)$$
(9.7)

and

$$e_{2b} = \frac{M}{M_0 l} = 1 + \frac{\alpha_2 bK}{2}(k-1) - \frac{\beta_2 b^2 K^2}{3}(k^2 - k + 1)$$
(9.8)

where

$$\alpha_2 = \frac{2(g_0)_2}{3 + (g_0)_2}$$
(9.9)

$$\beta_2 = \frac{1 - (g_0)_2}{3 + (g_0)_2}$$
(9.10)

The constants k and b are given as

$$k = \frac{4na^2(1-2n)}{4n^2a^2 + 1} + 1$$
(9.11)

and

$$b = \frac{1.1\,l^2}{8K(A + B + C - D)}$$
(9.12)

where

$$A = \frac{1}{2}\left(\frac{1}{1+k}\right)\left[\frac{l^2}{8n} - \frac{\left(\frac{L}{2}-nL\right)}{nL}\left((nL)^2 + \frac{l^2}{4}\right) - \frac{1}{3}\left(\frac{1}{1+k}\right)\left((nL)^2 + \frac{l^2}{4}\right)\right]$$

$$B = \frac{1}{2}\left(\frac{k^2}{1+k}\right)\left[\frac{nL^2}{2} - \frac{k}{3(1+k)}\left((nL)^2 + \frac{l^2}{4}\right)\right]$$

$$C = \frac{l^2}{16n}\,(k-1)$$

$$D = \left(\frac{L}{2}-nL\right)\left(\frac{L}{4}-\frac{nL}{2}\right)$$

(9.13)

9.3.5.1 Simplified design method: practicalities

The enhancement of yield line capacity due to tensile membrane action given in Equation (9.1) can be very easily set up on a standard spreadsheet, given the requisite data for the slab and the protection strategy adopted. The optimum small-deflection yield line capacity for an orthotropically reinforced slab, simply supported at its edges, is given (Park 1964a) by

$$p = \frac{24m_x\mu_s}{l^2\left[\sqrt{3\mu_s + \left(\frac{l}{L}\right)^2} - \frac{l}{L}\right]^2}$$

(9.14)

which, for the usual case of isotropic reinforcement (composite slabs very often contain only an anticracking mesh, with no additional bars), may be reduced to Wood's (1961) equation:

$$p = \frac{24m}{l^2\left[\sqrt{3 + \left(\frac{l}{L}\right)^2} - \frac{l}{L}\right]^2}$$

(9.15)

This applies to slabs without additional secondary beams attached within the unsupported area, which obviously does not represent the case for normal composite slabs cast on ribbed metal decking. However, the critical state considered in fire resistance calculations is when the unprotected

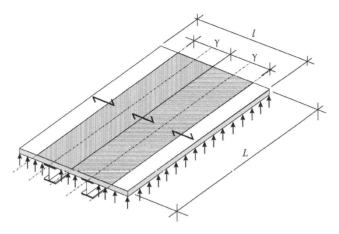

Figure 9.11 A composite slab showing the loaded areas assumed to be supported by
         unprotected beams.

steelwork is at very high temperature and can effectively be ignored as part
of the slab. It should be noted that, in the New Zealand version of the
method, which is generally identical to the original, the reduced strength of
the steel downstand beams is included in the slab's moment capacity in the
relevant direction but is not used in any other way.

In the BRE-Bailey method, the total load is assumed to be shared, at
the final temperature appropriate to the desired fire resistance period,
between any composite beams within the supported slab, acting in flexure and strictly in accordance with BS 5950 strength reduction factors
at elevated temperature and the slab's enhanced capacity due to tensile
membrane action. In the case of the composite beams, to ensure conservatism the beam with the highest load ratio is used to determine the
strength reduction of all the composite beams included. The loading
on each beam is apportioned according to the floor area it "supports",
as shown in Figure 9.11, a typical assumption taken from ambient-
temperature design.

The necessary contribution to the total load bearing in fire from the slab,
with enhancement from membrane action, is then calculated by subtract-
ing this composite beam grillage capacity from the total load carried by
the slab in the fire limit state (that is, according to the partial safety factors
used in fire). The process for calculating whether the slab is capable of providing this extra load capacity is as follows:

1. The small-deflection load capacity of the composite slab acting in
   flexure is calculated based on the lower-bound yield line mechanism,
   assuming that the beams have zero resistance.

2. The necessary enhancement factor is calculated by dividing this into the required contribution from the slab to total load capacity.
3. The slab deflection needed to provide this enhancement factor is calculated and checked for acceptability.

9.3.5.2 Contribution of the composite beams

Considering the simple example shown in Figure 9.11, consisting of a rectangular slab with two unprotected composite beams, the part of the slab load carried by the beams ($w_{beam\theta}$) is based on the lower-bound mechanism of one beam (the plastic hinge being at midspan), given by

$$w_{beam,\theta} = \frac{8M_{fi}}{\gamma L^2} \tag{9.16}$$

where γ defines the width of slab supported by one beam, which is taken as the secondary beam spacing, in this case $l/3$. This is not consistent with the yield line mechanism assumed for the slab but is conservative. Future research will reveal whether this conservatism is excessive or is justified by the real load sharing within the composite slab system.

The composite beams may, of course, be capable of carrying the whole slab load in fire, so that no contribution from tensile membrane action of the slab is required. This is checked initially by using a slightly modified form of the Eurocode 4 Part 1.2 (Committee of European Normalization [CEN] 1994–2005b) critical temperature method. Using the Eurocode symbols, the load level is calculated as

$$\eta_{fi,t} = \frac{f_{a\max,\theta cr}}{f_{ay,20°C}} \tag{9.17}$$

Table 3.2 from EN 1994-1-2 (CEN 1994–2005b) gives critical temperatures in terms of the load level η_{fi}. This is reproduced here as Table 9.1.

Only if the critical temperature for the secondary beams is below their design temperature at the desired fire resistance period is it necessary to use the tensile membrane action enhancement of the slab's load capacity.

If it is necessary to use the slab's membrane capacity, then it is necessary to calculate how much of the load is carried by the composite beams. The ambient-temperature plastic moment capacity is first calculated normally, by balancing the tension force in the steel section against the effective-width concrete compression block (Figure 9.12), ignoring the steel decking, and then multiplying by the lever arm between the centroids of these blocks.

Table 9.1 Critical temperatures for unprotected steel composite beams supporting a floor slab

Load level	0.7	0.6	0.5	0.4	0.3	0.2	0.1	0.08	0.06	0.04	0.02	0.01
Critical temperature, °C	526	558	590	629	671	725	820	860	900	1000	1100	1150

Although, strictly speaking, the same full calculation should be done again at elevated temperature, it is sufficient, and conservative, simply to reduce the plastic moment capacity in accordance with the appropriate strength reduction factor for steel. This can be done by interpolation from Table 9.1, using the design temperature of the steel section at the required fire resistance time.

9.3.5.3 Contribution of slabs in membrane action

As a preliminary to calculating the tensile membrane action enhancement, Wood's Equation (9.15) for small-deflection slab flexural load capacity using the lower-bound yield line mechanism gives the value of pl^2/m in terms of the slab aspect ratio (Table 9.2).

The isotropic moment capacity of the slab m is dependent on the position of the reinforcement, the ambient-temperature strengths of the materials used and the temperature distribution through the cross section.

It now remains to assess whether tensile membrane action can be used to provide an adequate increase in the strength of the composite slab. The enhancement factor relative to the yield line load capacity can be calculated using Equations (9.1) to (9.13), if acceptable slab deflections are known; a method of assessing these is given in the next section. Alternatively, design charts of the type shown in Figure 9.13 can be used. Each of these charts applies to a constant value of g_0, between 0.3 and 0.9, based on the proportion of the slab thickness that is used as a compressive layer to balance the net tension force in the steel downstand section, in pure flexure. It is acceptable to interpolate between different charts.

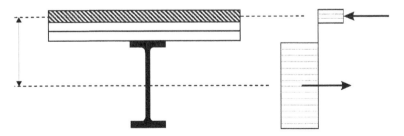

Figure 9.12 Schematic diagram showing plastic moment capacity assumptions for a composite beam.

Table 9.2 Yield line failure loads for slabs with isotropic reinforcement

$L/l =$	1.0	1.2	1.4	1.6	1.8	2.0	2.5	3.0	5.0	10	∞
$pl^2/m =$	24.0	20.3	17.9	16.2	15.0	14.1	12.6	11.7	10.1	9.0	8

9.3.5.4 Limiting slab deflection in membrane action

The equations and charts given in previous sections to represent the enhancement of slab yield line capacity due to tensile membrane action, provided that the materials used are infinitely ductile, imply that greater load capacity simply requires greater deflection. This neglects the fact that reinforcement will fracture eventually, and at this point the full-depth crack observed in tests will form, and the slab's integrity will be lost. If the slab's longer span is assumed to deflect as a parabolic curve and the end edges do not pull in, then the total strain in the long-span reinforcement is approximately

$$\varepsilon = \frac{8 v^2}{3L^2} \qquad (9.18)$$

This equation assumes that the extension of the reinforcement is averaged along the length of the slab as a constant strain, whereas in reality the strain is concentrated at crack locations. The strain in the reinforcement will increase significantly once the crack forms in the concrete, resulting in fracture of the reinforcement. Predicting the strain levels at which the crack

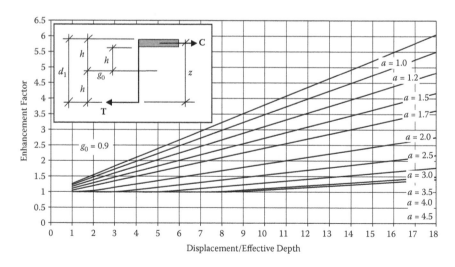

Figure 9.13 Enhancement factors for isotropically reinforced slab; example for $g_0 = 0.9$.

forms is complex, so a pragmatic approach is adopted in which a limit is defined for the average strain in the reinforcement, based on a maximum reinforcement stress of $0.5f_y$. This leads to a maximum allowable displacement of

$$v_{mech} = \sqrt{\left(\frac{0.5f_y}{E}\right)_{Reinf't} \frac{3L^2}{8}} \qquad (9.19)$$

A limiting deflection of $l/30$ is applied to this equation.

Equation (9.19) considers only mechanical straining, whereas thermal expansion due to the temperature of the reinforcement at any stage actually relieves its stress level.

Differential thermal expansions within the depth of the slab also cause it to deflect into double curvature. This can assist the membrane action of slabs since vertical displacements can be increased without an increase in mechanical strain. Restraint to thermal expansion can also cause the grillage of composite beams to buckle in fire, which will also increase the vertical displacement without any significant increase in mechanical strain. However, it is difficult to define how much restraint to expansion exists due to adjacent structure, so the effect of thermal buckling is conservatively ignored. To include the effects of thermal curvature, the temperature distribution through the slab is assumed to be linear, allowing the corresponding thermal displacement in one-way curvature to be estimated as

$$v_{therm} = \frac{\alpha(T_2 - T_1)l^2}{\psi\, 8h} \qquad (9.20)$$

in which

v_{therm} = thermally induced vertical displacement
α = coefficient of thermal expansion
T_2 = bottom temperature
T_1 = top temperature
h = depth of slab
l = length of shorter span of the slab
ψ = calibration factor = 2.4

The calibration factor ψ has been calculated by carrying out a comparison with the displacements obtained in the Cardington tests. This is based on a temperature difference of $T_2 - T_1 = 770°C$ for all the tests.

The maximum allowable deflection is given (Bailey and Moore 2000) by aggregating the mechanical and thermal effects, although strictly speaking the individual parts are calculated base on assumptions that are inconsistent in their treatment of the slab's boundary conditions:

$$v = \frac{\alpha(T_2 - T_1)l^2}{19.2h} + \sqrt{\left(\frac{0.5f_y}{E}\right)_{\text{Reinf}\,t_{20°C}} \frac{3L^2}{8}}$$ (9.21)

but

$$v = \frac{\alpha(T_2 - T_1)l^2}{19.2h} + l/30$$ (9.22)

Further research is still needed to obtain a better estimate of the vertical displacement at which tensile failure of the slab across the short midspan occurs. In its present form, the limit criterion often results in the familiar furnace test (BSI 1990a) lower-bound limiting deflection of *span*/30 eventually being applied in practical design cases. This is useful in indicating to building control departments and fire service authorities that, although a more complex structural interaction is being used here than in the familiar element-based design, limiting deflections are used that have been considered acceptable for some time in fire situations. It is also worth noting that in a normal composite-beam fire test the beam is close to runaway failure at a deflection of *span*/30, whereas in tensile membrane action the floor will remain structurally stable considerably beyond this deflection. Even in terms of observed behaviour, however, there is no real logic in conflating a deflection criterion used to indicate imminent runaway of simply supported beams with one for pure tensile fracture of lightly reinforced concrete slabs.

9.3.5.5 The process for design of composite floors using the BRE-Bailey method

The design sequence within the basic BRE-Bailey method is shown in Figure 9.14. This is quite applicable for most practical cases. The method has been developed subsequently (Bailey 2003) to allow for orthotropic reinforcement, where the mesh has different areas per unit length in each direction. For cases where the reinforcement area across the predicated short-span tension crack is high, the small areas of concrete at each side of the slab across the short span (whose force has to balance the net tensile force in the central zone to maintain equilibrium) may fail in crushing before the tensile crack can form. Bailey and Toh (2007a, 2007b) extended the basic method for such cases.

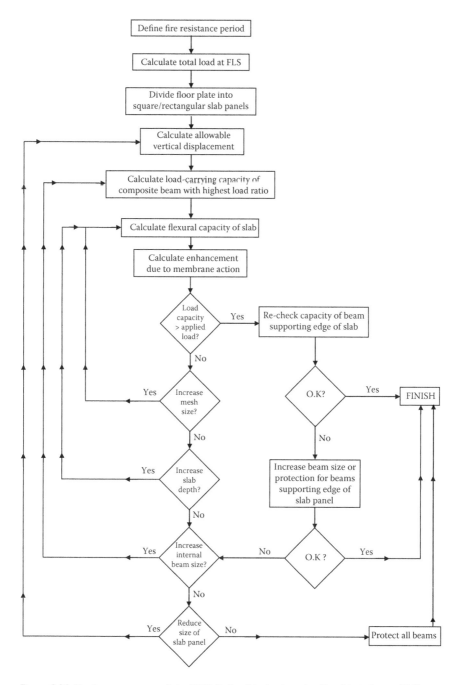

Figure 9.14 Design sequence of the BRE-Bailey Method at the Fire Limit State (FLS).

9.3.5.6 Further development of the BRE membrane action method: TSLAB

In addition to the design tables that are reproduced in full in the original design guide document P-288 (Newman et al. 2000) and cover the method as described, the SCI developed a Microsoft Excel-based spreadsheet called TSLAB, launched with the second edition of P-288 (Newman et al. 2006), which developed its ideas further. This tool determines whether the reinforcement preselected for particular slab panel geometries will be satisfactory and includes all the advances that have been incorporated into the method recently. TSLAB begins by performing one- (1-D) and 2-D thermal analyses on the unprotected intermediate beam and the composite slab. Whereas the basic method limits slab deflections using the assumption of nominal slab top and bottom temperatures based on the Cardington fire tests, the vertical deflection limit in TSLAB is calculated by using T_2 and T_1 values and reinforcement temperatures, obtained as weighted means from the 2-D thermal analysis of the slab cross section, which varies in thickness with the profile of its steel decking. Equations (9.21) and (9.22) are then used, with the reinforcement properties assessed at the mean reinforcement temperature appropriate to the fire resistance period required, assuming an ISO 834 Standard Fire regime, to calculate a maximum allowable deflection. The basic Equations (9.1) to (9.15) are then used to provide the enhanced load capacity of the slab. If this is inadequate to provide the required contribution from slab tensile membrane action, then the slab details must be changed; in general, this implies that a reinforcing mesh of higher area is specified. An increase of mesh area alone does not affect the limiting displacement. However, it must be remembered that the whole calculation is based on an underpinning assumption that perfect vertical support is provided by the protected composite beams around the slab's perimeter; as can be seen in the following section, this cannot be taken for granted.

9.3.5.7 Recent developments: plastic folding of edge beams

Abu (2009), Abu and Burgess (2010) and Abu et al. (2010, 2011) looked at the effect on failure of edge beams of using the tensile membrane action design method described with increased reinforcement mesh areas. If edge beams fold, together with the internal unprotected beams and the slab, then the integrity failure represented by the appearance of a through-depth crack is replaced by a structural (resistance) failure, which may remain local or may lead to progressive collapse. The folding failure mechanism for a slab panel is illustrated in Figure 9.15.

A simple way of considering this structural failure of a slab panel that includes plastic folding of opposing protected beams around the edge of

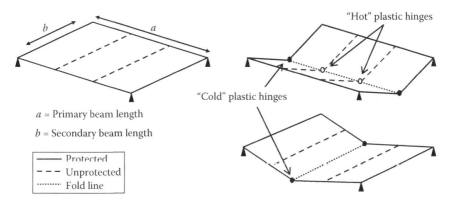

a = Primary beam length

b = Secondary beam length

———	Protected
- - -	Unprotected
··········	Fold line

Figure 9.15 Slab panel failure mechanisms.

a slab panel is to use a work balance equation similar to that used in the small-deflection yield line calculation. This can be used to predict when the parallel arrangements of primary or secondary (unprotected intermediate and protected secondary) composite beams simultaneously lose their ability to carry the applied fire limit state load due to their temperature-induced strength reductions.

The expressions for failure across either primary or secondary beams for a single slab panel are, respectively,

Primary beam failure:

$$\frac{wab}{4} - \left(\frac{4M_{pp}}{a}\right) \geq 0 \tag{9.23}$$

Secondary beam failure:

$$\frac{wab}{4} - \left(\frac{4M_{ps}}{b} + \frac{2n_{us}M_{us}}{b}\right) \geq 0 \tag{9.24}$$

In these equation, a and b are the lengths of the primary and secondary beams, respectively; w is the applied fire limit state floor loading; n_{us} is the number of unprotected secondary beams in the panel; and M_{us}, M_{ps} and M_{pp} are the temperature-dependent capacities of the unprotected, protected secondary and protected primary composite beams, respectively, at the temperatures corresponding to any given time. This can be extended across several adjacent panels folding along the same fold lines (possibly even across a complete width of a multibay floor slab). In this case, Equations (9.23) and (9.24) can be expressed more generally as

Primary beam failure:

$$\frac{wabn_{panels}}{4} - \left(\frac{2\sum M_{pp}}{a} \right) \geq 0 \qquad (9.25)$$

Secondary beam failure:

$$\frac{wabn_{panels}}{4} - \left(\frac{2\sum M_{ps}}{b} + \frac{2\sum M_{us}}{b} \right) \geq 0 \qquad (9.26)$$

in which n_{panels} is the number of adjacent panels (of equal size) taking part in the folding, and ΣM_{us}, ΣM_{ps} are the aggregates of the temperature-modified plastic moment capacities of the unprotected and protected secondary beams failing simultaneously on the same fold line. For primary beam failure, ΣM_{pp} is the aggregate of the plastic capacities of the parallel primary beams.

Abu et al. (2011) performed a series of studies on isolated composite slab panels with protected edge beams and unprotected internal secondary beams, looking at different panel dimensions and areas of mesh reinforcement. They compared the time-displacement behaviour of these panels, subjected to standard fire heating, obtained from numerical modelling using the software Vulcan with the simple folding failure predictions given by Equation (9.24). One such comparison considers a 9×9 m square panel with two unprotected secondary beams that achieve 963°C after one hour of the standard fire, at which time the "1-hour" protected edge secondary beams are at a temperature of 621°C. The comparison is shown in Figure 9.16b, which shows that for four mesh areas, from 142 mm²/m to 393 mm²/m, runaway deflection happens at almost the same time, which is well predicted by Equation (9.24). On the other hand, Figure 9.16a shows the tensile membrane action method predictions of strength with the four different mesh areas, together with the limiting deflections according to the generic method (marked BRE) and TSLAB. The maximum strength enhancement that can be used is at the intersection of the appropriate strength enhancement curve with whichever limiting deflection is being used. These suggest that increasing the reinforcement area is highly effective in obtaining increased fire resistance, in marked contrast to the implication from the numerical modelling and the plastic folding mechanism calculation. For example, A142 mesh produces only about 27 min of fire resistance according to either the original or the TSLAB criteria, but doubling the mesh area produces about 90 min according to the generic BRE

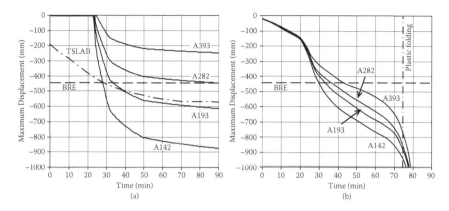

Figure 9.16 The 9 ´ 9 m slab panel results. (a) BRE tensile membrane method strength and deflection limit; (b) modelling displacement time and simple plastic folding limit.

method and clearly considerably more according to TSLAB. However, both numerical modelling of the actual time-deflection behaviour and the plastic folding calculation give runaway failure for all the mesh areas at approximately the same time. The key point here is that structural runaway failure involving collapse of protected edge beams is almost insensitive to mesh area and will take over from tensile cracking as the critical mode of failure as the mesh is increased.

As a result of this work the plastic folding calculation was included in a revision of TSLAB issued in 2010.

9.3.5.8 Other design documents

Two design methods that are very similar to the Bailey-BRE method described in this section have been launched since its appearance; these are not described in detail here. They differ in some details from this method but are broadly similar to it. The New Zealand slab panel method (Clifton et al. 2001) expanded the Bailey-BRE method to include the effects of continuity and additional reinforcement that may be present in the ribs of slabs. This method also includes the capacity of the heated unprotected composite beams in its yield line calculation. It imposes no limitation on slab size and checks individual components of a slab panel, such as protected beams and columns. More recently, the European-funded project FRACOF (SCI and CTICM 2009) has retested the Bailey-BRE method and has made some fairly minor amendments to it, including an additional limiting deflection criterion of $v< (L + l)/30$ in addition to Equation (9.22), which seems empirically based, and a version of the plastic folding calculation now included in the TSLAB software.

Chapter 10

Robustness of structures in fire

10.1 INTRODUCTION

In structural design, prevention of disproportionate collapse under accidental loading or malicious actions is an increasingly important design requirement. Progressive collapse is the ultimate form of disproportionate collapse, and its occurrence in the World Trade Center buildings on September 11, 2001, brought home, in the most tragic way, the devastating consequences of fire-induced progressive structural collapse. The prevention of disproportionate or progressive structural collapse is provided by structural robustness, which equates to the ability of a structure, or structural system, to accept a certain amount of damage without the structural failing to any great extent (Institution of Structural Engineers [ISE] 2010); it implies insensitivity to local failure. In Eurocode EN 1991-1-7 (Committee of European Normalization [CEN] 1991–2006), the formal definition of *robustness* is "the ability of a structure to withstand events like fire, explosions, impact or the consequences of human error without being damaged to an extent disproportionate to the original cause". This chapter is concerned with robustness of structures in fire. This is an active research topic, and there is still no definitive methodology of providing means of achieving adequate robustness for structures in fire. So, what this chapter intends is to define the context within which structural robustness in fire is considered, to identify a number of scenarios that may initiate fire-induced disproportionate collapse, and to provide a framework under which methods for achieving robust structural design for fire may be developed. Joint (connection) behaviour in fire is a critical issue affecting structural robustness in fire. Chapter 8 provided a detailed commentary on joint behaviour in fire. This chapter explains the implications of joint behaviour on structural robustness in fire.

Structural robustness is associated with accidental loading or malicious actions, in particular, structural response under unforeseen and unquantifiable loading. Although fire is a type of accident and its occurrence is rare, fire safety is an everyday requirement and is treated as a standard load case.

Design for structural fire safety should not be confused with design for structural robustness in fire. In fact, the Eurocode on structural robustness (EN 1991-1-7, CEN 1991–2006) includes such extreme loadings as impact and explosion, but not fire. This does not mean that it is not necessary to consider structural robustness in fire. It implies that dealing with the risk of disproportionate collapse under fire loading should consider situations that are not covered in standard fire limit state design. In other words, fire loading may be the initiating event leading to disproportionate structural failure when the effects of the actual fire action are more severe than those of the design fire action. Therefore, to find effective methods of ensuring structural robustness in fire, the initiating events associated with fire attack that may lead to disproportionate/progressive structural collapse should be systematically identified.

At the same time, it should be remembered that the effort spent on design for structural robustness should be consistent with the risk of disproportionate collapse. Since fire safety is already included in building design requirements, it is suggested that consideration of structural robustness in fire be limited to buildings of high risk. For example, Eurocode EN 1991-1-7 (CEN 1991–2006) classifies buildings into Classes 1, 2A, 2B and 3, with Class 1 buildings having low risk, Class 2 (2A and 2B) buildings having a normal level of risk and Class 3 buildings having high risk (e.g. hospitals, high-rise buildings, assembly buildings). It is recommended that, for buildings of Classes 1, 2A and 2B, no additional consideration is made other than performing the required normal fire limit state design. Additional consideration of structural robustness in fire should be limited to Class 3 buildings. However, since the building risk classification system in EN 1991-1-7 is necessarily a broad-brush approach, this does not suggest that the risk of disproportionate collapse be ignored for Classes 1 and 2 buildings. In these buildings, the risk of disproportionate collapse should still be assessed to identify specific high-risk concerns. For example, if fire protection to steelwork is applied in an area that is known to suffer abrasion, then specific measures are needed to ensure structural robustness.

When dealing with the structural robustness of Class 3 buildings, EN 1991-1-7 recommends a systematic risk assessment approach. This chapter attempts to help this risk assessment process by addressing structural robustness under fire attack.

10.2 CAUSES OF FIRE-INDUCED DISPROPORTIONATE COLLAPSE

Standard structural fire safety design (either prescriptive or engineered) ignores the risk of disproportionate or progressive collapse. Hence, design for robustness under fire loading should be carried out in a way similar to that under ambient-temperature loading. However, since there are a number of features unique to fire action, it is necessary to analyse the sources of risk of disproportionate collapse under fire attack. These are largely related to assumptions made in standard structural fire safety design. Fire engineering of structures encompasses decisions on fire compartment size, quantification of fire behaviour within the fire compartment, heat transfer analysis and structural behaviour. Fire-induced failure may occur when the actual fire condition is different from the design assumptions. This is explained next.

10.2.1 Fire compartment assumptions

One important means of controlling the extent of potential fire damage is the provision of fire-resistant compartments. Under this provision, fire is considered to be contained within the fire compartment of origin. In the majority of cases, this assumption is fulfilled. However, in rare cases, fire may spread out of the initial fire compartment. This is termed *compartment integrity failure*, as explained in Chapter 9. The reasons for fire compartment integrity failure can be many, including damage to the fire-resistant construction caused by other events, such as impact or explosion, earthquake, inadequate design/construction/maintenance, or abuse. In particular, the integrity of fire-resistant construction has been, and will continue to be, assessed under Standard Fire condition and on an individual-component basis. This may not be adequate because the real fire condition may be more severe, and interactions between the different individual fire-resistant construction components may have an adverse influence on some of these components, rendering them less effective than when assessed individually. Furthermore, inadequate fire-resistant compartment design and construction can never be ruled out.

Hence, the risk of fire compartment integrity failure is not negligible, and when considering structural robustness in fire, it should be treated as a potential initiating event. In fact, in structural design for robustness, if the cause of structural damage is unknown, one normal design scenario is removal of a structural member. Adapting this to fire safety design, a direct analogy is removal of a fire-resistant compartment boundary. The consequence of this is that fire spreads to adjacent fire-resistant compartments. Continuing the analogy with structural element removal at ambient temperature, fire-resistant compartment failure should consider individual component failures one at a time. Demonstrating this using the two-dimensional sketches in Figure 10.1, this would lead to fire exposure in adjacent fire-resistant compartments, either horizontally (Figure 10.1a) or vertically (Figure 10.1b).

Because design for structural robustness under fire attack requires consideration of exceptional cases, a balance should be sought so that the

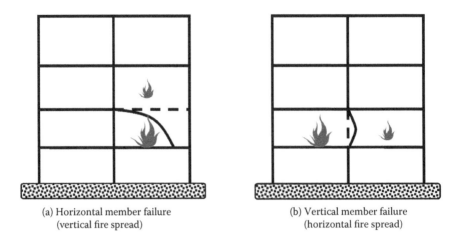

(a) Horizontal member failure   (b) Vertical member failure
(vertical fire spread)   (horizontal fire spread)

Figure 10.1 Fire spread due to fire compartment member failure.

design is not excessively punitive. Therefore, whilst progressive collapse of the World Trade Center Buildings 1 and 2 was a result of aeroplane impact causing integrity failure and simultaneous ignition on a number of floors, making this extreme loading condition a requirement for any fire-resistant design scenario would be prohibitively costly, even in most cases where explicit consideration of structural robustness is necessary. This scenario should only be considered in a very small number of exceptionally important buildings, as may be required by the client, in the case of extremely tall buildings. However, in most buildings that require systematic risk assessment of fire-induced disproportionate collapse, it can be assumed that any fire-resistant construction failure would result in the fire spreading to no more than two adjacent fire-resistant compartments.

10.2.2 Quality of information for fire safety design

Even if a fire is contained within its initial compartment, a design that has satisfied regulatory requirements may still be vulnerable to structural collapse. This is because there are many uncertainties in structural fire safety design, as explained here.

Fire loading is analogous to mechanical loading in ambient temperature structural design. However, the Standard Fire exposure condition still forms the basis of most structural fire design, even though it is unrepresentative of real fire conditions. Furthermore, even when more realistic fire conditions, such as the parametric fire curves (see Chapter 3), are used, there are significant uncertainties with regard to the input parameters, in particular, the magnitude of the fire load. Compared to ambient-temperature structural design, for which considerable research has been undertaken to collate loading data (to evaluate uncertainties in loading and in derivation of appropriate partial safety factors based on the framework of reliability theory), there has not yet been such comprehensive research directed at quantifying fire loading or the reliability of the design data. Therefore, it should not be a surprise that the probability of inadequate design input data is high.

In addition to uncertainties in fire behaviour, there are considerable uncertainties in material properties. Evaluation of fire resistance requires much more information on material properties than is required for calculating ambient-temperature load-carrying capacity. Despite this, the quality of information for fire resistance design is much poorer than that for ambient-temperature design. This comes about because (1) the probability of structural collapse under fire attack is considerably lower than under normal loading conditions; (2) obtaining high-quality data for fire-resistant design is inherently much more time consuming. Consider the stress-strain curve of steel as a simple example. It is a routine task to carry out tensile coupon testing at ambient temperature. Therefore, there has been an almost-exhaustive study of this property, including substantial testing and statistical analysis. In contrast, not only is obtaining steel tensile stress-strain relationships at elevated temperatures much more complicated, but also it is difficult to establish the exact condition under which the steel acts when it is affected. This casts doubt on the applicability of some existing material property data. Furthermore, many data may not even be collected during material testing if their importance is not recognised. For example, when establishing stress-strain relationships of steel at ambient temperature, the information on total elongation is not a priority provided there is sufficient elongation to allow steel to be treated as a ductile material for stress redistribution. However, for fire resistance calculation, due to the large deformations of the structure, how the material behaves at large strains becomes important; this information is usually not obtained. Therefore, there is insufficient mechanical property data to reflect accurately the material used under fire attack.

The shortcomings described for the simple quantity of steel stress-strain curve at elevated temperatures are relatively benign compared to more substantial problems with other material properties that are required for fire-resistant design, including the following:

- *Thermal properties of fire protection materials:* Whilst some data exist for their ambient-temperature properties, reliable data for elevated-temperature applications are virtually non-existent. In particular, the behaviour of intumescent coatings, which have become the dominant passive fire protection material for steel structures, at least in the United Kingdom, is poorly understood. Although there are now some published studies to aid understanding of intumescent coating behaviour in fire, the effects of time and environmental conditions on intumescent coating behaviour remain poorly understood.

- *Mechanical properties of structural materials:* Steel has traditionally been perceived to have poor fire resistance. Consequently, the majority of time and effort (although not perhaps as much as is required, as noted previously) has been devoted to understanding steel's mechanical properties at elevated temperatures. In contrast, the mechanical properties at elevated temperatures of other mainstream structural engineering materials, including concrete, masonry and timber, are still poorly understood.

Despite these shortcomings, it is not proposed that extensive efforts be made to obtain elevated-temperature material mechanical properties to a level similar to the data on ambient-temperature mechanical properties, recognising the diminishing returns in these efforts on improving structural safety. However, this does mean that the effects of these uncertainties in material properties should be accounted for in fire-resistant design. Designing for structural robustness offers a possible framework for this.

10.3 DESIGN STRATEGIES

As explained in the introduction to this chapter, when considering the risk of fire-induced disproportionate structural collapse, it is proposed that designing for structural robustness under fire be performed for Class 3 buildings, in addition to the normal structural fire safety design required for all building structures.

As part of a systematic risk assessment for Class 3 building structures, it is recommended that an assessment is made of the risk posed by the two significant causes identified in Section 10.2, which might induce disproportionate structural collapse: fire-resistant compartment failure and uncertainties in design information.

For uncertainties in information on material properties, some risk of fire-induced structural collapse may be effectively controlled by sensitivity analysis. This might involve using the most unfavourable thermal properties to give the highest structural temperatures and the most conservative estimate of mechanical properties to give the least structural resistance. Should this process not be sufficient, then it might be necessary to consider the effects of local structural failure on disproportionate failure of the structure. This may coincide with consideration of fire-resistant compartment failure.

Therefore, when designing for structural robustness under fire attack, it is proposed to include the effects of a fire-resistant compartment failure in Class 3 building structures, on the assumption that the accidental fire loading cannot be accurately identified. However, to ensure that the design does not become exceedingly costly, the consideration of fire-resistant compartment failure should in general be limited to any single component of the original compartment at one time, although the client of a building may want to impose more stringent conditions.

The remainder of this section explains different possible scenarios as a result of removing one of the fire-resistant compartment components and possible methods of design to control disproportionate collapse.

10.3.1 Loss of a horizontal component

In a structure, failure of the fire-resistant floor construction may result in the vertical members (columns, walls) losing lateral support, as illustrated by Figure 10.2. In this case, the vertical member (e.g. the column on grid-line C in Figure 10.2) should be designed based on an increased effective length. If failure of the fire-resistant floor construction does not affect the lateral support to the vertical member, then it would behave in the same way as if the fire were contained in separate fire-resistant compartments, and design of the vertical member (e.g. column on gridline B in Figure 10.2) is unchanged. In all cases, reduced partial safety factors should be used. During heating, the horizontal member may impose lateral loads on the affected vertical members. As explained in Section 2.3.3.2, concrete may suffer irreversible loss of strength due to the bending moments caused by the lateral loads. Therefore, these lateral loads should be included in the robustness design calculations for the affected columns. Sections 10.6.2 and 10.6.5 provide further discussion.

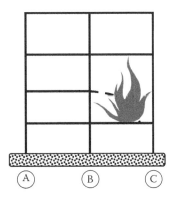

Figure 10.2 Effects of loss of compartment horizontal member on column.

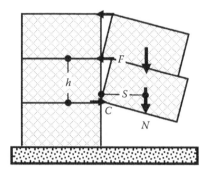

Figure 10.3 Tying action on loss of corner wall.

10.3.2 Loss of a vertical component

10.3.2.1 Panel construction with load-bearing walls

In the case of removal of a corner wall, which is exactly what happened in the Ronan Point incident that led to the current U.K. requirements for robustness in building structures, the tying resistance method may be applicable. As shown in Figure 10.3, the wall immediately above the damaged vertical support acts as a cantilevered deep beam. If there is sufficient tying resistance at the top of the panel wall immediately above the damaged vertical support, the vertical load (N in Figure 10.3) originally supported by the damaged vertical members may be resisted by the bending moment resistance developed by the tying resistance acting at a long lever arm (the height of the wall). The equilibrium condition is

$$F.h = N.s \tag{10.1}$$

It should be noted that the undamaged wall only has to resist the loads on the floor below because the walls on each of the storeys above can be assumed to develop the same load-carrying mechanism to resist the associated floor loads.

Additional design checks include checking the rest of the structure to resist the tensile (tying) force and the accompanying compressive force (at the bottom of the wall, C = F in Figure 10.3).

In the case of removal of an internal wall (Figure 10.4), the wall panel above becomes a deep beam, and any of the following two possible alternative load-carrying mechanisms may be sufficient:

1. Arching action within the plane of the wall immediately above the damaged wall;

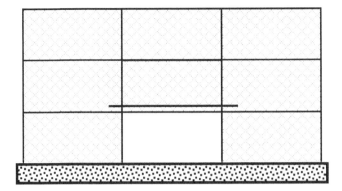

Figure 10.4 Loss of internal wall.

2. Simply supported deep beams in bending, with relatively high bending moment resistance at the centre. To enable this mechanism, tensile reinforcement (the thick line in Figure 10.4) should be provided at the bottom of the wall.

In summary, in structures with load-bearing walls, alternative load-bearing mechanisms can be readily identified provided that this initiating damage is envisaged during design.

10.3.2.2 Framed construction with non-load-bearing walls

10.3.2.2.1 Internal column removal

After losing an internal support column, the spans of the two connected beams in one direction are joined, and the effective beam span is much increased. It is unlikely that the original bending resistance of either beam will be sufficient to resist the applied load in fire. Alternative load-carrying mechanisms include arching action and catenary action if the horizontal members are beams (one-dimensional). If the horizontal member is a slab (two-dimensional), the alternative load-carrying mechanisms include compressive and tensile membrane actions in the floor slab.

Among these alternative load-carrying mechanisms, compressive membrane action in slabs or arching action in beams is unlikely to be effective. These load-carrying mechanisms work only if the structural deflection is small, typically less than half the depth of the structure. They also require significant horizontal restraint to resist the horizontal compression forces in the slab or beam. In a fire situation, slab or beam deflections are likely to be much higher, so tensile membrane action in the floor slab may work as a secondary load-carrying mechanism. However, this is based on the assumption that this alternative mechanism has not already been exploited in the structural fire-engineering design without member removal.

Catenary action has the potential to offer a reliable alternative load-carrying mechanism and is explained in the sections to follow. Catenary action is in fact the alternative load-carrying mechanism implied in the tying resistance approach when dealing with structural robustness design at ambient temperature. Its effectiveness in this respect at ambient temperature has been questioned due to a very high tying force that can be imposed on the connected structure and the large rotational capacities required of the connections (Byfield 2004). However, under fire conditions, the tensile force will be much lower because of the very large beam deflections involved. After explaining how catenary action works, the remaining part of this chapter explains the conditions necessary to enable catenary action to provide structural robustness in fire.

10.3.2.2.2 Corner column removal

If the removed vertical member is at the corner of a building, it will not be possible for the floor slab to develop either compressive or tensile membrane action if the beams are not able to provide edge vertical support. If the beams remained in place vertically, there would be no need for the floor slabs to develop any alternative load-carrying mechanism because the structural actions would not be changed from the situation when the column was in place. It would not be possible to utilise either arching or catenary action in the beams because there is no horizontal support mechanism for these axial forces in the beams to be resisted. The only alternative load-carrying mechanism is the bending resistance of the beams or floor slabs or a combination of both.

Figure 10.5 illustrates one alternative load-carrying mechanism, enhanced bending moment resistance of the floor slab along a diagonal.

A possible method for the beams to develop additional bending moment resistance is to use moment-resisting connections. However, for this to provide robustness, the bending moment resistance of the connections should not already have been used in ambient-temperature design, or there would not be sufficient reserve in connection bending moment resistance to be utilised for robustness. Therefore, this method is effective only when moment-resisting connections are treated as simple in ambient-temperature design; the reserve of real bending resistance of the connections can then be used for robustness.

Table 10.1 summarises the alternative load-carrying mechanisms that may be enabled in different situations. Among all the feasible ways, providing extra bending resistance (in slabs, walls, beams or connections) is familiar to engineers, causes no strong interactions between the different structural members, and is not pursued further. However, dealing with

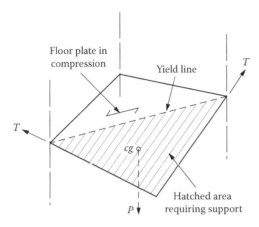

Figure 10.5 Load-carrying mechanism after removal of corner column. (From Khabbazan, M.M., *Progressive Collapse, The Structural Engineer*, 21 June, pp. 28–32, 2005. With permission.)

Table 10.1 Possible alternative load-carrying mechanisms on removal of a member

Structural type	Removed member	Alternative load-carrying mechanism	Feasibility	Additional consideration
Panelled	Corner wall	Tying	√	Compression at bottom of panel
	Edge internal/ interior wall	Bending	√	Provide bottom reinforcement
		Arching	√	Horizontal support to wall
	Horizontal (floor)	No change	√	Possible increased wall effective height
Framed	Horizontal (beam)	No change	√	Possible increased column effective length
	Edge internal/ interior column	Beam catenary action	√	See Section 11.4
		Beam arch action	X	Beam deflection too large
		Beam arch/ catenary action	X	No horizontal support
	Corner column	Rigid joints	√	Joint design as pinned at ambient condition
		Slab diagonal bending resistance	√	

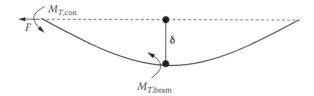

Figure 10.6 Beam with axial force.

catenary action requires consideration of the adjacent structure, in particular, the adequacy of connection performance. It is developed in more detail in Section 10.4.

10.4 CATENARY ACTION

Consider the beam shown in Figure 10.6. Its equilibrium equation is

$$M_{T,beam} + M_{T,con} + F\delta = M_{applied} \tag{10.2}$$

where $M_{T,beam}$ is the beam sagging bending moment, and $M_{T,con}$ is the connection hogging moment. T is the horizontal component of the beam axial force, and δ is its vertical deflection.

If the beam's vertical deflection is negligibly small, as assumed in normal ultimate limit state design, Equation (10.2) becomes the familiar bending moment equilibrium equation:

$$M_{T,beam} + M_{T,con} = M_{applied} \tag{10.3}$$

This equation is applied to normal fire limit state design.

If the beam and connection bending moment resistance is reduced further, the equilibrium condition expressed in Equation (10.3) can no longer be satisfied. It is this state of behaviour that is addressed in robustness design. In this design situation, catenary action is the alternative load-carrying mechanism, represented by $F\delta$ in Equation (10.2).

Although catenary action as a load-carrying mechanism can develop in both steel and reinforced concrete structures, its application to steel-framed structure has been researched to a much greater extent and is used to demonstrate the various features and implications.

Figure 10.7 shows typical vertical deflection and axial force curves in axially restrained steel beams with increasing temperature. The tensile (catenary) force in the beam is developed when its contraction is restrained. This happens at high temperatures, when the axial shortening between the

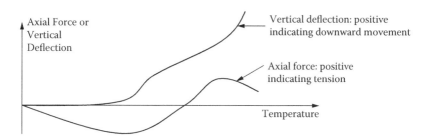

Figure 10.7 Typical behaviour of axially restrained beam.

beam's ends, as a result of large lateral deflections, overtakes the beam's thermal expansion. Owing to the extremely high ductility of steel, fracture of the steel beam is unlikely. Therefore, under catenary action, the limit state of this system depends mainly on the characteristics of the connections. Chapter 8 explained connection (joint) behaviour and different methods of quantifying connection behaviour. Section 10.5 explains the requirements for connection behaviour, to enable catenary action to develop fully, as well as possible methods of improving connection performance to prolong catenary action.

10.5 REQUIREMENTS FOR CONNECTIONS

Catenary action is required when the bending moment resistance of the beam (including any contribution from the connections) is insufficient to resist the applied load [the terms on the left-hand side of Equation (10.3) sum to less than the right-hand side]. As the beam temperature increases, the beam's bending contribution decreases, and from Equation (10.2), the contribution of catenary action ($F\delta$) must increase. Whether or not beam catenary action is sufficient will primarily depend on the connection's ability to rotate, thus allowing large beam deflections δ to occur, and on whether the resistance of the adjacent structure can sustain the tensile force (F) generated.

10.5.1 Required rotation capacity

The required amount of rotation capacity to enable catenary action to develop will be influenced by a large number of parameters, including the beam span, the level of applied load, the cross-section depth, and the structural temperatures. Analytical methods have yet to be developed to calculate the behaviour of a restrained beam in catenary action. However, it is possible to perform a rough order-of-magnitude analysis to obtain the required connection rotational capacity.

The pure catenary action form of Equation (10.2) (ignoring the contribution from bending moment resistance, which becomes orders of magnitude smaller during the catenary action stage) is

$$F\delta = M_{applied} \tag{10.4}$$

Assume that $M_{applied} = \mu M_{pl}$, where M_{pl} is the plastic bending moment capacity of the beam cross section at ambient temperature, and μ is the beam's load ratio in fire.

For a universal beam section, the flanges primarily contribute to the plastic bending moment capacity; therefore,

$$M_{pl} \approx A_f f_y d \tag{10.5}$$

where A_f is the flange area, f_y is the yield stress of steel and d is the beam depth.

Under pure catenary action, the catenary force reaches the beam's cross-sectional tensile capacity, giving

$$F \approx A.k_{y,\theta} f_y \tag{10.6}$$

where A is the cross-sectional area, and $k_{y,\theta}$ is the steel yield strength reduction factor at temperature θ.

For a universal steel beam section, $A \approx 3A_f$. Substituting the approximate Equations (10.5) and (10.6) into Equation (10.4) gives

$$3A_f k_{y,\theta} f_y.\delta \approx \mu A_f f_y h \tag{10.7}$$

or

$$\frac{\delta}{h} \approx \frac{\mu}{3k_{y,\theta}} \tag{10.8}$$

For a single-span beam, the load ratio is approximately $\mu = 0.5$. In the case of catenary action at very high temperatures (800–1000°C), $k_{y,\theta} = 0.05$–0.1. The approximate Equation (10.8) gives

$$\frac{\delta}{h} \approx \frac{5}{3} \sim \frac{10}{3}$$

Assuming an approximate span/depth ratio L/h of 20, then

$$\frac{\delta}{L} \approx \frac{5/3 \sim 10/3}{20} = \frac{1}{12} \sim \frac{1}{6}$$

During this derivation, many assumptions have been made, so the results from Equation (10.8) are approximate. Nevertheless, this range of deflection is close to those given by Yin and Wang (2004), who performed numerical simulations using the general software ABAQUS.

If the beam deflection profile is a half-sine wave, the deflection range corresponds to a connection rotation (beam-end rotation) of $180\left(\frac{1}{12} \sim \frac{1}{6}\right) = 15° \sim 30°$. If the catenary action deformation profile is linear, the end rotation is approximately $(15° \sim 30°) \times \frac{2}{\pi} \approx 10° \sim 20°$.

In the case of column removal, the beam span is often doubled. The applied bending moment will therefore quadruple. The beam-end rotation (the required connection rotational capacity) will double.

10.5.2 Available connection rotation capacity

The order-of-magnitude analysis presented in Section 10.5.1 requires connections to possess very high rotational capacities (10° or more). This is an order of magnitude higher than that required at ambient temperature (commonly taken as 30 mrad, about 1.7°) to allow for plastic bending moment distribution. Clearly, some connections will not be able to achieve sufficient rotational capacity to allow substantial development of catenary action. This section briefly examines the available rotation capacity of different types of steel beam/column connections and explores methods of improving connection rotation capacity.

Burgess and his research group (Yu et al., 2008, 2009a-d) have carried out elevated-temperature tests of connections under combined axial tension and bending moment. This loading condition was deemed to be representative of that in a connected beam under catenary action. Chapter 8 gave more background to this research and presented more detailed information. Figure 10.8 shows a summary of the load rotation characteristics of different types of common steel beam/column connections. Among the four types of connection examined, fin-plate, partial-depth end plate and flush end-plate connections can generally only achieve rotation capacities less than 10°. Whilst this level of rotational capacity is substantially higher than that required at ambient temperature, it is not likely to be sufficient to allow full development of catenary action. Also, the failure modes of these connections are typically brittle, involving weld fracture (partial-depth end plate), beam web shearing (fin plate), bolt tension and plate shearing (flush end plate). The inability of these connections to develop the required rotation capacity to allow the connected beam to develop substantial catenary action was also demonstrated in a series of restrained beam-column subassembly fire tests, conducted by Wang et al. (2011).

In contrast, web-cleat connections appeared to be able to develop much greater rotation capacity owing to the ability of the heel of the web cleat to straighten (Figure 10.9). Extended end-plate connections have also been

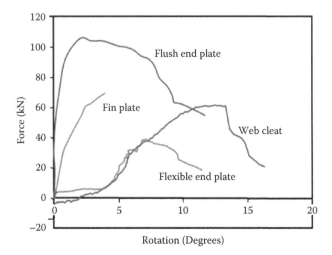

Figure 10.8 Indicative rotational capacity of different types of connection.

shown to achieve high rotation capacity owing to deformation of the end plate, as shown in Figure 10.10. The rotation capacity of web-cleat connection shown in Figure 10.9 is at the lower bound of that required for catenary action. Nevertheless, it is possible to enhance the rotation capacity of these connections.

The rotation capacity of a connection is mainly determined by deformational ductility of the connection component (bolt row) that is furthest from the compression (lower) flange of the beam, which acts as a fulcrum for rotation of the connection. Therefore, concentrating connecting bolts as near as possible to the compression flange of the beam should increase a connection's rotation capacity. It is possible to adopt this principle in detailing for fin plate, flexible end plate, and web-cleat connections.

For flush end plate and extended end-plate connections, the only way to obtain rotational ductility is by bending deformation of the end plate. Therefore, thin end plates with wide bolt spacing should be used wherever ductility is required. However, this may also reduce the tying capacity of the connection.

10.5.3 Practical implications for steel structures

The very large connection rotation capacity required to enable beams to develop catenary action fully may not be achievable in practice. However, even if this is the case, providing high rotation capacity will enable some catenary tension to develop, thus improving the structure's fire resistance in excess of that under pure bending. The approximate analysis in the previous

Figure 10.9 Deformation of web-cleat connection. (From Yu, H.X., Burgess, I.W., Davison, J.B., and Plank, R.J., Tying Capacity of Web Cleat Connections in Fire. Part I: Test and Finite Element Simulation, *Engineering Structures*, 31(3), pp. 651–663, 2009c. With permission from Elsevier.)

section has made the most conservative assumption, that the connected beam is the only structural element to resist the applied load in fire. In reality, other parts of the structure may participate in resisting fire-induced structural collapse. In addition, even if the temperature of the column that is assumed to be removed has exceeded its design limiting temperature, it will still develop some postbuckling strength to contribute to the equilibrium of the damaged structure (Wang 2004). Furthermore, because of fairly low loads in the fire limit state, the structure above the damaged column may have some reserve in strength and offer support to the structure originally supported by the damaged column. Figures 10.11a and 10.11b compare the most conservative assumptions for the load-carrying mechanism in catenary action with a more realistic mechanism.

In the case of unreliable fire protection to steelwork, the most desirable condition in the event of exceptional fire loading is when the alternative

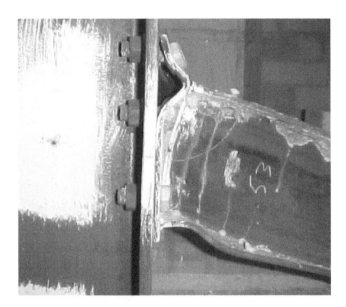

Figure 10.10 Deformation of extended end-plate connection. (From Wang, Y.C., Dai, X.H., and Bailey, C.G., An Experimental Study of Relative Structural Fire Behaviour and Robustness of Different Types of Steel Joint in Restrained Steel Frames, *Journal of Constructional Steel Research*, 67(7), pp. 1149–1163, 2011. With permission from Elsevier.)

load-carrying mechanism can allow the structure to survive temperatures above those that could ever be expected to be reached in unprotected steel. This temperature is usually on the order of 800–1000°C. If the building's importance is such that this should be considered in design, then connections with the highest rotation capacity should be used. However, should the steel temperature be expected to exceed the design limiting temperature by about 100°C, then the alternative load-carrying mechanism may be sufficient to give this reserve to steel structures in fire. The example in Section 10.5.4 can be used to demonstrate this statement.

10.5.4 Resistance of surrounding structure to catenary force

Should catenary action be used as the load-bearing mechanism for structural robustness in fire, the surrounding structure, in addition to the connections, must be designed to possess sufficient resistance to the combined catenary force and other applied forces in the structural member. This is different from the requirements of structural robustness design at ambient temperature using the tying method. At ambient temperature, the catenary

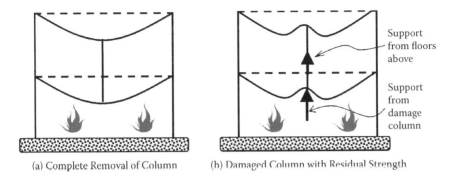

(a) Complete Removal of Column   (b) Damaged Column with Residual Strength

Figure 10.11 Two possible situations of the damaged column in fire.

action mechanism is implicitly assumed, but there is no need to check the resistance of the connected structure, other than the connections, to resist the tying force.

If catenary action is required, the catenary force has to be resisted by the surrounding structure, including the connections and the columns. Figure 10.12 shows that the catenary force increases and reaches a maximum and then decreases to follow the reduction in steel strength at high temperatures. To allow catenary action to occur, the surrounding structure must be designed to resist the maximum catenary force.

No accurate manual calculation method is currently available. Based on extensive simulation results on catenary action in steel beams by Yin and Wang (2004), the following method is suggested.

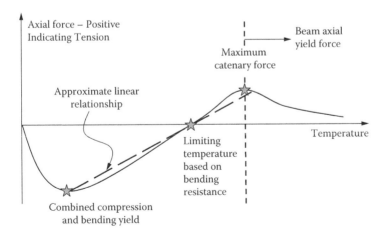

Figure 10.12 Different key stages of beam axial force-temperature relationship.

Figure 10.12 shows a few key stages of the restrained beam axial force-temperature relationship. The beam increases its compressive force due to restrained thermal expansion. The compressive force starts to decrease when the beam has reached its load-carrying capacity under combined bending and compression. This phase of reduction in beam compressive force will last until the beam reaches its conventional limiting temperature (based on bending resistance) at zero axial load. When the beam reaches its conventional limiting temperature, it indicates that the applied force in the beam is resisted by pure bending, and it experiences rapid vertical deflection. This rapid increase in vertical deflection causes the beam shortening to overtake its thermal expansion, and it now enters the catenary action phase. Due to decreasing bending moment resistance at temperatures above its limiting temperature, catenary action contributes to the load-carrying capacity of the beam. This continues until catenary action dominates. At this stage, the axial tensile force in the beam approaches the tensile yield capacity of the beam. The exact variation in the axial force can be complex, but based on extensive numerical simulations by Yin and Wang (2004), it appears that this variation may be represented by a linear relationship (indicated by the broken thick line in Figure 10.12). The axial force at the intersection of this line and the beam tensile yield capacity-temperature relationship gives a conservative approximation of the maximum axial tensile force in the beam.

If the temperature distribution in the beam cross section is uniform, the full beam tensile yield capacity will be reached. Most realistic beams have non-uniform temperature distributions in their cross sections, and the full cross-sectional tensile yield capacity is not reached. However, the beam reaches the full tensile yield capacity of the cross section based on the temperature of the lower flange.

Example 10.1

Implication of catenary action on column design

Figure 10.13 shows the structural steel beam arrangement of part of a simple regular floor structure. Since reliability of the fire protection on the steel beams cannot be guaranteed, the steel beam temperatures may be higher than the beam's limiting temperature in bending. The design requirement for robustness is to ensure that the columns remain stable by being able to resist the additional maximum catenary force that will be developed in the attached beams. Consider beam B1-B2 in Figure 10.13.

DESIGN DATA

Permanent action: $G = 3.5$ kN/m^2
Variable action: $Q = 5$ kN/m^2

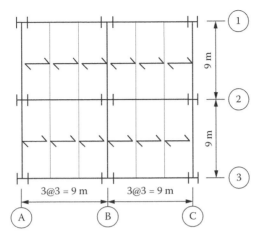

Figure 10.13 Floor plan for example.

Partial safety factors at ambient temperature: $\gamma_G = 1.35$, $\gamma_Q = 1.5$
Partial safety factors for accidental limit state: $\gamma_{A,G} = 1$, $\gamma_{A,Q} = 1/3$
Steel grade: S275 (yield stress = 275 N/mm²)
Floor height = 4 m

SUGGESTED SOLUTION

Limiting temperature of the beam

UDL on beam = $3*(1.35*3.5 + 1.5*5) = 36.675$ kN/m
$M_{max} = 1/8*36.675*9^2 = 372$ kN
Use U.K. steel section size UKB 457 × 191 × 67
$M_{pl} = 0.275*1471 = 404.5$ kN.m
For accidental fire limit state:

UDL on beam = $3*(3.5 + 5/3) = 15.5$ kN/m

$M_{fire} = 1/8*15.5*9^2 = 157$ kN.m

Load Ratio = $157/404.5 = 0.388$

Limiting temperature $T_{lim} = 685°C$

Axial failure load of beam

$A_b = 8550$ mm², $I = 29380$ cm⁴

$N_{pl} = 0.275*8550 = 2351$ kN

$N_{cr} = \pi^2\, EI/L_b{}^2 = \pi^2 * 200000 * 293800000/(9000)^2/1000 = 7160$ kN

$\lambda = \sqrt{2351/7160} = 0.573$

Using column buckling curve b according to EN 1993-1-1 (CEN 1993–2005a):

Column axial resistance $N_{b,Rd} = 1998$ kN

Assuming linear interaction between bending moment and axial force, the column axial resistance in the presence of bending is

$N_{Sd} = (1 - 0.388)*1998 = 1223$ kN

Maximum catenary force in beam

Axial stiffness of beam $K_b = EA/L_b = 200000*8550/9000 = 190000$ kN/m

Consider column size UKC 305 × 406 × 340 and assume the beam is connected to the minor axis of the column. $I_z = 46850$ cm⁴.

Column restraint stiffness to beam axial deformation:

$$K_c = \frac{48EI_{c,z}}{(2L_c)^3} = \frac{48*200000*46850*10000}{8000^3} = 8784\text{kN/m},$$

giving

$K_c/K_b = 0.046$

Note that double floor height is used.

Assuming elastic behaviour, the temperature rise ΔT at which the beam axial compression resistance is reached can be obtained from

$$\frac{K_b K_c}{K_b + K_c}(0.000012 * \Delta T)\, L_b = 1223\text{kN},$$

giving $\Delta T = 114°C$, $T = 134°C$ if the ambient temperature is 20°C.

In the calculation, 0.000012 is the thermal expansion coefficient of steel.

According to the approximation introduced in Figure 10.12, the maximum beam catenary force is 258 kN, which is reached at a steel maximum temperature of 800°C. The complete approximate beam axial force-temperature relationship is shown in Figure 10.14.

Required rotational capacity

Suppose the beam is required to survive until the lower flange has reached 800°C, which is 115°C higher than its limiting temperature. Assume

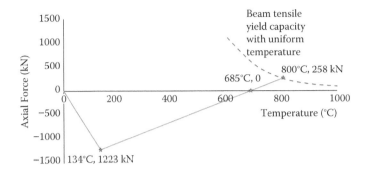

Figure 10.14 Approximate axial force-temperature relationship for example in Section 10.5.4.

non-uniform temperature distribution in the beam cross section, with temperature in the web and the lower flange being the same at 800°C, and the upper flange temperature is 600°C. The steel strength retention factors at 600°C and 800°C are 0.47 and 0.11, respectively. The flange area is 2412 mm².

The beam's bending moment capacity is approximately (EN 1993-1-2, CEN 1993–2005b)

$$\frac{0.11 \times 404.5}{0.7} = 63.6 \text{kN} \cdot \text{m}$$

The total tensile yield capacity of the beam cross section with non-uniform temperature is

258 + (0.47− 0.11)*0.275*2412 = 497 kN

Assuming a linear interaction curve between bending moment and axial force, the residual bending moment capacity of the beam is

(1 − 258/497)*63.6 = 30.5 kN.m

According to Equation (10.2), the maximum beam vertical deflection is

δ = (157 − 30.5)/258 = 0.49 m

Assuming a half-sine shape for beam vertical deflection, the required joint rotational capacity is

180*(0.49/9) = 9.8°C

This level of connection rotation is achievable by some common steel connections.

Implication for column design

If the beam catenary action is to be resisted by the column, the column size has to be increased above that without considering catenary action, or the column limiting temperature must be lowered.

For the column size used, assume plastic load-carrying capacity (stocky column) and linear bending moment-axial force interaction curve.

$A_c = 43300$ mm^2, $W_{pl,z} = 3544$ cm^3

$N_{pl} = 0.275*43300 = 11907$ kN, $M_{pl,Rd} = 0.275*3544 = 975$ kN.m

Assume this column supports 10 storeys and supports $6*9 = 54$ m^2 area on each floor. The column axial load is fire is $10*54*(3.5 + 5/3) = 2790$ kN.

If the column does not resist beam catenary force, the column load ratio is $2790/11907 = 0.234$, and the column limiting temperature according to BS 5950 Part 8 is 691°C.

The beam catenary force (258 kN) generates a bending moment of $258*8/4 = 516$ kN in the column. If the column resists the additional catenary action force, the column load ratio is $2790/11907 + 516/975 = 0.763$. The column limiting temperature is about 490°C.

10.6 OTHER CONSIDERATIONS

This chapter has explained in some detail issues that should be considered when designing for structural robustness under fire attack, using steel-framed structures during their heating phase as an example to demonstrate the main points. It is not the intention, and it is not possible, to cover every situation. However, it is hoped that the logic embedded in this process can be adopted when considering other issues. The main step in the process is to identify possible alternative load-carrying mechanisms in the event of fire-induced local structural failure and check their suitability to enable the damaged structure to resist the applied loads in fire. The following paragraphs explain additional considerations that may arise under different situations.

10.6.1 Other structural forms

Other alternative load-carrying mechanisms may exist in other types of structure. For example, in steel-framed tall structures, floor-height trusses are often used. These floor trusses would be supported by a number of columns in normal design. As a strategy for structural robustness in fire, the truss may be checked for sufficient load-carrying capacity on the assumption that it has lost some column supports.

10.6.2 Thermal expansion

When thermal expansion is restrained, compression forces are generated in the structure. In a concrete structure, restrained thermal expansion in a beam may result in the connected column deforming laterally and being under high bending. For example, Figure 2.7c shows an example taken from the Cardington concrete structure fire testing. If this additional bending moment in the Reinforced Concrete (RC) column has not been considered during fire-engineering design, there will be an increased risk of RC column failure, possibly leading to more extensive or disproportionate damage to the structure.

The compression force in the beam due to restrained thermal expansion depends on the axial restraint of the surrounding structure, which depends on the bending stiffness of the connected columns. When a steel column is approaching failure, its bending stiffness reduces. Therefore, in steel structures, it is possible for the connected steel columns to self-adjust the compression force in the connected beam so as not to be the cause of progressive collapse. In contrast, when a reinforced concrete column is subjected to bending due to compressive load in the connected beam, irreversible failure in the form of cracking may develop.

Most important, design for robustness is about dealing with unforeseen, unexpected situations. When considering structural robustness in fire, the issue of restrained thermal expansion can be identified easily. If it is effectively dealt with, it ceases to be a source of disproportionate collapse.

10.6.3 Cooling

During the cooling stage, tensile forces may be induced in beams. Cooling-induced tensile forces have been observed to cause connection fracture, but the consequences of this fracture during the cooling phase are not always unfavourable. By fracturing parts of the connection, the cooling-induced tension loads can be relieved, and this can be favourable to the connected column. Furthermore, if the connected columns have been designed to be safe over two floors, connection fracture is not important if (1) fracture happens when the fire has died out, so there is no danger of further fire spread; (2) any debris loading from the fractured connection does not cause a cascading failure of the floors below.

If cooling-induced connection fracture needs be prevented, one design strategy for steel-framed structures may be to slightly lower the limiting temperature of the connected beams. Ding and Wang (2007) carried out a limited experimental study. Figure 10.15 compares the axial force development in two almost-identical beams, but with a small difference in the temperature at which cooling started. Due to the rapid change in beam axial force near the beam limiting temperature (at which its axial force is zero),

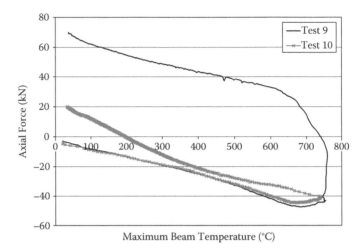

Figure 10.15 Effects of cooling temperature on restrained beam axial force. (From Ding, J., and Wang Y.C., Experimental Study of Structural Fire Behaviour of Steel Beams to Concrete Filled Tubular Column Assemblies with Different Types of Joints, *Engineering Structures*, 29, pp. 3485–3502, 2007. With permission from Elsevier.).

by starting cooling at a temperature a few degrees lower than the beam's limiting temperature, the residual tensile force in the connection was substantially lower. It is recommended to design for a limiting temperature slightly lower than that based on pure bending (e.g. by 50°C). For reinforced concrete structures, a small increase in cover to the reinforcement in beams (say by 5 mm) would be sufficient to reduce the steel reinforcement temperature rise.

10.6.4 Dynamic loading

The total collapse of the World Trade Center buildings was clearly a result of dynamic loading caused by falling structure because the lower-storey columns were designed to resist static loads. However, it is not recommended that the effects of dynamic amplification be considered when designing structures for robustness under fire attack. This is on the assumption that, if fire-induced disproportionate collapse is an explicit design requirement and the design achieves the objective of a stable structure after fire damage, any load redistribution process would be very slow compared to the free fall experienced in the World Trade Center buildings.

10.6.5 Reinforced concrete structures

Although the same issues exist when dealing with reinforced concrete structures, little information is available to allow detailed quantification of their risk of disproportionate collapse. Nevertheless, reinforced concrete structures have one important advantage over steel structures: The problem of unreliable external fire protection material does not exist. It is likely that the risk of fire-induced progressive collapse in reinforced concrete structures will be lower than in steel-framed structures. Behaviour of the Madrid Windsor Tower may be offered as partial evidence of this statement. Nevertheless, as mentioned in Section 10.6.2, since concrete cracking is irreversible, the effect of thermal expansion will be more severe in reinforced concrete structures than in steel structures. However, if design for thermal expansion is included in normal fire safety design, which would be expected for important structures requiring explicit assessment of the risk of fire-induced disproportionate collapse, the risk of restrained thermal expansion inducing disproportionate collapse will be low.

10.6.6 Combination with other accidental conditions

Designing for structural robustness is intended to deal with accidental situations. Each accidental situation is already a rare event with a very small probability of occurrence. Combination of two or more accidental conditions will be exceedingly punitive to the structure and prohibitively expensive. World Trade Center Buildings 1 and 2 suffered a sequence of airplane impact followed by fire, and it is their collapse that has brought fire-induced progressive collapse into prominence. However, including this sequence of events into design consideration can only be appropriate for structures of the utmost significance (such as nuclear facilities) or for structures where such combined accidental conditions have been identified to have a sufficiently high probability of occurrence (e.g. in offshore oil platforms or other petrochemical plants). However, these are structures requiring specialist treatment and already have well-established systematic risk assessment procedures to identify accidental loading conditions for robustness design. For the vast majority of structures that fall into the Class 3 building type, each accidental loading condition should be considered separately.

10.7 SUMMARY

This chapter has explained structural robustness in the context of fire attack. It has argued that, although fire is an accidental event, normal structural fire safety design does not cover fire-induced disproportionate collapse. The main issue is that there are many uncertainties associated with structural fire safety design assumptions, in particular the reliability of fire protection materials and fire-resistant compartmentation. Dealing with robustness in fire, as presented in this chapter, is about mitigating the risk of fire-induced disproportionate collapse when faced with such uncertainties. The central argument of this chapter is that, if the consequence of structural collapse is very severe (e.g. Class 3 buildings according to U.K. and European building classification), the requirement for structural robustness design should be over and above that required for normal structural fire safety design.

To provide sufficient robustness, this chapter recommends that the vertical members of structures should be designed to survive vertical fire spread resulting from possible integrity failure of one horizontal fire-resistant construction. In addition, the vertical members should be designed to resist any additional force that may be generated in the attached horizontal members due to additional design considerations for robustness. For example, the maximum catenary force developed in the beam should be capable of being resisted by the attached columns.

Very large deformations and rotations may develop as a result of alternative load-carrying mechanisms such as catenary action. Depending on the eventual fire exposure condition, existing connection technologies may not be able to provide sufficient deformation capacity to allow full development of such load-carrying mechanisms. For example, if a steel beam behaves as if unprotected due to the total destruction of the applied fire protection, existing connection technology will not be sufficient to allow the beam to survive, even with the development of catenary action. However, existing connections may still have substantial rotational capacity to allow some development of catenary action. Whether or not this is sufficient will depend on the acceptable extent of damage. This issue has not been addressed in this chapter. However, if safety of the vertical members can be ensured, the risk of disproportionate collapse will be lowered.

Chapter 11

The practical application of structural fire engineering for a retail development in the United Kingdom[1]

11.1 INTRODUCTION

This chapter describes how a performance-based approach has been adopted to optimise the structural fire protection for a composite-steel-framed structure in the United Kingdom. The adopted solution involves designing the structure so that fire protection can be omitted from selected beams. Various methods and design tools described in previous chapters are used in the design, so the methodologies are not new, but this chapter explains how they have been adopted on a real project and examines the differences between four design methods when applied to a real structure.

In the United Kingdom, buildings are required to comply with the requirements of the Building Regulations. Part B of the Building Regulations (Department of Communities and Local Government [DCLG] 2006) relates to fire. In 1984, a new Building Act was passed, and in 1985, the Building Regulations were revised. Since 1985, all versions of Part B of the Building Regulations have stipulated functional requirements that have to be achieved by a building. For structural stability in fire, the Building Regulation requirement is that, "The building shall be designed and constructed so that, in the event of fire, its stability will be maintained for a reasonable period". Designers and engineers are able to demonstrate compliance with these functional requirements by a variety of methods. The simplest and most common approach is to follow the prescriptive recommendations of guidance documents and design standards. For structural fire resistance, the prescriptive standards recommend that elements of a structure be tested in accordance with a Standard Fire test, and that they achieve a certain standard depending on the purpose group and height of the building in question. However, as described, alternative methods can be used to demonstrate compliance with the Building Regulations. Typically, where non-prescriptive methods are adopted, they will either demonstrate

---

[1] Chapter contributed by Dr. Florian M. Block of BuroHappold Limited.

that the proposed solution achieves a standard that is at least as good as that which would be achieved by adopting a prescriptive solution ("equivalency") or will adopt a performance-based approach to demonstrate that the proposed solution will achieve a suitable performance. In terms of structural fire resistance, a typical performance-based approach will be to demonstrate that the structure can maintain its stability for the entire duration of a realistic, worst-case fire.

11.2 METHODOLOGY

The section explains the five steps followed to justify the omission of passive fire protection on selected beams in the structure considered.

11.2.1 Obtain stakeholder agreement on methodology

At the outset of a project, it is important that the relevant stakeholders agree with the approach. These stakeholders are likely to include the client, the insurers, the design team and the building control authorities. It is necessary to ensure that

- there is value to be gained from conducting a performance-based approach;
- the proposed approach does not increase the life-safety or insurance risk;
- sufficient information and time will be available; and
- the control authorities are likely to grant approval.

Doing this before embarking on the analysis raises any potential problems at an early stage and reduces the project risk.

11.2.2 Develop design fires (including cooling)

The temperature rise within the different structural elements of a fully protected structure will be similar for a given fire exposure. As the structure becomes hotter, deflections will increase, and the load-bearing capacity will reduce. Therefore, when comparing the relative performance of different fully protected structures, it is possible to use a Standard Fire curve. However, in a partially protected structure, the unprotected beams will heat and cool significantly quicker than the protected elements. This differential heating can have a significant impact on the performance of the structure and its survival period. Therefore, when designing such structures, it is important to consider different fire scenarios and heating regimes. For example, short, hot fires can result in large differential expansion between protected and unprotected beams, which in turn leads to large differential deflections and high connection forces; a long-duration fire can have a larger effect on the overall load-bearing capacity of the structural system.

11.2.3 Develop assessment criteria

It is good practice to agree on the assessment criteria to be used to assess appropriate performance in fire prior to conducting analyses. The selection of appropriate assessment criteria should consider the functional requirements of the structure at fire limit state. For example, some floors are required to be compartment floors, but others are not. If a floor is a compartment floor, it is required to maintain overall stability, prevent smoke and flames passing from the fire side to the non-fire side (integrity), and to prevent excessive temperature rise on the unexposed surface (insulation). If a floor is not a compartment floor, it is only required to maintain its overall stability. Similarly, in some instances it is necessary for a structure to meet a particular assessment criterion for the entire duration of the fire, but in others this might only be for the heating phase or for the evacuation period.

11.2.4 Build geometry of the subframes and analyse for different fires

When assessing the performance of the structure at the fire limit state, it is necessary to determine how much of the structure should be analysed to be representative. Simple design methods, such as BS 5950: Part 8 (British Standards Institution [BSI] 1990b), only consider individual elements. Others, such as VulcanLite (Huang et al. 2003a, 2003b), TSLAB (Newman et al. 2006) and SCI P288 (Newman et al. 2006) consider individual bays. When adopting finite element analysis, it is often beneficial, and more realistic, to consider large portions of the structure. The extent of the subframe adopted for finite element analysis is often a compromise between absolute accuracy and the time available; it is typically impossible to analyse a complete structure, so a representative subframe will be selected, and appropriate boundary conditions will be applied.

11.2.5 Assess connection forces

The differential heating that occurs in a partially protected floor system can result in high connection forces. Therefore, it is important to assess these connection forces and to design appropriately. The connections can be designed to resist the forces, ductile connections can be used to reduce the forces, or the structure can be designed such that, if the connections fail, overall stability will be maintained.

Example 11.1

The example used in this chapter to show the practical application of structural fire-engineering methods is a large retail, leisure and cinema complex in northwestern England. It occupies a triangular site, and the building footprint is approximately 150 × 85 m in plan and comprises two storeys plus a mezzanine floor. For design purposes, the building can be considered to act as two separate structures due to the placement of a movement joint running north to south through the building. On the ground floor level, the space is subdivided into entrance, circulation and retail spaces. A service yard occupies the centre of the site, splitting the building at ground floor level. The mezzanine level has similar use to the ground floor, but with the addition of leisure facilities along the southern edge of the site. Above the mezzanine level, the usage of the building is primarily for leisure facilities. Double-height bowling and cinema spaces are separated by a concourse area, with the cinema spanning across the service yard located below.

The building is constructed as a steel-composite structure with normal downstand composite beams supporting an in situ composite floor on metal decking. This type of structure is ideal to utilise the benefits of tensile membrane action during a fire, which can be used to omit fire protection from off-grid secondary beams. Due to the size, and the multiple usage and changing floor construction, of the buildings, five different subframes have been analysed. According to the Approved Document B (DCLG 2006), the required Standard Fire resistance period of the building is 60 min, based on the height of the top occupied floor above ground.

11.3 DESIGN FIRES

Normal buildings should be designed for postflashover compartment fires. These are traditionally described by a compartment temperature development with respect to time. However, in any given building, there are an infinite number of time-temperature histories that could occur, depending on the fire load and compartment characteristics. Clearly, it is not possible to assess the impact of all possible fires; therefore, an important step of a structural fire-engineering assessment is to determine and agree on the appropriate design fires. As mentioned, partially protected structures should be exposed to a number of fires with different durations and temperature development, including the cooling phase. The parametric fire curves, as given in EN 1991 Part 1.2 (Committee of European Normalization [CEN] 1991–2002), have been used here with different ventilation conditions and are shown in Figure 11.1.

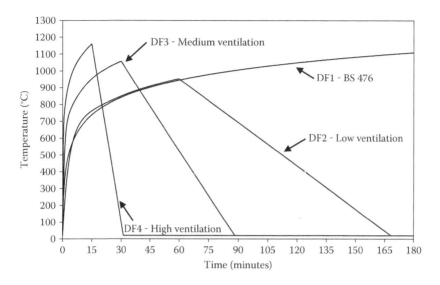

Figure 11.1 Design fires. (From Block, F.M., Yu, C., and Butterworth, N.A., The Practical Application of Structural Fire Engineering on a Retail Development in the UK, *Journal of Structural Fire Engineering*, 1(4), 205–218, 2010. With permission.)

11.4 ACCEPTANCE CRITERIA

Fire resistance is generally defined by three criteria: resistance (or stability), integrity and insulation. Therefore, each of these parameters has to be assessed.

11.4.1 Resistance (stability)

To ensure overall stability, it is necessary to ensure that protected beams will continue to perform as required. Limiting the deflection of protected beams to the limits that are imposed in Standard Fire tests suggests that the protection on protected beams will continue to perform as required. Therefore, the deflections of the protected beams are limited to *span*/20. In postflashover fires, unprotected beams become hot, lose strength and ultimately contribute very little to the overall stability of the structure. Therefore, it is not necessary to impose a similar deflection limitation on the unprotected beams. The overall performance of the structure (in terms of its overall stability) does not rely on the performance of the slab as this will still act as a diaphragm if deformed. Providing that the performance of the stabilising members (protected beams and column) can be ensured, overall stability will be maintained. Failure of the slab will only lead to local failure. However, in the Cardington fire tests, slab deflections of approximately *span*/10 were experienced without loss of local load-carrying capacity. Therefore, if deflections are limited to span/10, not only will the overall stability be maintained, but also it is unlikely that any local collapse will occur.

11.4.2 Integrity

The requirement for integrity is achieved by ensuring that large cracks do not develop in the floor slab. Most finite element programs are unable to predict integrity failures; therefore, when it is necessary to maintain the integrity of the floor slab, checks should be put in place to ensure that integrity failures do not occur. This is most commonly achieved by limiting deflections, as the deflection of the slab is an indication of its curvature (and the strain that is induced in hogging over protected beams), and if the curvature is too great, the reinforcement may fracture over the protected beams. Furthermore, excessive deflections within the slab could lead to a tension crack forming in its middle region. In either of these scenarios, integrity failures of the slab might occur. In reality, curvature is a function of the differential deflection across the slab; therefore, the differential deflection between the slab midregion and protected beams should be assessed. Sufficient research and testing have not been conducted to allow specific guidance on deflection limitations to be developed. Therefore, typically, deflections are limited to those that have been demonstrated to be acceptable in tests. The BS 476 Standard Furnace test (BSI 1990a) limits deflections to span/20. Therefore, it is likely that limiting deflections to span/20 will ensure that integrity failures do not occur.

11.4.3 Insulation

Insulation is not affected by deflections (even if it cracks or spalls, the concrete material will be retained by the metal decking), and integrity failures in the slab will occur before insulation is lost. Therefore, the requirements for insulation will be ensured by complying with the prescriptive requirements for slab thickness.

11.5 FINITE ELEMENT ANALYSES

The software used to analyse the building is Vulcan (2008), a three-dimensional frame analysis program that has been developed mainly to model the behaviour of skeletal steel and composite frames, including their floor slabs, under fire conditions. The first step in the finite element analysis process is to determine the relevant subframes.

11.5.1 Numerical modelling

An initial scoping study on individual bays using VulcanLite (Huang et al. 2003a, 2003b) was carried out to determine the required reinforcement mesh type, as well as to conduct a cost-benefit analysis showing the economical viability of the project. Subsequently, a detailed assessment was conducted on four representative subframes for the different parts of the building. The building geometry, the proposed fire protection regime and the location of the subframes are shown in Figure 11.2.

In this chapter, models F2 and F3 are discussed. Figure 11.3 shows the steel section reference numbers and the assessed fire protection regime. The steel section sizes of the floor beams are shown in Table 11.1. The main characteristics of the models are described next.

11.5.1.1 Concrete slabs

The concrete slabs were modelled using the non-linear, temperature-dependent, layered, nine-node slab element provided in Vulcan. The complexity of the slab element allows relatively large element sizes of up to 1×1 m without losing significant accuracy in the representation of the floor deflections in fire. In the area of F2, the slab is 150 mm thick, cast on Ribdeck E60 trapezoidal deck. As a result of the stage 1 analyses, the U.K. reinforcement mesh size A252 (8-mm bars at 200 centres in both directions, giving 252 mm^2/m) was proposed for the mezzanine floor. In the area of F3, the floor slab is 200 mm thick cast on a Holorib reentrant deck. The analyses in stage 1 suggested that the U.K. reinforcement mesh size A393 (10-mm bars at 200 centres in both directions, giving 393 mm^2/m) would be required to reduce the slab deflection in the compartment floor. Grade 35 normal-weight concrete is used. According to BS5950: Part 8 (BSI 1990b), a material strength factor 1.10 needs to be applied to the calculations for concrete at elevated temperatures. Therefore, in the numerical analyses, the concrete strength at ambient temperature was taken as 32 N/mm^2. As in F2, a trapezoidal deck was used, so only the continuous part of the slab was modelled to capture the worst-case temperature distribution of the reinforcement. In F3, however, the full depth of the slab was used considering different stiffnesses parallel and perpendicular to the ribs as a reentrant deck heats up more uniformly.

11.5.1.2 Boundary conditions

Where possible, the advantage of symmetry in the structure was used to reduce computing time. On the lines of symmetry, the translations perpendicular to these lines and the rotations across them were restrained. This is to account for the adjacent structure, which was not modelled, but would provide lateral restraint.

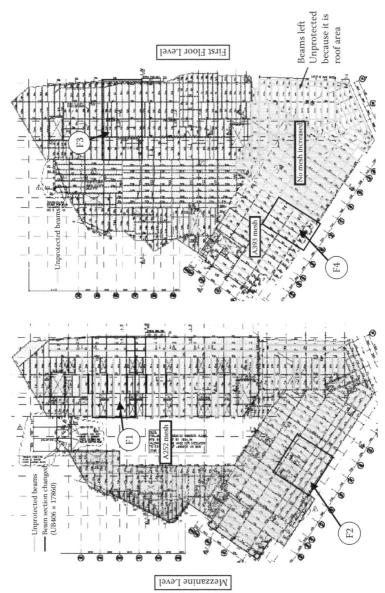

Figure 11.2 Geometry, proposed fire protection regime and the location of the subframes. (From Block, F.M., Yu, C., and Butterworth, N.A., The Practical Application of Structural Fire Engineering on a Retail Development in the UK, *Journal of Structural Fire Engineering*, I(4), 205–218, 2010. With permission.)

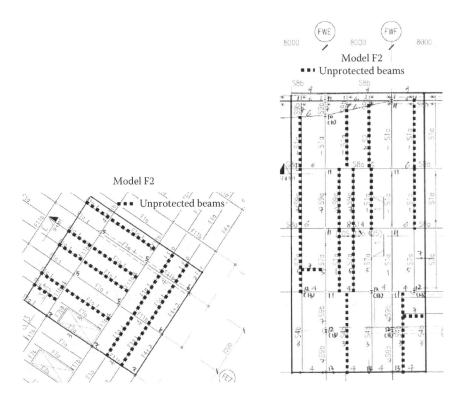

Figure 11.3 Section sizes and unprotected beam layout. (From Block, F.M., Yu, C., and Butterworth, N.A., The Practical Application of Structural Fire Engineering on a Retail Development in the UK, *Journal of Structural Fire Engineering*, 1(4), 205–218, 2010. With permission.)

11.5.1.3 Loadings

Characteristic design loads were provided, and the fire limit state loads were calculated in accordance with BS 5950: Part 8, using partial safety factors of 1.0 for dead load and permanent imposed load and 0.8 for non-permanent imposed load. The two Vulcan models are shown in Figure 11.4. The line loads and the applied boundary conditions are also shown. The loads introduced by the superstructure were represented by point loads at the top of the columns.

11.5.2 Material temperatures and behaviour

To determine the fire protection level for the protected elements, it was assumed that temperatures of 620°C and 550°C were reached after 60 min in the beams and columns, respectively. A heat-transfer analysis of the beams and columns was then undertaken in accordance with EN 1993

Table 11.1 Steel section sizes for models F2 and F3

	Beam reference	Section size
Model F2	F1a	406 × 140 × 39 UB
	F4	350 × 350 × 12.5 SHS
	F7a/F7b	457 × 152 × 52 UB
	F11b	457 × 191 × 89 UB
	S1a	457 × 191 × 67 UB
	S2	457 × 191 × 74 UB
	S4b	533 × 210 × 82 UB
Model F3	S5	533 × 210 × 92 UB
	S7a	533 × 210 × 109 UB
	S8a/S8b	533 × 210 × 122 UB
	S9b/S9c	406 × 140 × 39 UB
	S14	610 × 305 × 238 UB
	S25	254 × 146 × 31 UB

Source: Block, F.M., Yu, C., and Butterworth, N.A., The Practical Application of Structural Fire Engineering on a Retail Development in the UK, *Journal of Structural Fire Engineering*, 1(4), 205–218, 2010. With permission.

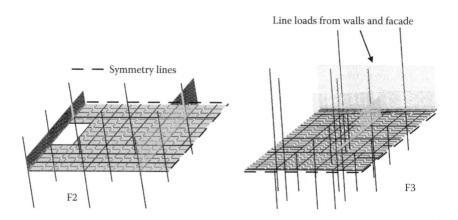

Figure 11.4 Finite element meshes of the Vulcan models F2 and F3. (From Block, F.M., Yu, C., and Butterworth, N.A., The Practical Application of Structural Fire Engineering on a Retail Development in the UK, *Journal of Structural Fire Engineering*, 1(4), 205–218, 2010. With permission.)

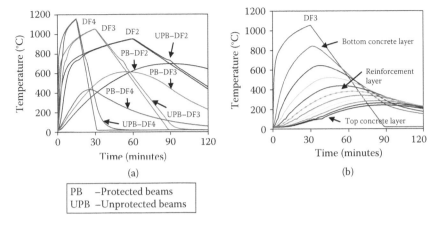

Figure 11.5 Typical material temperatures: (a) steel and (b) concrete (layer temperatures). (From Block, F.M., Yu, C., and Butterworth, N.A., The Practical Application of Structural Fire Engineering on a Retail Development in the UK, *Journal of Structural Fire Engineering*, 1(4), 205–218, 2010. With permission.)

Part 1.2 (CEN 1993–2005b). To include the effects of thermal curvature, the web and each flange were analysed separately. For the concrete slabs, a one-dimensional heat-transfer calculation was conducted that included the effects of moisture in the concrete. For the three parametric fires, the resulting temperature histories for protected and unprotected typical steel elements, as well as for the slabs, are shown in Figure 11.5.

It can be seen from Figure 11.5 that the slow fire (design fire 2) is critical for both protected beams and columns. The fast fire (design fire 4) causes the largest temperature differences between protected and unprotected beams, as well as the hottest temperatures of the unprotected beams.

To represent the effect provided by the increasing material temperatures on the material behaviour, the temperature-dependent constitutive relations given in EN 1992-1-2 (CEN 1992–2005b) and EN 1993-1-2 (CEN 1993–2005b) were used.

11.5.3 Results of the numerical analyses

The results of the non-linear analyses, including the deflected shapes, deflections and forces, were analysed and compared with the assessment criteria. The deflected shapes from the two models are shown in Figure 11.6, from which it can be seen that the deflection of the protected beams and the floor slab remains above the relevant deflection limit in the parametric fire. The results for model F2 are shown in Figure 11.7, and it can be seen that the protected beams exceed this limit in the International Organisation for Standardisation (ISO) fire (DF1) but well after the prescriptively required

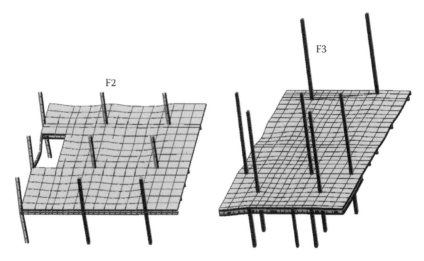

Figure 11.6 Maximum deflections of the two Vulcan models. (From Block, F.M., Yu, C., and Butterworth, N.A., The Practical Application of Structural Fire Engineering on a Retail Development in the UK, *Journal of Structural Fire Engineering*, 1(4), 205–218, 2010. With permission.)

fire resistance period of 60 min. Therefore, it can be said that the omission of fire protection to the off-grid secondary beams is acceptable.

The analyses of the model F3 in design fire 2 and design fire 4 were carried out. As the floor that is modelled in F3 is a compartment floor, span/20 should be taken as the deflection limit for both the slab and the

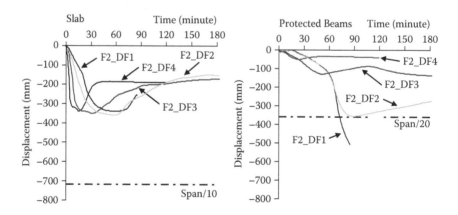

Figure 11.7 Vertical deflections of the protected beams and the slab in F2. (From Block, F.M., Yu, C., and Butterworth, N.A., The Practical Application of Structural Fire Engineering on a Retail Development in the UK, *Journal of Structural Fire Engineering*, 1(4), 205–218, 2010. With permission.)

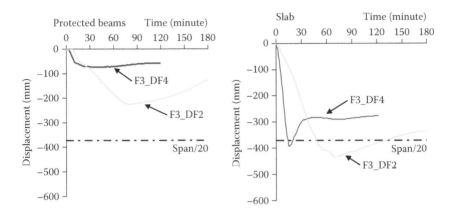

Figure 11.8 Vertical deflections of the slab and the protected beams in F3. (From Block, F.M., Yu, C., and Butterworth, N.A., The Practical Application of Structural Fire Engineering on a Retail Development in the UK, *Journal of Structural Fire Engineering*, 1(4), 205–218, 2010. With permission.)

protected beams. The results are shown in Figure 11.8, and it can be observed that the maximum differential deflections of the slab in the two design fires are both larger than span/20 after 60 min. Since the deflections of the protected beams are still much less than span/20, it is clear that the load-carrying capacity of the floor slab is not sufficient to limit the deflections to the assessment criteria. An increase of the protected beams will therefore not lead to a significant reduction of the floor deflections, and the only alternatives are either to increase the reinforcement mesh in the slab or to fire protect the currently unprotected secondary beams. As an A393 reinforcement mesh is already specified for the slab, all beams and columns supporting the first cinema floor are protected to maintain the integrity of the compartment floor.

11.6 STEEL BEAM CONNECTIONS

Generally, connection forces in partially protected structures are dominated by the differential thermal movements between the protected and the unprotected beams, as well as the restrained thermal expansion of the columns. As thermal stresses depend on the level of restraint, the forces in the structure reduce with increasing flexibility between the floor slab and beams as well as in the connections. Vulcan currently does not include the local behaviour of connections or the effects of local buckling of the beams. Therefore, any connection forces directly extracted from the software will be on the high side as the structure is modelled as rigidly

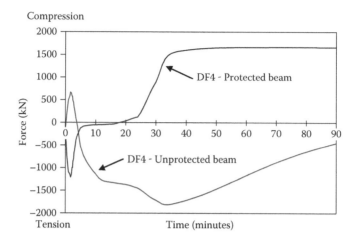

Figure 11.9 Typical connection forces in the fast design fire (DF4). (From Block, F.M., Yu, C., and Butterworth, N.A., The Practical Application of Structural Fire Engineering on a Retail Development in the UK, *Journal of Structural Fire Engineering*, 1(4), 205–218, 2010. With permission.)

connected. However, the software can give a realistic trend of the forces during the course of a fire. For the fast fire, such a typical curve is shown in Figure 11.9. It can be seen that, in the first few minutes of the fire, a peak in tensile forces in the protected beams and compressive forces in the unprotected beams is developed. This peak is caused by the rapid heating, and therefore thermal expansion, of the unprotected beams. It is very likely that this peak will cause local buckling in the bottom flange of the unprotected beams, as could be seen in the full-scale fire tests at Cardington. The buckling will release several thermal forces in the whole system. The next occurrence of high connection forces will be during the cooling phase of the fire and is caused by the shortening of the plastically deformed unprotected beams. It is this phase of a fire in which connection failure is most likely. One example of this was seen during the fire tests at Cardington, in which the fin-plate connections failed by bolt shearing (see Figure 2.7b). To prevent such a failure affecting larger parts of the structure, three options can be given:

1. The connections can be designed to resist the applied forces.
2. Sufficient ductility can be provided in the beam-to-beam connections to allow the thermal movements of the unprotected beams without breaking. Tests at the University of Sheffield (Yu et al. 2009d) have shown that this can be delivered by double web-cleat

or partial-depth end-plate connections. These connections would only have to maintain their vertical shear capacity during and after the fire, as they are ductile enough to follow the thermal movements without breaking.

3. It can be ensured that the unprotected beams can be supported after their connections have failed. For example, the bottom flange of the primary beam could be designed to support the shear loading transferred by the unprotected beam in the fire limit state by overlapping the unprotected secondary beam and the bottom flange of the primary beam. For this option, the tying resistance of the connections should be reduced to the structurally allowable minimum to minimise the forces introduced into the structure before the connection fails.

For the connections between protected beams and columns in bays with unprotected beams, end-plate connections should be provided. This type of connection has good fire performance and is particularly good at transferring compression forces due to the full contact between the end plate and the column flange whilst being ductile enough to deal with the initial tension forces until the unprotected beams have buckled.

11.7 SIMPLIFIED METHODS

In the United Kingdom, simplified methods are currently applied to justify partially protected steel structures. Each of these methods has its advantages and disadvantages.

11.7.1 SCI P288

The SCI publication P288 (Newman et al. 2006) contains design tables that are used to determine whether the partially protected structure will have sufficient load-bearing capacity at fire limit state. The method is applied to a single bay of structure, and the tables are quick to use. The method is restricted to rectangular geometries, certain bay sizes and a maximum fire resistance period of 60 min. It is not possible to assess likely deflections or connection forces. The tables only consider exposures to the BS 476 Standard Fire curve and as such do not consider the impact of the differential heating between protected and unprotected elements. Therefore, it is our opinion that SCI P288 should not be used in isolation to justify partially protected structures. However, they can be used as part of a scoping study at the beginning of a project that is to be assessed in much more detail during the detailed design stage.

11.7.2 TSLAB

TSLAB (Newman et al. 2006) is a software program that uses the same theoretical background as the tables in SCI P288 and is therefore similar in scope and approach to SCI P288 but has additional features. It can be used for any rectangular geometry. It can adopt both standard and parametric fires but is limited to fire resistance periods of up to 120 min. The methodology considers individual bays. It assumes that there is no continuity over protected beams, which is a conservative assumption, but also assumes that the protected beams do not deflect, which is a non-conservative assumption. Evidence suggests that within a practical range these assumptions compensate to give acceptable solutions, but there is no means of knowing whether this is true for all conditions. This is particularly true for edge or corner bays, where there is no continuity over some of the projected beams anyway. Therefore, it is our opinion that TSLAB should only be used for structures that fall within a certain range (the definition of this range is unclear and is not the subject of this chapter), and that it should be used with great care. Like the tables in SCI P288, TSLAB can be a useful tool during the initial stages of a project.

11.7.3 VulcanLite

VulcanLite (Huang et al. 2003a, 2003b) is a bespoke interface for Vulcan (2008) that uses a "wizard" to generate models for single bays of composite-steel-framed structures. Modelling a single bay is sufficient for most buildings where the objective of the analysis is to omit fire protection from beams that do not frame into columns. VulcanLite is not limited to rectangular bays or uniformly distributed loads. It models standard or parametric fires and is not limited to any fire resistance period. Importantly, VulcanLite is based on fundamental mechanics and predicts actual structural performance in fire conditions, including deflections and forces. With the appropriate use of boundary conditions, it can model the effects of continuity over the protected beams. As such, it can be used to deliver economic, safe and reliable solutions for relatively simple structures. More complicated geometries require a detailed finite element assessment using a validated finite element code.

11.8 COMPARISON BETWEEN VULCAN AND SIMPLIFIED METHODS

As part of the study, the behaviour of the large subframe models has been compared with the results of individual-bay analysis methods. The tables in SCI P288 could not be used for the study building as they are based on a

Table 11.2 Comparison of different analysis approaches

Model	Criterion	T-Slab	VulcanLite	Vulcan
F2	Slab	Pass (A393)	Pass (A252)	Pass (A252)
	Perimeter beams	Failed	Failed	Pass
F3	Slab	Pass (A393)	Pass (A393)	Pass (A393)
	Perimeter beams	Pass	Pass	Pass

Source: Block, F.M., Yu, C., and Butterworth, N.A., The Practical Application of Structural Fire Engineering on a Retail Development in the UK, *Journal of Structural Fire Engineering*, 1(4), 205–218, 2010. With permission.

partial safety factor for imposed load of 0.5 rather than the 0.8 recommend for all buildings other than offices.

As a comparison, two different typical bays have been analysed within the models F2 and F3. The results of this comparison are given in Table 11.2.

One can see the benefit in running more advanced analyses. Although the original perimeter beams in model F2 failed in both simplified methods, it would be possible to increase the perimeter beam size and then to justify a reduced mesh thickness using VulcanLite. However, if more of the structure is analysed, the columns and the surrounding structure provide sufficient longitudinal restraint to the perimeter beams to enable them to maintain their stability. For model F3, all approaches show that the chosen mesh size and the original perimeter beams are sufficient to withstand the loads at fire limit state (FLS) under all design fires. The final solution for F3 was fully protected, as the cinema seating, which is supported on this floor, introduces significant point loads on most intermediate beams, causing failure of the floor slab.

What could not be shown in this comparison was that in model F1 a number of beams between columns had to be increased in size due to their additional axial loads and $P\delta$ effects. This could only be picked up by the Vulcan analyses and shows the benefit of including columns in the model. However, if one only wants to conduct an initial study or has to analyse a very simple building, VulcanLite would be sufficient. In addition, it is our opinion that VulcanLite is preferable to TSLAB. This is not only primarily due to the fact that it predicts actual performance using fundamental mechanics but also because of its ability to apply different boundary conditions, to include point loads, to model non-regular geometries, and to assess real deflections and forces.

11.9 CONCLUSIONS

This chapter summarises the structural fire engineering assessment of a large retail and leisure development in the United Kingdom. After discussing the regulatory background in the United Kingdom, the methodology of a performance-based assessment was described. As an important part of this methodology, assessment criteria have been specified to ensure that the requirements of the Building Regulations are fulfilled after omitting fire protection on selected beams. Furthermore, the importance of analysing a partially protected structure exposed to a series of design fires with different ventilation conditions, including the cooling phase was highlighted.

On the example of two large subframe models, it was shown how finite element analyses could be used to calculate temperatures, determine the structural response and process the results. The behaviour of connections in partially protected structures was discussed, and three different options were given to ensure that local connection behaviour had no negative effect on the overall performance during and after the fire.

Finally, a comparison was made between the various available simplified methods to justify partially protected steel-concrete composite frames. In this instance, it was concluded that best value could be attained by analysing a large subframe using finite element analysis, but that using TSLAB or VulcanLite would result in a conservative design. Attention was drawn to some of the deficiencies of the simple methods, and VulcanLite was identified as a viable compromise between approximate methods and finite element analysis of whole subframes.

References

ABAQUS (2007), *Abaqus Users' Manual v6.7*, Simulia, RI, USA.

Abecassis-Empis, C., Reszka, P., Steinhaus, T., Cowlard, A., Bieau, H., Welch, S., Rein, G., and Torero, J.L. (2008), Characterisation of Dalmarnock Test One, *Experimental and Thermal Fluid Science*, 32, pp. 1334–1343.

Abu, A.K. (2009), Behaviour of Composite Floor Systems in Fire, PhD thesis, University of Sheffield, UK.

Abu, A.K., and Burgess, I.W. (2010), The Effect of Edge Support on Tensile Membrane Action of Composite Slabs in Fire, keynote paper, *Proceedings of SDSS 2010*, Rio de Janeiro, Brazil, pp. 21–32.

Abu, A.K., Burgess, I.W., and Plank, R.J. (2012), Reinforcement Ratios and Their Effects on Composite Slab Panel Failure in Fire, *Proceedings of the Institution of Civil Engineers and Buildings*, 165(SB).

Abu, A.K., Ramanitrarivo, V., and Burgess, I.W. (2010), Collapse Mechanisms of Composite Slab Panels in Fire, *Proceedings of the Structures in Fire Conference*, East Lansing, MI, USA, pp. 382–389.

ADAPTIC (2008), *Adaptic Users' Manual v1.2*, http //www3.imperial.ac.uk/pls/portallive/docs/1/25733696.PDF.

Al-Jabri, K.S. (1999), The Behaviour of Steel and Composite Beam-to-Column Connections in Fire, PhD thesis, Department of Civil and Structural Engineering, University of Sheffield, UK.

Al-Jabri, K.S., Burgess, I.W., and Plank, R.J. (2002), Prediction of the Degradation of Connection Characteristics at Elevated Temperature, *Proceedings of 3rd European Conference on Steel Structures*, Coimbra, Portugal, pp. 1391–1400.

Alpert, R.L. (1972), Calculation of Response Time of Ceiling-mounted Fire Detectors, *Fire Technology*, 8, pp. 182–195.

American Society for Testing and Materials (ASTM) (1997), *ASTM C 177-97: Standard Test Method for Steady-State Heat Flux Measurements and Thermal Transmission Properties by Means of the Guarded-Hot-Plate Apparatus*, American Society for Testing and Materials, West Conshohocken, PA, USA.

Anderberg, Y., and Thelandersson, S. (1976), *Stress and Deformation of Concrete at High Temperatures. 2 Experimental Investigation and Material Behaviour*, Bulletin 54, Lund Institute of Technology, Sweden.

Ang, C.N., and Wang, Y.C. (2004), The Effect of Water Movement on Specific Heat of Gypsum Plasterboard in Heat Transfer Analysis under Natural Fire Exposure, *Construction and Building Materials*, 18, pp. 505–515.

Ansys (2007), *Ansys Release 11.0, Help System*, Ansys, Canonsburg, PA, USA.

Association for Specialist Fire Protection (ASFP) (2002), *Fire Protection for Structural Steel in Buildings*, Steel Construction Institute and Association for Specialist Fire Protection, London.

Bailey, C.G. (2000), *Design of Steel Structures with Composite Slabs at the Fire Limit State*, Final Report prepared for the Department of the Environment, Transport and the Regions, and the Steel Construction Institute, Report No. 81415, Building Research Establishment, Garston, Watford, UK.

Bailey, C.G. (2001), Membrane Action of Unrestrained Lightly Reinforced Concrete Slabs at Large Displacements, *Engineering Structures*, 23, pp. 470–483.

Bailey, C.G. (2002a), Holistic Behaviour of Concrete Buildings in Fire, *Proceedings of the Institution of Civil Engineers*, 152(3), pp. 199–212.

Bailey, C.G. (2002b), Structural Fire Design of Unprotected Steel Beams Supporting Composite Floor Slabs, Presented at II International Conference on Steel Construction—II CICOM, São Paulo, Brazil.

Bailey, C.G. (2003), Efficient Arrangement of Reinforcement for Membrane Behaviour of Composite Floor Slabs in Fire Conditions, *Journal of Constructional Steel Research*, 59, pp. 931–949.

Bailey, C.G., and Moore D.B. (2000), The Structural Behaviour of Steel Frames with Composite Floor Slabs Subjected to Fire: Part 1: Theory, Part 2: Design, *The Structural Engineer*, 78(11), pp. 19–27, 28–33.

Bailey, C.G., and Toh, W.S. (2007a), Behaviour of Concrete Floor Slabs at Ambient and Elevated Temperatures, *Fire Safety Journal*, 42, pp. 425–436.

Bailey, C.G., and Toh, W.S. (2007b), Small-scale Concrete Slab Tests at Ambient and Elevated Temperatures *Engineering Structures*, 29, pp. 2775–2791.

Bailey, C.G., Burgess, I.W., and Plank, R.J. (1996), Computer Simulation of a Full-Scale Structural Fire Test, *The Structural Engineer*, 74(6), pp. 93–100.

Bailey, C.G., Lennon, T., and Moore, D.B. (1999), The Behaviour of Full-Scale Steel Framed Buildings Subjected to Compartment Fires, *The Structural Engineer*, 77(8), pp. 15–21.

Benichou, N., and Sultan M.A. (2005), Thermal Properties of Components of Lightweight Wood-Framed Assemblies at Elevated Temperatures, *Fire and Materials*, 25, pp. 165–179.

Block, F.M. (2006), Development of a Component-Based Finite Element for Steel Beam-to-Column Connections at Elevated Temperatures, PhD thesis, University of Sheffield, UK.

Bresler, B., and Pister, K.S. (1958), Strength of Concrete under Combined Stresses, *Journal of American Concrete Institute*, September, pp. 321–345.

British Steel (1998), *The Fire Resistance of Steel Buildings*, British Steel, London.

British Steel (1999), *The Behaviour of Multi-Storey Steel Framed Buildings in Fire: A European Joint Research Programme*, British Steel Swinden Technology Centre, Rotherham, UK.

British Standards Institution (BSI) (1987), *British Standard 476, Fire Tests on Building Materials and Structures, Part 20: Method for Determination of the Fire Resistance of Elements of Construction (General Principles)*, British Standards Institution, London.

British Standards Institution (BSI) (1990a), *BS 476: Method for Determination of the Fire Resistance of Elements of Construction: Part 20*, British Standards Institution, London.

British Standards Institution (BSI) (1990b), *BS 5950: Structural Use of Steelwork in Buildings. Part 8: Code of Practice for Fire Resistant Design*, British Standards Institution, London.

British Standards Institution (BSI) (2007), *PD 6688-1-2: Background Paper to the UK National Annex to BS EN 1991-1-2*, British Standards Institution, London.

Brotchie, J.F., and Holley, M.J. (1971), *Membrane Action in Slabs: Cracking, Deflection and Ultimate Load of Concrete Slab Systems*, Publication SP-30, American Concrete Institute, Detroit, Paper 30-16, pp. 345–377.

Byfield, M.P. (2004), Design of Steel Framed Buildings at Risk from Terrorist Attack, *The Structural Engineer*, 16 November, pp. 31–38.

Carslaw, H.S., and Jaeger, J.C. (1959), *Conduction of Heat in Solids*, 2nd ed., Oxford University Press, Oxford, UK.

Chana, P., and Price, B. (2003), The Cardington Fire Test, *Concrete*, 37(1), pp. 28–33.

Clifton, C. (1996), *Fire Models for Large Firecells*, Hera Report R4-83, Heavy Engineering Research Association, Auckland, New Zealand.

Clifton, G.C., Hinderhofer, M.D., and Schmid, R. (2001), *Design of Multi-Storey Steel Framed Buildings with Unprotected Secondary Beams or Joists for Dependable Inelastic Response in Severe Fires*, HERA Steel Design and Construction Bulletin, 60, HERA, Manukau City, New Zealand.

Committee of European Normalisation (CEN) (1990–2002), *EN 1990-2002 and Amendment A1:2005, Eurocode 0: Basis of Structural Design*, CEN, Brussels.

Committee of European Normalisation (CEN) (1991–2002), *EN 1991-1-2-2002, Eurocode 1: Actions of Structures, Part 1-2: Actions of Structures Exposed to Fire*, CEN, Brussels.

Committee of European Normalisation (CEN) (1991–2006), *EN 1991-1-7: 2006, Actions on Structures Part 1-7: General Actions—Accidental Actions*, CEN, Brussels.

Committee of European Normalisation (CEN) (1992–2005a), *EN 1992-1-1-2005, Eurocode 2, Design of Concrete Structures, Part 1-1: General Rules and Rules for Buildings*, CEN, Brussels.

Committee of European Normalisation (CEN) (1992–2005b), *EN 1992-1-2-2005, Eurocode 2, Design of Concrete Structures, Part 1-2, General Rules, Structural Fire Design*, CEN, Brussels.

Committee of European Normalisation (CEN) (1993–1995), *ENV 1993-1-2, Eurocode 3: Design of Steel Structures, Part 1.2: General Rules—Structural Fire Design*, CEN, Brussels.

Committee of European Normalisation (CEN) (1993–2005a), *EN 1993-1-1-2005, Eurocode 3, Design of Steel Structures, Part 1-1, General Rules and Rules for Buildings*, CEN, Brussels.

Committee of European Normalisation (CEN) (1993–2005b), *EN 1993-1-2-2005, Eurocode 3: Design of Steel Structures, Part 1-2: Structural Fire Design*, CEN, Brussels.

Committee of European Normalisation (CEN) (1993–2005c), *EN 1993-1-8-2005, Eurocode 3, Design of Steel Structures, Part 1-8, Design of Joints*, CEN, Brussels.

Committee of European Normalisation (CEN) (1994–2005a), *EN 1994-1-1-2005, Eurocode 4, Design of Composite Steel and Concrete Structures, Part 1-1, General Rules and Rules for Buildings*, CEN, Brussels.

Committee of European Normalisation (CEN) (1994–2005b), *EN 1994-1-2-2005, Eurocode 4: Design of Composite Steel and Concrete Structures: Part 1.2 General Rules, Structural Fire Design*, CEN, Brussels.

Committee of European Normalisation (CEN) (1995–2005a), *EN 1995-1-1-2005, Eurocode 5, Design of Timber Structures, Part 1-1, General Rules and Rules for Buildings*, CEN, Brussels.

Committee of European Normalisation (CEN) (1995–2005b), *EN 1995-1-2-2005, Eurocode 5, Design of Timber Structures, Part 1-2, General Rules, Structural Fire Design*, CEN, Brussels.

Committee of European Normalisation (CEN) (1996–2005a), *EN 1996-1-1-2005, Eurocode 6, Design of Masonry Structures, Part 1-1, General Rules and Rules for Buildings*, CEN, Brussels.

Committee of European Normalisation (CEN) (1996–2005b), *EN 1996-1-2-2005, Eurocode 6, Design of Masonry Structures, Part 1-2, General Rules, Structural Fire Design*, CEN, Brussels.

Committee of European Normalisation (CEN) (1999–2007a), *EN 1999-1-1-2007, Eurocode 9, Design of Aluminium Structures, Part 1-1, General Rules and Rules for Buildings*, CEN, Brussels.

Committee of European Normalisation (CEN) (1999–2007b), *EN 1999-1-2-2007, Eurocode 9, Design of Aluminium Structures, Part 1-2, General Rules, Structural Fire Design*, CEN, Brussels.

Committee of European Normalisation (CEN) (2010), *EN 13381-8-2010: Test Methods for Determining the Contribution to the Fire Resistance of Structural Members, Part 8: Applied Reactive Protection to Steel Members*, CEN, Brussels.

Cooke, G.M.E. (1998), *Tests to Determine the Behaviour of Fully Developed Natural Fires in a Large Compartment*, Fire Note 4, Fire Research Station, Building Research Establishment, Watford, UK.

Coulomb, C.A. (1776). Essai sur une Application des Regles des Maximis et Minimis a Quelquels Problemes de Statique Relatifs, a la Architecture, *Mémoires de Mathématique et de Physique présentés a i Académie Royale des Sciences par Divers Savans, et lûs dans ses Assemblées.*, 7, pp. 343–387.

Dai, X.H., Wang, Y.C., and Bailey, C.G. (2007), Temperature Distributions in Unprotected Steel Connections in Fire, in *Proceedings of the International Conference on Steel and Composite Structures*, ed. Wang, Y.C. and Choi, C.K. Taylor & Francis, London.

Dai, X.H., Wang, Y.C., and Bailey, C.G. (2009), Effects of Partial Fire Protection on Temperature Developments in Steel Joints Protected by Intumescent Coating, *Fire Safety Journal*, 44, pp. 376–386.

Dai, X.H., Wang, Y.C., and Bailey, C.G. (2010a), Numerical Modelling of Structural Fire Behaviour of Restrained Steel Beam-Column Assemblies Using Typical Joint Types, *Engineering Structures*, 32, pp. 2337–2351.

Dai, X.H., Wang, Y.C., and Bailey, C.G. (2010b), A Simple Method to Predict Temperatures in Steel Joints with Partial Intumescent Coating Fire Protection, *Fire Technology*, 46, pp. 19–35.

Department of Communities and Local Government (DCLG) (2006), *The Building Regulations 2000, Fire Safety, Approved Document B*, Her Majesty's Stationary Office, London.

DiNenno, P. (2002), *The SFPE Handbook of Fire Protection Engineering*, 3rd ed., Society of Fire Protection Engineers, Bethesda, MD, USA.

Ding, J., and Wang Y.C. (2007), Experimental Study of Structural Fire Behaviour of Steel Beams to Concrete Filled Tubular Column Assemblies with Different Types of Joints, *Engineering Structures*, 29, pp. 3485–3502.

Ding, J., and Wang, Y.C. (2009), Temperatures in Unprotected Joints between Steel Beams and Concrete-Filled Tubular Columns in Fire, *Fire Safety Journal*, 44, pp. 16–32.

Do, C.T., Bentz, D.P., and Stutzman, P.E. (2007), Microstructure and Thermal Conductivity of Hydrated Calcium Silicate Board Materials, *Journal of Building Physics*, 31, pp. 55–67.

Drucker, D.C., and Prager, W. (1952), Soil Mechanics and Plastic Analysis or Limit Design, *Quarterly of Applied Mathematics*, 10, pp. 157–165.

Drysdale, D. (1998), *An Introduction to Fire Dynamics*, 2nd ed., Wiley, New York.

Elghazouli, A.Y., Izzuddin, B.A., and Richardson, A.J. (2000), Numerical Modelling of Structural Fire Behaviour of Composite Buildings, *Fire Safety Journal*, 35(4), pp. 279–297.

El-Rimawi, J.A. (1989), *The Behaviour of Flexural Members under Fire Conditions*, PhD thesis, University of Sheffield, UK.

Elsawaf, S., Wang, Y.C., and Mandal, P. (2011), Numerical Modelling of Restrained Structural Subassemblies of Steel Beam and CFT Columns Connected Using Reverse Channels in Fire, *Engineering Structures*, 31, pp. 1217–1231.

Federal Emergency Management Agency (FEMA) (2002), *World Trade Center Building Performance Study: Data Collection, Preliminary Observations and Recommendations*, FEMA Technical Report 403, Federal Emergency Management Agency, Washington, DC, USA.

Fletcher, I., Borg, A., Hitchen, N., and Welch, S. (2006), Performance of Concrete in Fire: A Review of the State of the Art with a Case Study of the Windsor Tower, *Proceedings of the Fourth International Workshop on Structures in Fire*, Aveiro, Portugal.

Flint, G. (2005), Fire Induced Collapse of Tall Buildings, PhD thesis, University of Edinburgh, UK.

Franssen, J.M. (2005), SAFIR A Thermal/Structural Program Modelling Structures under Fire, *Engineering Journal of American Institution of Steel Construction*, 42(3), pp. 123–158.

Franssen, J.-M. (2006), Calculation of Temperature in Fire-Exposed Bare Steel Structures: Comparison between ENV 1993-1-2 and EN 1993-1-2, *Fire Safety Journal*, 41(2), pp. 139–143.

Franssen, J.M., and Real P.V. (2010), *Fire Design of Steel Structures*, ECCS Eurocode Manuals, Ernst and Sohn, Berlin.

Franssen, J.M., and Zaharia, R. (2005), *Design of Steel Structures Subject to Fire*, Les Éditions de l'Universite de Liege, Belgium.

Gillie, M. (2009), Analysis of Heated Structures: Nature and Modelling Benchmarks, *Fire Safety Journal*, 44(5), pp. 673–680.

Gillie, M., and Stratford, T. (2007), *The Dalmarnock Fire Tests: Experiments and Modelling: Chapter 8 Behaviour of the Structure During the Fire*, School of Engineering and Electronics, University of Edinburgh, UK.

Gillie, M., Usmani, A.S., and Rotter, J.M. (2001a), A Structural Analysis of the First Cardington Test, *Journal of Constructional Steel Research*, 56(6), pp. 581–601.

Gillie, M., Usmani, A.S., Rotter, J.M., and O'Connor, M. (2001b), Modelling of Heated Composite Floor Slabs with Reference to the Cardington Experiments, *Fire Safety Journal*, 36(8), pp. 745–767.

Hayes, B. (1968), Allowing for Membrane Action in the Plastic Analysis of Rectangular Reinforced Concrete Slabs, *Magazine of Concrete Research*, 20(65), pp. 205–212.

Hayes, B., and Taylor R. (1969), Some Tests on Reinforced Concrete Beam-Slab Panels. *Magazine of Concrete Research*, 21(67), pp. 113–120.

Hu, Y., Burgess, I.W., Davison, J.B., and Plank, R.J. (2008), Modelling of Flexible End Plate Connections in Fire Using Cohesive Elements, *Proceedings of the Structures in Fire Workshop, Singapore*, pp. 127–138.

Hu, Y., Davison, J.B., Burgess, I.W., and Plank, R.J. (2009), Component Modelling of Flexible End-plate Connections in Fire, *International Journal of Steel Structures*, 9, pp. 29–38.

Huang, Z., Burgess, I.W., and Plank, R.J. (2001a), Non-linear Structural Modelling of a Fire Test Subject to High Restraint, *Fire Safety Journal*, 36(8), pp. 795–814.

Huang, Z., Burgess, I.W., and Plank, R.J. (2002), Modelling of Six Full-Scale Fire Tests on a Composite Building, *The Structural Engineer*, 80(19), pp. 30–37.

Huang, Z., Burgess, I.W., Plank, R.J., and Bailey, C.G. (2001b), Strategies for Fire Protection of Large Composite Buildings, *Proceedings of Interflam 2001*, Edinburgh, pp. 395–406.

Huang, Z., Burgess, I.W., and Plank, R.J. (2003a), Modelling Membrane Action of Concrete Slabs in Composite Buildings in Fire. Part I: Theoretical Development, *Journal of Structural Engineering, ASCE*, 129(8), pp. 1093–1102.

Huang, Z., Burgess, I.W., and Plank, R.J. (2003b), Modelling Membrane Action of Concrete Slabs in Composite Buildings in Fire. Part II: Validations, *Journal of Structural Engineering, ASCE*, 129(8), pp. 1103–1112.

Huang, Z., Burgess, I.W., and Plank, R.J. (2004), Fire Resistance of Composite Floors Subject to Compartment Fires, *Journal of Constructional Steel Research*. 60(2), pp. 339–360.

Huang, Z.-F., Tan, K.-H., and Ting, S.-K. (2006), Heating Rate and Boundary Restraint Effects on Fire Resistance of Steel Columns with Creep, *Engineering Structures*, 28(6), pp. 805–817.

Institution of Structural Engineers (ISE) (2010), *Practical Guide to Structural Robustness and Disproportionate Collapse in Buildings*, Institution of Structural Engineers, London.

Instituto Tecnico De Materiales Y Construcciones (Intemac) (2005), *Fire in the Windsor Building, Madrid: Survey of the Fire Resistance and Residual Bearing Capacity of the Structure after the Fire*, Intemac, Madrid, Spain.

International Organisation for Standardisation (ISO) (1975), *ISO 834: Fire Resistance Tests, Elements of Building Construction,* International Organization for Standardization, Geneva.

Kawagoe, K. (1958), *Fire Behaviour in Rooms*, Report No. 27, Building Research Institute, Tokyo.

Khabbazan, M.M. (2005), Progressive Collapse, *The Structural Engineer*, 21 June, pp. 28–32.

Khoury, G.A. (2000), Effect of Fire on Concrete and Concrete Structures, *Progress in Structural Engineering and Materials*, 2(4), pp. 429–447.

Khoury, G.A., Grainger, B.N., and Sullivan, P.J.E. (1985), Transient Thermal Strain of Concrete: Literature Review, Conditions within Specimen and Behaviour of Individual Constituents, *Magazine of Concrete Research*, 37(132), pp. 131–144.

Kirby, B.R., and Preston, R.R. (1988), High Temperature Properties of Hot-Rolled Structural Steels for Use in Fire Engineering Studies, *Fire Safety Journal*, 13(1), pp. 27–37.

Koronthalyova, O., and Matiasovsky, P. (2003), Thermal Conductivity of Fibre Reinforced Porous Calcium Silicate Hydrate-based Composites, *Journal of Thermal Envelope and Building Science*, 27(1), pp. 71–89.

Kruppa, J., Joyeux, D., and Zhao, B. (2005), Scientific Background to the Harmonization of Structural Eurocodes, *Heron*, 50(4), pp. 219–235.

Kupfer, H.B., Hilsdorf, H.K., and Rusch, R. (1969), Behavior of Concrete under Biaxial Stresses, *Journal of American Concrete Institute*, 66, pp. 119–128.

Lamont, S., Lane, B., Flint, G., and Usmani, A.S. (2006), Behavior of Structures in Fire and Real Design—A Case Study, *Journal of Fire Protection Engineering*, 16(1), pp. 5–35.

Law, A. (2010), The Assessment and Response of Concrete Structures Subject to Fire, PhD thesis, University of Edinburgh, UK.

Lawson, R.M. (1990a), Behaviour of Steel Beam-to-Column Connections in Fire, *The Structural Engineer*, 68(14), pp. 263–271.

Lawson, R.M. (1990b), *Enhancement of Steel Beam Capacity in Fire Utilizing Connections*, Steel Construction Institute Publication P-086, Steel Construction Institute, Ascot, UK.

Lawson, R.M., and Newman, G.M. (1996), *Structural Fire Design to EC3 & EC4, and Comparison with BS 5950*, Technical Report, SCI Publication 159, Steel Construction Institute, Ascot, UK.Lennon, T. (1996), *Cardington Fire Tests: Instrumentation Locations for Large Compartment Fire Test*, Building Research Establishment Report N100/98, Building Research Establishment, Watford, UK.

Lennon, T., and Moore, D. (2003), The Natural Fire Safety Concept—Full-Scale Tests at Cardington, *Journal of Fire Safety Engineering*, 38, pp. 623–643.

Lennon, T., Moore, D., Wang, Y.C., and Bailey, C.G. (2007), *Designers' Guide to EN 1991-1-2, EN 1992-1-2, EN 1993-1-2 and EN 1994-1-2 (Designers' Guides to the Eurocodes)*, Thomas Telford, London.

Leston-Jones, L.C. (1997), The Influence of Semi-Rigid Connections on the Performance of Steel Framed Structures in Fire, PhD thesis, Department of Civil and Structural Engineering, University of Sheffield, UK.

Liu, T.C.H. (1996), Finite Element Modelling of Behaviour of Steel Beams and Connections in Fire, *Journal of Constructional Steel Research*, 36(3), pp. 181–199.

Loeb, A.L. (1954), A Theory of Thermal Conductivity of Porous Materials, *Journal of American Ceramics Society*, 37, p. 96.

Luo, M.C., Yin, Y.Z., Lamont, S., and Lane, B. (2005), Eastern Solutions, Focus: Building Design, *Fire Engineering Journal and Fire Protection (FEJ & FP)*, June, pp. 29–31.

Maljaars, J., Twilt, L., Fellinger, J.H.H., Snijder, H.H., and Soetens, F. (2010), Aluminium Structures Exposed to Fire Conditions—An Overview, *Heron*, 55(2), pp. 85–122.

Mehaffey, J.R., Cuerrier, P., and Carisse, G.A. (1994), A Model for Predicting Heat Transfer through Gypsum Board/Wood-Stud Walls Exposed to Fire, *Fire and Materials*, 18, pp. 297–305.

National Institute of Standards and Technology (NIST) (2005), *Federal Building and Fire Safety Investigation of the World Trade Center Disaster: Final Report of the National Construction Safety Team on the Collapses of the World Trade Center Towers (Draft)*, NIST NCSTAR 1 (Draft), National Institute of Standards and Technology, Gaithersburg, MD, USA.

National Institute of Standards and Technology (NIST) (2008), *Federal Building and Fire Safety Investigation of the World Trade Center Disaster: Structural Response and Probable Collapse Sequence of World Trade Center Building 7*, NIST Report NCSTAR 1-9(2), National Institute of Standards and Technology, Gaithersburg, MD, USA.

Newman, G.M., Robinson, J.T., and Bailey, C.G. (2006), *Fire Safe Design: A New Approach to Multi-Storey Steel-Framed Buildings*, 2nd ed., SCI Publication P288, Steel Construction Institute, Ascot, UK.

Park, R. (1964a), Tensile Membrane Behaviour of Uniformly Loaded Rectangular Reinforced Concrete Slabs with Full Restrained Edges, *Magazine of Concrete Research*, 16(46), pp. 39–44.

Park, R. (1964b), Ultimate Strength of Rectangular Concrete Slabs under Short-Term Uniform Loading with Edges Restrained against Lateral Movement, *Proceedings of the Institution of Civil Engineers*, 28, pp. 125–150.

Pettersson, O., Magnuson, S.E., and Thor, J. (1976), *Fire Engineering Design of Structures*, Publication 50, Swedish Institute of Steel Construction, Stockholm.

Pope, N., and Bailey, C.G. (2006), Quantitative Comparison of FDS and Parametric Fire Curves with Post-Flashover Compartment Fire Test Data, *Fire Safety Journal*, 41, pp. 99–110.

Rahamanian, I., and Wang, Y.C. (2012), A Combined Experimental and Numerical Method for Extracting Temperature-Dependent Thermal Conductivity of Gypsum Boards, *Construction and Building Materials*, 26, pp. 707–722.

Ramberg, W., and Osgood, W.R. (1943), *Description of Stress-Strain Curves by 3 Parameters*, Technical Report 902, National Advisory Committee for Aeronautics, Hampton, VA, USA.

Rein, G., et al. (2007a), *The Dalmarnock Fire Tests: Experiments and Modelling: Chapter 10: A Priori Modelling of Fire Test One*, School of Engineering and Electronics, University of Edinburgh, UK.

Rein, G., et al. (2007b), *The Dalmarnock Fire Tests: Experiments and Modelling: Chapter 3: A Priori Modelling of Fire Test One*, School of Engineering and Electronics, University of Edinburgh, UK.

Rein, G., et al. (2007c), Multi-story Fire Analysis for High-Rise Buildings, *Proceedings of the 11th International Interflam Conference*, London, pp. 605–615.

Röben, C. (2010), The Effect of Cooling and Non-Uniform Fires on Structural Behaviour, PhD thesis, University of Edinburgh, UK.

Röben, C., Gillie, M., and Torero, J.L. (2010), Structural Behaviour during a Vertically Travelling Fire, *Fire Safety Journal*, 66, pp. 191–197.

Russell, H.W. (1935), Principles of Heat Flow in Porous Insulators, *Journal of American Ceramic Society*, 18:1.

Sanad, A.M., Rotter, J.M., Usmani, A.S., and O'Connor, M.A. (2000), Composite Beams in Large Buildings under Fire—Numerical Modelling and Structural Behaviour, *Fire Safety Journal*, 35(3), pp. 165–188.

Sawczuk, A., and Winnicki, L. (1965), Plastic Behaviour of Simply Supported Reinforced Concrete Plates at Moderately Large Deflections, *International Journal of Solids Structures*, 1, pp. 97–111.

Schneider, U., and Horvath, J. (2003), *Behaviour of Ordinary Concrete at High Temperatures*, Research Reports of Vienna University of Technology, Institute of Building Materials, Building Physics and Fire Protection 9, Vienna.

Smith, J.M. (1981), *Chemical Engineering Kinetics*, McGraw-Hill, New York.

Spyrou, S., Davison, J.B., Burgess, I.W., and Plank, R.J. (2004a), Experimental and Analytical Investigation of the "Compression Zone" Component within a Steel Joint at Elevated Temperatures, *Journal of Constructional Steel Research*, 60(6), pp. 841–865.

Spyrou, S., Davison, J.B., Burgess, I.W., and Plank, R.J. (2004b), Experimental and Analytical Investigation of the "Tension Zone" Component within a Steel Joint at Elevated Temperatures, *Journal of Constructional Steel Research*, 60(6), pp. 867–896.

Steel Construction Industry Forum (SCIF) (1991), *Investigation of Broadgate Phase 8 Fire*, Steel Construction Institute, Ascot, UK.

Steel Construction Institute (SCI) (1992), *Joints in Simple Construction Vol. 1: Design Methods*, SCI Publication P-105, Steel Construction Institute and BCSA, Ascot, UK.

Steel Construction Institute (SCI) (1993), *Joints in Simple Construction Vol. 2: Practical Application*, SCI Publication P-105, Steel Construction Institute and BCSA, Ascot, UK.

Steel Construction Institute (SCI) and CTICM (2009), *Fire Resistance Assessment of Partially Protected Composite Floors (FRACOF [Centre Technique Industriel des Conservations Metalliques]): Engineering Background*, Technical Report, Steel Construction Institute, Ascot, UK.

Stern-Gottfried, J., Rein, G., Lane, B., Torero, J.L. (2009), An Innovative Approach to Design Fires for Structural Analysis of Non-Conventional Buildings, a Case Study, *Proceedings of International Conference on Applications of Structural Fire Engineering*, Czech Technical University, Prague, Czech Republic.

Sukumar, N. Moës, N., Moran, B., and Belytschko, T. (2000), Extended Finite Element Method for Three-Dimensional Crack Modelling, *International Journal for Numerical Methods in Engineering*, 48(11), pp. 1549–1570.

Suvorov, S.A., and Skurikhin, V.V. (2002), High-Temperature Heat-Insulating Materials Based on Vermiculite, *Refractories and Industrial Ceramics*, 43(11–12), pp. 383–389.

Suvorov, S.A., and Skurikhin, V.V. (2003), Vermiculite—a Promising Material for High-Temperature Heat Insulators, *Refractories and Industrial Ceramics*, 44(3), pp. 186–193.

Tan, K.H., Vimonsatit, V., and Qian, Z.H. (2004), Testing of Plate Girder Web Panel Loaded in Shear at Elevated Temperature, *Proceedings of the Third Structures in Fire Conference*, Ottawa, Canada, pp. 89–97.

TF (2000), Grantham R, Enjily v, Milner M, Bullock M, G Pitts. Multi-story Timber Frame Buildings: A Design Guide, BREPress, Garston.

Thomas, G. (2002), Thermal Properties of Gypsum Plasterboard at High Temperatures, *Fire and Materials*, 26, pp. 37–45.

Tresca, H. (1864). Mémoire sur L'écoulement des Corps Solides Soumis à de Fortes Pressions, *Comptes Rendu de l' Academie des Sciences Paris*, 59, pp. 754.

Usmani, A.S., Chung Y.C., and Torero, J.L (2003), How Did the WTC Towers Collapse: A New Theory, *Fire Safe Journal*, 38, pp. 501–533.

Usmani, A.S., Rotter, J.M., Lamont, S., Sanad, A.M., and Gillie, M. (2001), Fundamental Principles of Structural Bbehaviour under Thermal Effects, *Fire Safety Journal*, 36(8), pp. 721–744.

Varma, A.H., Agarwal, A., Hong, S., and Prasad, K. (2008), Behavior of Steel Building Structures with Perimeter MRFs under Fire Loading Effects, *Proceedings of the Fifth International Conference on Structures in Fire*, Singapore, pp. 266–276.

von Mises, R. (1913), Mechanik der festen Körper im plastisch deformablen Zustand, *Göttingem Nachrichten Mathematisce Physikalische*, 1, pp. 582–592.

Vulcan (2008), Vulcan Solutions Ltd., http://www.vulcan-solutions.com/index.html.

Wald, F., Simöes da Silva L., Moore, D., Lennon, T., and Chladná, M. (2006), Experimental Behaviour of a Steel Structure under Natural Fire, *Fire Safety Journal*, 41(7), pp. 509–522.

Wang Y.C. (2002), *Steel and Composite Structures, Behaviour and Design for Fire Safety*, Spon Press, London.

Wang, Y.C. (2004), Post-Buckling Behaviour of Axially Restrained and Axially Loaded Steel Columns under Fire Conditions, *Journal of Structural Engineering*, ASCE, 130(3), pp. 371–380.

Wang, Y.C., Lennon, T., and Moore, D.B. (1995), The Behaviour of Steel Frames Subject to Fire, *Journal of Constructional Steel Research*, 35, pp. 291–322.

Wang, Y.C., Dai, X.H., and Bailey, C.G. (2011), An Experimental Study of Relative Structural Fire Behaviour and Robustness of Different Types of Steel Joint in Restrained Steel Frames, *Journal of Constructional Steel Research*, 67(7), pp. 1149–1163.

Welch, S., Jowsey, A., Deeny, S., Morgan, R., and Torero, J.L. (2007), BRE Large Compartment Fire Tests—Characterising Post-Flashover Fires for Model Validation, *Fire Safety Journal*, 42, pp. 548–567.

Wickstrom, U. (1982), Temperature Calculation of Insulated Steel Columns Exposed to Natural Fire, *Fire Safety Journal*, 4(4), pp. 219–225.

Wickstrom, U. (1985), Temperature Analysis of Heavily-Insulated Steel Structures Exposed to Fire, *Fire Safety Journal*, 9, pp. 281–285.

Wickstrom, U. (2005), Comments on Calculation of Temperature in Fire-Exposed Bare Steel Structures in prEN 1993-1-2: Eurocode 3—Design of Steel Structures—Part 1-2: General Rules—Structural Fire Design, *Fire Safety Journal*, 40(2), pp. 191–192.

Wong, M.B., and Ghojel, J.I. (2003), Sensitivity Analysis of Heat Transfer Formulations for Insulated Structural Steel Components, *Fire Safety Journal*, 38, pp. 187–201.

Wood, R.H. (1961), *Plastic and Elastic Design of Slabs and Plates, with Particular Reference to Reinforced Concrete Floor Slabs*, Thames and Hudson, London.

Yin, Y.Z., and Wang Y.C. (2004), A Numerical Study of Large Deflection Behaviour of Restrained Steel Beams at Elevated Temperatures, *Journal of Constructional Steel Research*, 60, pp. 1029–1047.

Yu, H.X., Burgess, I.W., Davison, J.B., and Plank, R.J. (2008), Numerical Simulation of Bolted Steel Connections in Fire Using Explicit Dynamic Analysis, *Journal of Constructional Steel Research*, 64, pp. 515–525.

Yu, H.X., Burgess, I.W., Davison, J.B., and Plank, R.J. (2009a), Experimental Investigation of the Behaviour of Fin Plate Connections in Fire, *Journal of Constructional Steel Research*, 65, pp. 723–736.

Yu, H.X., Burgess, I.W., Davison, J.B., and Plank, R.J. (2009b), Development of a Yield-Line Model for Endplate Connections in Fire, *Journal of Constructional Steel Research*, 65(6), pp. 1279–1289.

Yu, H.X., Burgess, I.W., Davison, J.B., and Plank, R.J. (2009c), Tying Capacity of Web Cleat Connections in Fire. Part 1: Test and Finite Element Simulation, *Engineering Structures*, 31(3), pp. 651–663.

Yu, H.X., Burgess, I.W., Davison, J.B., and Plank, R.J. (2009d), Tying Capacity of Web Cleat Connections in Fire. Part 2: Development of Component-Based Model, *Engineering Structures*, 31(3), pp. 697–708.

Yu, H.X., Burgess, I.W., Davison, J.B., and Plank, R.J. (2011), Experimental and Numerical Investigations of the Behaviour of Flush Endplate Connections at Elevated Temperatures, *Journal of Structural Engineering, ASCE*, 137(1), pp. 80–87.

Yu, X., Huang, Z., and Burgess, I.W. (2008b), Nonlinear Analysis of Orthotropic Composite Slabs in Fire, *Engineering Structures*, 30(1), pp. 67–80.

Yuan, J. (2009), Fire Protection Performance of Intumescent Coating under Realistic Fire Conditions, PhD thesis, School of Mechanical, Aerospace and Civil Engineering, University of Manchester, UK.

Zienkiewicz, O.C., and Taylor, R.L. (2000), *The Finite Element Method, Vol. 2: Solid Mechanics*, 5th ed., Butterworth-Heinemann, Oxford, UK.

Zonnoni, M. (2008), Brand Bij Bouwkunde. Cot Instit Voor Veilingheids, *Crisismanagement*.

Index